Neuromuscular Aspects of Physical Activity

Phillip F. Gardiner, PhD
Université de Montréal

Human Kinetics

MT

Library of Congress Cataloging-in-Publication Data

Gardiner, Phillip F., 1949-
    Neuromuscular aspects of physical activity / Phillip F. Gardiner.
        p. cm.
    Includes bibliographical references and index.
    ISBN 0-7360-0126-3
        1. Exercise--Physiological aspects.   2. Neurophysiology.   3. Motor neurons.   4. Muscles.
    I. Title.
    QP301 .G367   2001
    612'.049--dc21   00-054203

ISBN: 0-7360-0126-3

Acquisitions Editor: Michael S. Bahrke, PhD
Developmental Editor: Rebecca Crist
Assistant Editor: M. Edward Zulauf
Copyeditor: Karen Bojda
Proofreader: Myla Smith
Permission Manager: Courtney Astle
Graphic Designer: Fred Starbird
Graphic Artist: Sandra Meier
Photo Manager: Clark Brooks
Cover Designer: Jack W. Davis
Art Manager: Craig Newsom
Illustrator: Mic Greenberg
Printer: Sheridan Books

Printed in the United States of America   10 9 8 7 6 5 4 3 2 1

Human Kinetics
Web site: www.humankinetics.com

United States: Human Kinetics
P.O. Box 5076
Champaign, IL 61825-5076
800-747-4457
e-mail: humank@hkusa.com

Canada: Human Kinetics
475 Devonshire Road Unit 100
Windsor, ON N8Y 2L5
800-465-7301 (in Canada only)
e-mail: hkcan@mnsi.net

Europe: Human Kinetics, Units C2/C3 Wira Business Park
Leeds LS16 6EB, United Kingdom
+44 (0) 113 278 1708
e-mail: hk@hkeurope.com

Australia: Human Kinetics
57A Price Avenue
Lower Mitcham, South Australia 5062
08 8277 1555
e-mail: liahka@senet.com.au

New Zealand: Human Kinetics
P.O. Box 105-231, Auckland Central
09-523-3462
e-mail: hkp@ihug.co.nz

3/23/04

I dedicate this book, with love, to my mother and father,
Verlie and Charles Gardiner.

Contents

Preface   vii

Acknowledgments   ix

Credits   xi

Chapter 1   Muscle Fiber Types   1

Grouping Fibers by Myosin Heavy-Chain (MHC) Composition   3

Functional Properties of Fibers Containing Different
   Myosin Heavy-Chain Profiles   8

Fiber Types and Performance   32

Summary   36

Chapter 2   Motoneurons and the Muscle   37
               Units They Innervate

The Muscle Unit and Muscle Unit Types   39

The Motoneuron Component of the Motor Unit   47

The Heckman–Binder Model of Motor Unit Recruitment   63

Motor Unit Recruitment During Different Types
   of Voluntary Contractions   66

Summary   81

Chapter 3   Neuromuscular Fatigue   83

Two Basic Fatigue Mechanisms Involving the Nervous System   85

Reduced Motoneuron Activity During Various Types of Contractions   89

Evidence From Reduced Animal Preparations on Mechanisms
   of Neuromuscular Fatigue   107

Summary   110

Chapter 4 Endurance Training of the Neuromuscular System   111

Muscle Adaptations   112
The Neuromuscular Junction   128
Motoneuron Adaptations to Endurance Training   133
Spinal Cord Adaptations to Endurance Training   137
Summary   141

Chapter 5 Strength Training   143

Acute Effects of Strength Training on Protein Synthesis
    and Degradation   145
The Chronic Effect of Resistance Overload on Muscle Phenotype   154
Neural Effects of Resistance Training   161
Summary   169

Chapter 6 Neuromuscular Responses to Decrease in Normal Activity   171

General Principles Underlying Neuromuscular Responses
    to Reduced Activity   173
Models of Decreased Neuromuscular Usage   176
Summary   202

References   203
Index   231
About the Author   237

Preface

During my career of over 20 years in the Department of Physical Education at the Université de Montréal, I have had the opportunity to develop and teach senior undergraduate and graduate courses in my area of expertise. The subject matter would be best described as *aspects neuro-musculaires de l'exercice,* or neuromuscular aspects of exercise. One of my goals in the undergraduate course has been to introduce students to the most current research findings available in the burgeoning fields of the neurosciences and molecular biology and their impact on our understanding of the physiology of exercise. The graduate course, a small-group reading and discussion format, has over the years attracted students from kinesiology, physical education, physical therapy, occupational therapy, biomedical sciences, physiology, and biology. The route that this course has taken has been largely dictated by the rapidly evolving trends in the field, although the metamorphosis of my own interests, which have also taken a rather tortuous route during this time, has certainly been a factor.

In this book, my purpose is to provide a balanced overview of neuromuscular involvement in physical activity, including how the neuromuscular system is used and how it responds to fatiguing exercise and to changes (increases and reductions) in chronic activation levels. As a result of the proliferation of our knowledge in the associated parent disciplines and the inevitable specialization that it breeds, books written on this particular subject vary considerably, reflecting the interests and expertise of the authors. This book is thereby destined to be unique.

The order of presentation of the six chapters of the book seems to me to be the most logical. To understand the neuromuscular responses to increased and decreased neuromuscular activity, one must have a grasp of the heterogeneity of muscle fibers and motor units, their properties, and how they are used during different types of movement. The first part of the book proceeds from the histochemical and functional properties of single fibers (chapter 1) to the patterns with which motor units are recruited (chapter 2). I have put particular emphasis on the latter, which constitutes the largest chapter of the book, based on my impression that research on motor unit recruitment, essential for understanding exercise, is not given sufficient emphasis in other exercise physiology texts. In chapter 2, I demonstrate an application of the fine motor unit recruitment model of C.J. Heckman and M. Binder, which, in my experience, helps students better understand the complexities involved in motor unit recruitment during exercise. The variables in the model can also be manipulated by the students themselves, using a simple computer program, to model the effects of training, fatigue, and inactivity and to determine recruitment results in muscles that differ in motor unit number and size, as well as in dependence on recruitment versus rate coding for voluntary force generation.

In chapter 3, I present a unique discussion of the central and peripheral components of fatigue during maximal and submaximal exercise in humans and give support for these interpretations from experimental results that are forthcoming from reduced animal preparations. Of particular interest in this chapter is the current debate as to the source and significance of fatigue at different levels of the central nervous system.

Chapters 4 and 5 explore the components of neuromuscular endurance and strength by examining general principles of the molecular stimuli for their development at the muscle level (such as the roles of stretch and muscle biochemical stimuli and the contributions of transcriptional, translational, and posttranslational mechanisms in the final phenotype) and also by discussing the neural mechanisms involved. The molecular biological aspects are presented in both chapters as general principles with examples. Chapter 4 emphasizes the chronic stimulation model, which provides us with considerable insight into the stimuli and phenotypic consequences of prolonged muscle activation, with the understanding that the model is in some respects different from, and in others much simpler than, exercise training. Also in chapter 4, I present as much of the sparse literature as is currently available on the effects of endurance training on motoneurons and spinal cord connectivity, with some reference to the current efforts to demonstrate the "learning" capacity of the spinal cord. In chapter 5, I discuss mechanisms for strength development at levels of the muscular and nervous systems, including a discussion of so-called neural aspects of training.

Finally, in chapter 6 I present the effects of neuromuscular inactivity, using a format similar to that used in the two previous chapters. Evidence of involvement of the nervous system in the inactivity-induced impairment of function is included, along with an analysis of the muscular changes resulting from reduced neuromuscular activity due to spinal cord transection and isolation, pharmacological motor-nerve blockade, bed rest, weightlessness, and limb immobilization.

In presenting the material as I do, I assume that the reader has followed a first-year university-level course in general physiology or exercise physiology and has knowledge of anatomy, biochemistry, and a bit of molecular biology. The text is heavily referenced because I believe it useful for students to have access to the original material if they do not like my interpretation. I have also relied heavily on figures, of which there are over 150, taken directly or adapted from the research literature.

A text based primarily on the research literature has from the start the shortcoming that not all of the available and pertinent information can be included. If you think that a specific research paper should have been cited but was not, it is because (1) I considered it redundant with another citation, (2) I did not know about it, or (3) I disagree with you as to its significance for the discussion. Please accept my apologies in the cases of 1 and 2. As for 3, well, perhaps we will have a chance to discuss it at a scientific meeting sometime soon.

Acknowledgments

Appreciation is extended to my friend, colleague, and wife, Kalan, not only for her love, but also for her technical help, incredible listening capacity, tolerance, and understanding during the writing. Thanks also to my sons Patrick and Matthew for putting up with my odd schedules and moods during the preparation of the book and to my graduate students for their patience in putting up with my days away from the office and lab.

Credits

Figure 1.2 Reprinted, with kind permission from Kluwer Academic Publishers, from J.A.A. Sant'ana Pereira et al., 1995, "New method for the accurate characterization of single human skeletal muscle fibres demonstrates a relation between mATPase and MyHC expression in pure and hybrid fibre types," *Journal of Muscle Research and Cell Motility 16*: 30, figure 7.

Figure 1.3 Reprinted, with permission from Blackwell Science Ltd., from L. Larsson and G. Salviati, 1992, "A technique for studies of the contractile apparatus in single human muscle fibre segments obtained by percutaneous biopsy," *Acta Physiologica Scandinavica 146*: 491.

Figure 1.4 Reprinted, by permission, from P.J. Reiser, R.L. Moss, G.G. Giulian, and M.L. Greaser, 1985, "Shortening velocity in single fibers from adult rabbit soleus muscles in correlated with myosin heavy chain composition, *Journal of Biological Chemistry 260(16):* 9078.

Figure 1.5 Reprinted, by permission, from L. Larsson and R.L. Moss, 1993, "Maximum velocity of shortening in relation to myosin isoform composition in single fibres from human skeletal muscles," *Journal of Physiology 472: 604.*

Figure 1.6 Data are from rats (Li and Larsson 1996; Bottinelli et al. 1994a, b, c; Galler et al. 1994; Rome et al. 1990), rabbits (Sweeney et al. 1988; Rome et al. 1990), monkeys (Fitts et al. 1998; Widrick et al. 1997a), humans (Larsson and Moss 1993; Widrick et al. 1996a; Harridge et al. 1996; Bottinelli et al. 1996) and horses (Rome et al. 1990).

Figure 1.7 Reprinted, by permission, from L. Larsson and R.L. Moss, 1993, "Maximum velocity of shortening in relation to myosin isoform composition in single fibres from human skeletal muscles," *Journal of Physiologyy 472*: 607.

Figure 1.8 Adapted, by permission, from R. Bottinelli, R. Betto, S. Schiaffino, and C. Reggiani, 1994, "Unloaded shortening velocity and myosin heavy chain and alkali light chain isoform composition in rat skeletal muscle fibres," *Journal of Physiology 478:* 346.

Figure 1.9 Reprinted, by permission, from J.J. Widrick, S.W. Trappe, C.A. Blaser, D.L. Costill and R.H. Fitts, 1996, "Isometric force and maximal shortening velocity of single muscle fibers from elite master runners, *American Journal of Physiology- Cell Physiology 271:* C672.

Figure 1.10 Reprinted, by permission, from B. Wohlfart and K.A.P. Edman, 1994, " Rectangular hyperbola fitted to muscle force-velocity data using three-dimensional regression analysis," *Experimental Physiology 79*: 236.

Figure 1.11 Adapted, by permission, from R. Bottinelli, M. Canepari, M.A. Pellegrino, and C. Reggiani, 1996, "Force-velocity properties of human skeletal muscle fibres: Myosin heavy chain isoform and temperature dependence," *Journal of Physiology 495:* 579.

Figure 1.12 Adapted, by permission, from R. Bottinelli, M. Canepari, M.A. Pellegrino, and C. Reggiani,1996, "Force-velocity properties of human skeletal muscle fibres: Myosin heavy chain isoform and temperature dependence," *Journal of Physiology 495:* 581.

Figure 1.13 Adapted, by permission, from R. Bottinelli, M. Canepari, M.A. Pellegrino, and C. Reggiani, 1996, "Force-velocity properties of human skeletal muscle fibres: Myosin heavy chain isoform and temperature dependence, *Journal of Physiology 495*: 581.

Figure 1.14 Adapted, by permission, from R. Bottinelli, M. Canepari, M.A. Pellegrino, and C. Reggiani, 1996, "Force-velocity properties of human skeletal muscle fibres: Myosin heavy chain isoform and temperature dependence," *Journal of Physiology 495*: 579.

Figure 1.15 Adapted, by permission, from R. Bottinelli, M. Caneparii, M.A. Pellegrino, and C. Reggiani, 1996, "Force-velocity properties of human skeletal muscle fibres: Myosin heavy chain isoform and temperature dependence," *Journal of Physiology 495*: 581.

Figure 1.16 Data are from: (1) Larsson and Moss 1993; (2) Larsson et al. 1996; (3) Widrick et al. 1996b; (4) Stienen et al. 1996; (5) Harridge et al. 1998; (6) Harridge et al. 1996; (7) Bottinelli et al. 1996.

Figure 1.17 Reprinted, by permission, from R. Bottinelli, M. Canepari, C. Reggiani, and G.J.M. Stienen, 1994, "Myofibrillar ATPase activity during isometric contraction and isomyosin composition in rat single skinned muscle fibres," *Journal of Physiology 481*: 671.

Figure 1.18 Reprinted, by permission, from G.J.M. Stienen, J.L. Kiers, R. Bottinelli, and C. Reggiani, 1994, "Myofibrillar ATPase activity in skinned human skeletal muscle fibres: Fibre type and temperature dependence," *Journal of Physiology 493:* 302.

Figure 1.19 Reprinted, by permission, from S. Schiaffino and C. Reggiani, 1996, "Molecular diversity of myofibrillar proteins: Gene regulation and functional significance," *Physiological Reviews 76*: 377.

Figure 1.20 Reprinted, by permission, from S. Lowey, G.S. Waller, and K.M. Trybus, 1993, "Skeletal muscle myosin light chains are essential for physiological speeds of shortening," *Nature 365:* 455.

Figure 2.43 Reprinted, by permission, from A. Nardone, C. Romano, and M. Schieppati, 1984, "Selective recruitment of high-threshold human motor units during voluntary isotonic lengthening of active muscles," *Journal of Physiology 409*: 456.

Figure 2.44 Reprinted, by permission, from J.N. Howell, A.J. Fuglevand, M. Walsh, and B.R. Bigland-Ritchie, 1995, "Motor unit activity during isometric and concentric-eccentric contractions of the human first dorsal interosseus muscle," *Journal of Neurophysiology 74*: 903.

Figure 2.45 Reprinted, with permission from Blackwell Science Ltd., from N. Vollestad, O. Vaage, and L. Hermansen, 1984, "Muscle glycogen depletion patterns in type I and subgroups of type II fibres during prolonged severe exercise in man," *Acta Physiologica Scandinavica 122*: 437.

Figure 2.46 Reprinted, with permission from Blackwell Science Ltd., from N.K. Vollestad and P.C.S. Blom, 1985, "Effect of varying exercise intensity on glycogen depletion in human muscle fibres," *Acta Physiologica Scandinavica 125*: 397, 398.

Figure 3.1 Reprinted, by permission, from A.J. Fuglevand, K.M. Zackowski, K.A. Huey, and R.M. Enoka, 1993, "Impairment of neuromuscular propagation during human fatiguing contractions at submaximal forces," *Journal of Physiology 460*: 555.

Figure 3.2a Reprinted, by permission, from J.H. Kuei, R. Shadmehr, and G.C. Sieck, 1990, "Relative contribution of neurotransmission failure to diaphragm fatigue," *Journal of Applied Physiology 68*: 176.

Figure 3.2b Reprinted, by permission, from J.H. Kuei, R. Shadmehr, and G.C. Sieck, 1990, "Relative contribution of neurotransmission failure to diaphragm fatigue," *Journal of Applied Physiology 68*: 177.

Figure 3.2c Reprinted, by permission, from B.D. Johnson and G.C. Sieck, 1993, "Differential susceptibility of diaphragm muscle fibers to neuromuscular transmission failure," *Journal of Applied Physiology 75*: 343.

Figure 3.3 Reprinted, by permission, from J.E. Desmedt, 1983, *Motor control mechanisms in health and disease* (New York: Raven Press), 187, 188.

Figure 3.4 Reprinted, by permission, from L. Grimby, J. Hannerz, and B. Hedman, 1981, "The fatigue and voluntary discharge properties of single motor units in man," *Journal of Physiology 316*: 548.

Figure 3.5 Reprinted, by permission, from K.-E. Hagbarth, E. Kunesch, M. Nordin, R. Schmidt, and E. Wallin, 1986, "Gamma loop contributing to maximal voluntary contractions in man," *Journal of Physiology 380*: 579.

Figure 3.6 Reprinted, by permission, from L.G. Bongiovanni and K.-E. Hagbarth, 1990, "Tonic vibration reflexes elicited during fatigue from maximal voluntary contractions in man," *Journal of Physiology 423*: 5, 6.

Figure 3.7 Reprinted, by permission, from L.G. Bongiovanni and K.-E. Hagbarth, 1990, "Tonic vibration reflexes elicited during fatigue from maximal voluntary contractions in man," *Journal of Physiology 423*: 8.

Figure 3.8 Reprinted, by permission, from V.G. Macefield, S.C. Gandevia, B. Bigland-Ritchie, R.B. Gorman, and D. Burke, 1993, "The firing rates of human motoneurons voluntarily activated in the absence of muscle afferent feedback," *Journal of Physiology 471*: 434, 436.

Figure 3.9 Reprinted, with permission from Blackwell Science Ltd., from S.C. Gandevia, 1998, "Neural control in human muscle fatigue," *Acta Physiologica Scandinavica 162*: 278.

Figure 3.10 Reprinted, by permission, from J. Woods, F. Furbush, and B. Bigland-Ritchie, 1987, "Evidence for a fatigue-induced reflex inhibition of motoneuron firing rates," *Journal of Neurophysiology 58*: 127, 131.

Figure 3.11 Reprinted, by permission, from J.L. Taylor, J.E. Butler, G.M. Allen, and S.C. Gandevia, 1996, "Changes in motor cortical excitability during human muscle fatigue," *J. Physiol. (Lond.) 490*: 520, 521.

Figure 3.12 Reprinted, by permission, from J.L. Taylor, J.E. Butler, G.M. Allen and S.C. Gandevia, 1996, "Changes in motor cortical excitability during human muscle fatigue," *Journal of Physiology 490*: 552, 523.

Figure 3.13 Reprinted, by permission, from S.C. Gandevia, G.M. Allen, J.E. Butler, and J.L. Taylor, 1996, "Supraspinal factors in human muscle fatigue: Evidence for suboptimal output from the motor cortex," *Journal of Physiology 490*: 531.

Figure 3.14 Reprinted, by permission, from W.N. Löscher, A.G. Cresswell, and A. Thorstensson, 1996, "Excitatory drive to the alpha-motoneuron pool during a fatiguing submaximal contraction in man," *Journal of Physiology 491*: 275.

Figure 3.15 Reprinted, by permission, from S.J. Garland, L. Griffin, and T. Ivanova, 1997, "Motor unit discharge rate is not associated with muscle relaxation time in sustained submaximal contractions in humans," *Neuroscience Letters 239*: 26.

Figure 3.16 Reprinted, by permission, from G. Macefield, K.-E. Hagbarth, R. Gorman, S.C. Gandevia, and D. Burke, 1991, "Decline in spindle support to alpha-motoneurones during sustained voluntary contractions," *Journal of Physiology 440*: 502, 504.

Figure 3.17 Reprinted, by permission, from J.A. Psek and E. Cafarelli, 1987, "Behavior of coactive muscles during fatigue," *Journal of Applied Physiology 74*: 172, 173.

Figure 3.18 Reprinted, by permission, from I. Zijdewind, D. Kernell, and C.G. Kukulka, 1995, "Spatial differences in fatigue-associated electromyographic behaviour of the human first dorsal interosseus muscle," *Journal of Physiology 483*: 504, 505.

Figure 4.1 Reprinted, by permission, from P. Hu, C. Yin, K.M. Zhang, L.D. Wright, T.E. Nixon, A.S. Wechsler, J.A. Spratt, and F.N. Briggs, 1995, "Transcriptional regulation of phospholamban gene and translational regulation of SERCA2 gene produces coordinate expression of these two sarcoplasmic reticulum proteins during skeletal muscle phenotype switching," *Journal of Biological Chemistry 270*: 11621.

Figure 4.2 Reprinted, by permission, from F. Jaschinski et al., 1998, "Changes in myosin heavy chain mRNA and protein isoforms of rat muscle during forced contractile activity," *Am. J. Physiol. Cell Physiol. 274*: C367.

Figure 4.3 Reproduced, with permission, from U. Seedorf et al., 1986, "Neural control of gene expression in skeletal muscle: Effects of chronic stimulation on lactate dehydrogenase isoenzymes and citrate synthase," *Biochemical Journal 239:* 115-120. © 1986 the Biochemical Society.

Figure 4.4 Reprinted, by permission, from P. Hu, C. Yin, K.M. Zhang, L.D. Wright, T.E. Nixon, A.S. Wechsler, J.A. Spratt, and F.N. Briggs, 1995, "Transcriptional regulation of phospholamban gene and translational regulation of SERCA2 gene produces coordinate expression of these two sarcoplasmic reticulum proteins during skeletal muscle phenotype switching," *Journal of Biological Chemistry 270*: 11621.

Figure 4.5 Reprinted, by permission, from A. Termin and D. Pette, 1996, "Dynamics of parvalbumin expression in low-frequency-stimulated fast-twitch rat muscle," *European Journal of Biochemistry 236*: 816,817.

Figure 4.6 Reprinted, by permission, from A. Termin and D. Pette, 1992, "Changes in myosin heavy-chain isoform synthesis of chronically stimulated rat fast-twitch muscle," *European Journal of Biochemistry 204*: 571.

Figure 4.7 Reprinted, by permission, from J. Henriksson, M.M.-Y. Chi, C.S. Hintz, D.A. Young, K.K. Kaiser, S. Salmons, and O.H. Lowry, 1986, "Chronic stimulation of mammalian muscle: changes in enzymes of six metabolic pathways," *American Journal of Physiology 251*: C622.

Figure 4.11 Reprinted, by permission, from H. Green et. al, 1992, "Metabolite patterns related to exhaustion, recovery, and the transformation of chronically stimulated rabbit fast muscle," *Pflügers Archive 420*: 363. © 1992 Springer-Verlag.

Figure 4.13 Reprinted, by permission, from P.V. Nguyen and H.L. Atwood, 1990, "Expression of long-term adaptation of synaptic transmission requires a critical period of protein synthesis," *J. Neurosci. 10:* 1105. Copyright 1990 by the Society for Neuroscience.

Figure 4.14 Reprinted, by permission, from V. Gisiger, M. Bélisle, and P.F. Gardiner, 1994, "Acetylcholinesterase adaptation to voluntary wheel running is proportional to the volume of activity in fast, but now slow, rat hindlimb muscles," *European Journal of Neuroscience 6*: 675.

Figure 4.15 Reprinted, by permission, from R. Panenic and P.F. Gardiner, 1998, "The case for adaptability of the neuromuscular junction to endurance exercise training," *Canadian Journal of Applied Physiology 23*: 341.

Figure 4.16 Reprinted, by permission, from M. Dorlöchter, A. Irintchev, M. Brinkers, and A. Wernig, 1991, "Effects of enhanced activity on synaptic transmission in mouse extensor digitorum longus muscle," *Journal of Physiology 436*: 289.

Figure 4.17 Reprinted, by permission, from C.-M. Kang et al., 1995, "Chronic exercise increases SNAP-25 abundance in fast-transported proteins of rat motoneurones," *Neuroreport 6*(3): 549-553.

Figure 4.18 Reprinted, with permission from Elsevier Science, from R. Gharakhanlou et al., 1999, "Increased activity in the form of endurance training increases calcitonin gene-related peptide content in lumbar motoneuron cell bodies and in sciatic nerve in the rat," *Neuroscience 89:*1236.

Figure 4.19 Reprinted, by permission, from J.B. Munson, R.C. Foehring, L.M. Mendell, and T. Gordon, 1997, "Fast-to-slow conversion following chronic low-frequency activation of medial gastrocnemius muscle in cats," *Journal of Neurophysiology 77*: 2608.

Figure 4.20 Reprinted, with permission from Elsevier Science, from J.R. Wolpaw and J.S. Carp, 1990, "Memory traces in spinal cord," *Trends in Neurosciences 13:* 138-140.

Figure 4.21 Reprinted, by permission, from R.D. De Leon, J.A. Hodgson, R.R. Roy, and V.R. Edgerton, 1998, "Locomotor capacity attributable to step training versus spontaneous recovery after spinalization in adult cats," *Journal of Neurophysiology 79:* 1333.

Figure 4.22 Reprinted, by permission, from J.A. Hodgson et al., 1994, "Can the mammalian lumbar spinal cord learn a motor task?" *Med. & Sci. Sports Exerc. 26:* 1494.

Table 4.2 Reprinted, by permission, from A. Adam, C.J. De Luca, and Z. Erim, 1998, "Hand dominance and motor unit firing behavior," *Journal of Neurophysiology 80*.

Figure 5.1 Reprinted, by permission, from S.M. Phillips, K.D. Tipton, A. Aarsland, S.E. Wolf, and R.R. Wolfe, 1997, "Mixed muscle protein synthesis and breakdown after resistance exercise in humans," *American Journal of Physiology, Endocrinology, and Metabolism 273*: E103.

Figure 5.3 Reprinted, by permission, from D. Aronson, S.D. Dufresne, and L.J. Goodyear, 1997, "Contractile activity stimulates the c-Jun NH$_2$-terminal kinase pathway in rat skeletal muscle," *J. Biol. Chem. 272*: 25638.

Figure 5.4 Reprinted, by permission, from N.J. Osbaldeston et al., 1995, "The temporal and cellular expression of *c-fos* and *c-jun* in mechanically stimulated rabbit latissimus dorsi muscle," *Biochemical Journal 308*: 465-471. © 1995 the Biochemical Society.

Figure 5.5 Reprinted, by permission, from N.J. Osbaldeston et al., 1995, "The temporal and cellular expression of *c-fos* and *c-jun* in mechanically stimulated rabbit latissimus dorsi muscle," *Biochemical Journal 308:* 465-471. © 1995 the Biochemical Society.

Figure 5.6 Reprinted, by permission, from K. Baar and K. Esser, 1999, "Phosphorylation of p70^{s6k} correlates with increased skeletal muscle mass following resistance exercise," *American Journal of Physiology-Cell Physiology 276*: C124, C125.

Figure 5.7 Reprinted, by permission, from H. Fridén and R.L. Lieber, 1998, "Segmental muscle fiber lesions after repetitive eccentric contractions," *Cell and Tissue Research 293*: 170. © 1998 Springer-Verlag.

Figure 5.8 Reprinted, by permission, from G.R. Adams, B.M. Hather, K.M. Baldwin, and G.A. Dudley, 1993, "Skeletal muscle myosin heavy chain composition and resistance training," *Journal of Applied Physiology 74*: 913.

Figure 5.9 Reprinted, by permission, from J.D. MacDougall, D.G. Sale, S.E. Alway, and J.R. Sutton, 1984, "Muscle fiber number in biceps brachii in bodybuilders and control subjects," *Journal of Applied Physiology 57*: 1402.

Figure 5.10 Reprinted, by permission, from D.L. Allen, S.R. Monke, R.J. Talmadge, R.R. Roy, and V.R. Edgerton, 1995, "Plasticity of myonuclear number in hypertrophied and atrophied mammalian skeletal muscle fibers," *Journal of Applied Physiology 78*: 1975.

Figure 5.11 Reprinted, by permission, from J. Duchateau and K. Hainaut, 1984, "Isometric or dynamic training, differential effects on mechanical properties of a human muscle," *Journal of Applied Physiology 56*: 297, 298.

Figure 5.12 Reprinted, by permission, from M.V. Narici, G.S. Roi, L. Landoni, A.E. Minetti, and P. Cerretelli, 1989, "Changes in force, cross-sectional area, and neural activation during strength training and detraining of the human quadriceps," *Eur. J. Appl. Physiol. 59:* 314. © 1989 Springer-Verlag.

Figure 5.13 Reprinted, by permission, from L.L. Ploutz, P.A. Tesch, R.L. Biro, and G.A. Dudley, 1994, "Effect of resistance training on muscle use during exercise," *Journal of Applied Physiology 76*: 1678.

Figure 5.14 Adapted, by permission, from T. Hortobágyi, J.P. Hill, J.A. Houmard, D.D. Fraser, N.J. Lambert, and R.G. Israel, 1996, "Adaptive responses to muscle lengthening and shortening in humans," *Journal of Applied Physiology 80*: 768.

Figure 5.15 Reprinted, by permission, from G. Yue and K. Cole, 1992, "Strength increases from the motor program: comparison of training with maximal voluntary and imagined muscle contractions," *Journal of Neurophysiology 67*: 1117.

Figure 5.16 Reprinted, by permission, from B. Carolan and E. Cafarelli, 1992, "Adaptations in coactivation after isometric resistance training," *Journal of Applied Physiology 73*: 914.

Figure 5.17 Reprinted, by permission, from M. Van Cutsem, J. Duchateau, and K. Hainaut, 1998, "Changes in single motor unit behaviour contribute to the increase in contraction speed after dynamic training in humans," *Journal of Physiology 513:* 298, 300.

Figure 6.1 Reprinted, by permission, from F.W. Booth and C.R. Kirby, 1992, "Changes in skeletal muscle gene expression consequent to altered weight bearing," *American Journal of Physiology. Regulatory, Integrative and Comparative Physiology 262:* 330

Figure 6.2 Reprinted, by permission, from R.N. Michel, G. Cowper, M.M-Y Chi, J.K. Manchester, H. Falter, and O.H. Lowry, 1994, "Effects of tetrodotoxin-induced neural inactivation on single muscle fiber metabolic enzymes," *American Journal of Physiology- Cell Physiology 267*: C58, C60.

Figure 6.3 Reprinted, by permission, from B. Cormery, F. Pons, J.-F. Marini, and P.F. Gardiner, 2000, "Myosin heavy chains in fibers of TTX-paralyzed rat soleus and medial gastrocnemius muscles," *Journal of Applied Physiology 88*: 71.

Figure 6.4 Reprinted, by permission, from P.F. Gardiner, M. Favron, and P. Corriveau, 1992, "Histochemical and contractile responses of rat medial gastrocnemius to 2 weeks of complete disuse," *Can. J. Physiol. Pharmacol. 70*: 1075-1081.

Figure 6.5 Reprinted, by permission, from B. Jiang, R.R. Roy, C. Navarro, Q. Nguyen, D. Pierotti, and V.R. Edgerton, 1991, "Enzymatic responses of cat medial gastrocnemius fibers to chronic inactivity," *Journal of Applied Physiology 70*: 236.

Figure 6.6 Reprinted, by permission, from R.R. Roy, D.J. Pierotti, V. Flores, W. Rudolph, and V.R. Edgerton, 1992, "Fibre size and type adaptations to spinal isolation and cyclical passive stretch in cat hindlimb," *Journal of Anatomy 180*: 493.

Figure 6.7 Reprinted, by permission, from R.R. Roy, D.J. Pierotti, V. Flores, W. Rudolph, and V.R. Edgerton, 1992, "Fibre size and type adaptations to spinal isolation and cyclical passive stretch in cat hindlimb," *Journal of Anatomy 180:* 496.

Figure 6.10 Reprinted, by permission, from R.J. Talmadge, R.R. Roy, and V.R. Edgerton, 1996, "Distribution of myosin heavy chain isoforms in non-weight-bearing rat soleus muscle fibers," *Journal of Applied Physiology 81*: 2545.

Figure 6.11 Reprinted, by permission, from H.E. Berg, G.A. Dudley, T. Häggmark, H. Ohlsén and P.A. Tesch, 1991, "Effects of lower limb unloading on skeletal muscle mass and function in humans," *Journal of Applied Physiology 70*: 1883.

Figure 6.12 Reprinted, by permission, from L.L. Ploutz-Snyder, P.A. Tesch, D.J. Crittenden, and G.A. Dudley, 1995, "Effect of unweighting on skeletal muscle use during exercise," *Journal of Applied Physiology 79*: 170.

Figure 6.13 Reprinted, by permission, from A.A. Ferrando, H.W. Lane, C.A. Stuart, J. Davis-Street, and R.R. Wolfe, 1996, "Prolonged bed rest decreases skeletal muscle and whole body protein synthesis," *American Journal of Physiology, Endocrinology, and Metabolism 270*: E630.

Figure 6.14 Reprinted, by permission, from L. Larsson, X.P. Li, H.E. Berg, and W.R. Frontera, 1996, "Effects of removal of weight-bearing function on contractility and myosin isoform composition in single human skeletal muscle cells," *Pflügers Archive 432*: 323.

Figure 6.15 Reprinted, by permission, from J. Duchateau, 1995, "Bed rest induces neural and contractile adaptations in triceps surae," *Med. Sci. Sports Exerc. 27*: 1583.

Figure 6.16 Reprinted, by permission, from J. Duchateau, 1995, "Bed rest induces neural and contractile adaptations in triceps surae," *Med. Sci. Sports Exerc. 27*: 1586.

Figure 6.17 Reprinted, by permission of the publisher, from S.A. Spector et al., 1982, "Architectural alterations of rat hind-limb skeletal muscles immobilized at different lengths," *Experimental Neurology 76:* 104. © 1982 by Academic Press.

Figure 6.18 Reprinted, by permission, from H. Jänkälä, V.P. Harjola, N.E. Petersen, and M. Härkönen, 1997, "Myosin heavy chain mRNA transform to faster isoforms in immobilized skeletal muscle: A quantitative PCR study," *Journal of Applied Physiology 82*: 980.

Figure 6.19 Reprinted, by permission, from C.T.M. Davies, L.C. Rutherford, and D.O. Thomas, 1987, "Electrically evoked contractions of the triceps surae during and following 21 days of voluntary leg immobilization," *European Journal of Applied Physiology 56:* 308, 309. © 1987 Springer-Verlag.

Figure 6.20 Reprinted, by permission, from J. Duchateau and K. Hainaut, 1987, "Electrical and mechanical changes in immobilized human muscle," *Journal of Applied Physiology 62*: 2171.

Figure 6.21 Reprinted, by permission, from J. Duchateau and K. Hainaut, 1987, "Electrical and mechanical changes in immobilized human muscle," *Journal of Applied Physiology 62*: 2169.

Table 6.1 Reprinted, by permission, from P.F. Gardiner, M. Favron, and P. Corriveau, 1992, "Histochemical and contractile responses of rat medial gastrocnemius to 2 weeks of complete disuse," *Can. J. Physiol. Pharmacol. 70:* 1075-1081.

Table 6.2 Reprinted, by permission, from G.M. Diffee, V.J. Caiozzo, R.E. Herrick, and K.M. Baldwin, 1991, "Contractile and biochemical properties of rat soleus and plantaris after hindlimb suspension," *American Journal of Physiology- Cell Physiology 260*: C350.

Table 6.4 Reprinted, by permission, from B.M. Hather, G.R. Adams, P.A. Tesch, and G.A. Dudley, 1992, "Skeletal muscle responses to lower limb suspension in humans," *Journal of Applied Physiology 72*: 1495.

Table 6.5 Adapted, by permission, from J. Duchateau and K. Hainaut, 1990, "Effects of immobilization on contractile properties, recruitment and firing rates of human motor units," *Journal of Physiology 422*: 59.

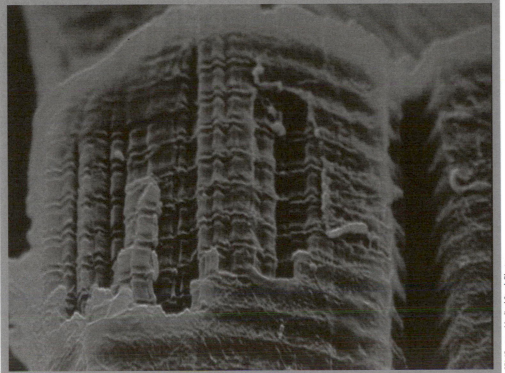

Muscle Fiber Types

"**M**uscular power is (other circumstances being equal) proportioned to the size of the muscle; but it often happens that great power is required where bulk of muscle would be inconvenient or cumbersome. In such cases, the muscle is supplied with an increased endowment of nervous filaments, which compensate, by the strength of stimulus, for what it wants in bulk of fibre."

A. Combe, *The Principles of Physiology Applied to the Preservation of Health and to the Improvement of Physical and Mental Education*, 1843

Most Important Concepts From This Chapter

1 Myosin heavy-chain (MHC) type composition is used most often to distinguish the type of muscle fiber. MHCs found in human limb muscle fibers include types I, IIa, and IIx.

2 Fiber properties that are determined primarily by the MHC include maximal shortening velocity, the shape of the force–velocity curve, maximal tension per unit of cross section, and the energetic cost of contraction.

3 The composition of isoforms of other contractile proteins (myosin light chains, troponin, tropomyosin) exerts modulatory effects on contractile function.

4 Many combinations of the sarcomeric proteins are present among fibers within a muscle, resulting in a significant variability in contractile characteristics among fibers.

5 Correlations between measures of muscle fiber composition and athletic performance are generally modest.

We accomplish movements of our limbs by contraction of muscles. The gross anatomy of muscles obviously is destined to play a major role in the patterns of movement of which the crossed joints are capable. Factors such as muscle size and the point of origin and insertion of the muscle relative to the axis of rotation of the joint affect the strength and speed of movement. Since muscles contain many individual muscle fibers (for example, there are approximately 200,000 fibers in the human biceps brachii), the geometrical relationships between their individual vectors of force generation during shortening and the whole-muscle line of pull provide an additional variable affecting movement. For example, muscles that have a pennate arrangement (arranged at an angle not completely in line with the angle of force generation between the proximal and distal attachments of the muscle) of short muscle fibers are potentially stronger, but slightly slower in shortening speed, than muscles of the same weight in which fibers are fusiform (run the entire length of the muscle along its line of force generation). This is because there are more, shorter fibers in a pennate muscle, thus more fibers in parallel for force generation, which are each slower in shortening distance per unit time

because of their shorter length. Human muscles show a wide range of pennateness of structure, from the fusiform (e.g., the sartorius) to the highly pennate (e.g., the quadriceps femoris, which is multipennate).

These considerations of gross muscle anatomy are informative regarding the potential for the performance of strong or fast movements. However, voluntary movements generally involve the activation of muscle fibers via the nervous system such that the resulting patterns of contractile activity are often submaximal, both in the numbers of fibers recruited and in the forces generated, and the individual forces generated by the fibers are relatively unsynchronized. While the potential for movement can be estimated by examining muscle structure, the actual characteristics of the movements are dictated by such neuromuscular considerations as the variability among the properties of individual fibers, the numbers and sizes of the individual motor units (an alpha-motoneuron and the muscle fibers it innervates), and the patterns with which these units are activated during the movements. For this reason, the performance of voluntary movements is more often than not a function of the characteristics of the subvolume of fibers participating in the movement and the patterns with which the nervous system recruits them. It is thus informative to study the variability that exists among the individual fibers of a muscle and how the innervating motor nerves orchestrate the recruitment of these different fibers to generate movements.

It is most likely that no two muscle fibers are exactly the same within the same muscle. If we extend this to a comparison of fibers from different skeletal muscles, including those muscles involved in chewing (masticatory muscles), seeing (eye muscles), and breathing (intercostals and diaphragm), as well as among individuals of various ages, we would soon be overwhelmed by the variety of morphological, biochemical, and functional properties of fibers that are present in the human body. Because our concern is chiefly with the primary muscles involved in exercise, thus those of the limbs, our task is a bit more manageable, since we thereby eliminate some of the more "exotic" muscle fiber phenotypes. Similarly, we are going to limit our discussion to adult muscle fibers, since embryonic and neonatal muscle fibers have unique protein profiles.

We find it simpler to understand variability in nature if we can find a basis upon which to categorize or "pigeonhole" large amounts of data, as did Carolus Linnaeus in the 1750s, when publication of his *Systema Naturae* described his system for classifying plants depending on shared characteristics. Classic ways in which muscle fibers have been grouped include fast versus slow, red versus white, and tonic versus phasic. During the last 30 years, histochemical and, more recently, immunological and immunohistochemical techniques have permitted researchers to classify fibers based on the presence or absence of specific proteins that determine specific functional fiber properties.

The following discussion is not meant to detail all the differences that exist among various fibers resident within a muscle. Rather, I have attempted to highlight structural and functional properties and their interrelationships, which are major players in creating functional heterogeneity among muscle fibers, which in turn allows us the plasticity of movement schemes that typify physical activity.

A list of proteins that vary among different fibers is presented in table 1.1. Fibers can be classified according to qualitative and quantitative differences in the proteins that are important in determining their functional properties. The most important property for grouping fiber types at this time appears to be the myosin heavy-chain isoform, and we thus begin our analysis with a look at how variability in this property among fibers is functionally important. The functional implications of variability in the other proteins listed in table 1.1 and the extent to which they covary with the myosin heavy-chain isoform are also considered further on in this chapter.

Grouping Fibers by Myosin Heavy-Chain (MHC) Composition

CONCEPT

1 Traditionally, and probably most appropriately, muscle fibers have been classified according to the myosin heavy-chain (MHC) protein that they possess. As of this writing, 10 MHC species have been identified in mammalian extrafusal muscle fibers (Moss, Diffee, and Greaser 1995; Pette and Staron 1997). In the limb muscles of adult mammals, including humans, myosin heavy-chain species include types I, IIa, IIx (also termed IId), and IIb. It has now been determined, based on cDNA nucleotide sequencing, that the fiber type normally classified as IIB in human limb muscles, and thus presumably containing the MHC IIb isoform, actually contains an isoform more closely resembling the MHC IIx present in rats (Smerdu et al. 1994). For these reasons, all references throughout the book to IIB fibers in human limb muscles assume that they contain MHC IIx and therefore are equivalent to the IIX fiber seen in other mammals.

The type I isoform is also referred to as the beta-cardiac isoform, which is identical to that expressed in ventricle (Schiaffino and Reggiani 1996). In addition, there probably exist in normal adult human muscle fibers more than one type I myosin heavy chain that have escaped detection using standard myofibrillar ATPase–based and immunohistochemical techniques (Sant'ana Pereira, Wessels, et al. 1995). These include an alpha-cardiac-like MHC (MHC I) and MHC Ia, both of which have been found in rabbit skeletal muscles (Galler, Hilber, et al. 1997; Peuker, Conjard, and Pette 1998).

Table 1.1

Major Protein Isoforms of Thick and Thin Filaments and Sarcoplasmic Reticulum

Protein	Type II fibers	Type I fibers
Myosin heavy chain	MHC IIa MHC IIx MHC IIb	MHC Iβ MHC Iα MHC Ia
Myosin light chains		
Regulatory	MLC 2f (fast)	MLC 2s (slow)
Alkaline (essential)	MLC 1f MLC 3f	MLC 1sa MLC 1sb
Tropomyosin	Tm αfast Tm β	Tm αslow/cardiac Tm β
Troponin C	TnC fast	TnC slow/cardiac
Troponin I	TnI fast	TnI slow
Troponin T	TnT 1f TnT 2f TnT 3f TnT 4f	TnT 1s TnT 2s
Myosin-binding protein C	MyBP-C$_{fast}$	MyBP-C$_{slow}$
Sarcoplasmic reticulum ATPase	SERCA 1a	SERCA 2a

Pette and Staron 1993, 1997; Schiaffino and Reggiani 1996; Pette 1998; Russell, Motlagh, and Ashley 2000.

These various isoforms of the myosin heavy chain are the consequence of their encoding by a multigene family. It is highly likely that not all MHCs have been identified, since this multigene family probably includes many more genes than the 10 MHC isoforms that have been identified. In the chicken, for example, the MHC family includes at least 31 genes (Moss, Diffee, and Greaser 1995; Pette 1998). The genes that encode for these myosin heavy chains show a high degree of similarity (80–95%) in nucleotide sequence across species (Reggiani, Bottinelli, and Stienen 2000), such that monoclonal antibodies produced by one species against a given myosin heavy chain can often, but not always, be used successfully for immunohistochemistry in another species.

The differences in fiber types among different species must be appreciated, however, in order to understand the extent to which animal experimentation can be extrapolated to humans. Since the classic work of Barany (1967), who showed an inverse relationship between the ATPase of myosin and body size across several species, we have understood that a slow fiber in a mouse is not the same as a slow fiber in a cat or a human. While four MHCs have been found in limb muscles of smaller mammals, only three have so far been found in humans. Differences in the myosin molecules of the same fiber types in different species are evident from several techniques. For example, when separation of MHC is accomplished using electrophoretic gels, rat MHC IIb and IIx migrate faster than IIa, while the reverse is true for humans (Sant'ana Pereira, Wessels, et al. 1995; Bottinelli et al. 1996). These differences express themselves when classic myofibrillar ATPase (mATPase)–based histochemistry is used. In rat muscle, for example, IIB fibers stain lighter than IIX after alkaline preincubation, while the reverse is true for rabbit muscles (Hämäläinen and Pette 1995). In addition, these fibers can be separated after acid preincubation in the rabbit but not in the rat. Finally, these species-specific responses are in evidence when antibody techniques are used as well. For example, the monoclonal antibody that reacts with all but IIX fibers in the rat reacts with 50% of IIX fibers in humans (Smerdu et al. 1994). These authors contend that there still may exist additional uncharacterized myosin heavy chains in human muscle to explain this type of irregularity.

Fibers as Hybrids Containing More Than One Myosin Heavy Chain

Fibers exist either as "pure," containing only one type of myosin heavy chain, or as hybrids, containing multiple forms. In humans, for example, the following combinations of myosin heavy chains have been found in the same fiber: I + IIa, I + IIa + IIx, and IIa + IIx. Whether or not these combinations represent the presence of different homodimers (myosin molecules formed by two MHCs of the same isoform) or heterodimers, or both, is not known at present.

As we shall see in a later chapter, the distribution of these hybrids in a muscle can tell us something about the sequence of adaptations that occur when a fiber is changing its MHC composition as a result of changes in its chronic activity level. For example, the fact that the combination I + IIb is very rarely found suggests that the transition of MHC composition from IIb to I (or the reverse) does not take place in the absence of an intermediate step, probably IIa or IIx MHC expression or both (referred to as the "next-neighbor rule"; see Stevens, Sultan et al. 1999). However, under certain circumstances, odd combinations of MHC proteins and their corresponding mRNAs, such as I + IIb, can be found. For example, this combination of MHC protein is found in the rat soleus after a period of no weight bearing under hyperthyroid conditions (Caiozzo, Baker, and Baldwin 1998).

Histochemistry-Based Fiber Typing Using Myofibrillar ATPase

Knowing the MHC profile within individual fibers has proven extremely helpful in the interpretation of physiological and pathological muscle adaptations and in the development of principles of motor unit recruitment during voluntary movement. The initial technical approach used to distinguish muscle fibers in cross sections according to their MHC complement used an enzymatic reaction involving myofibrillar ATPase coupled with a reaction producing a color change in the fiber.

The initial studies of fiber types in humans used the differences in acid and alkaline lability of the enzyme myofibrillar ATPase, which is an integral part of the myosin molecule residing in the S1 head region. Staining characteristics depend on heavy-chain composition and are not influenced by myosin light-chain complement (Billeter et al. 1981). The procedures technically involve preincubation of muscle serial sections at various acid and alkaline levels before submitting sections to incubation for the demonstration of myofibrillar ATPase activity (figure 1.1). Fibers are named according to the primary heavy chain present: type I, IIA, IIAB, and IIB. (In fact, MHCs were originally given their respective names according to the type of fiber in which they were found. Thus, the nomenclature of fiber type proposed by Brooke and Kaiser in the early 1970s preceded, and is the original basis of, that of the MHC.) A fiber type designated as *C* according to the original nomenclature scheme has since been shown to possess varying proportions of the MHCs I and IIa. Thus, types IC and IIC contain a predominance of type I and IIa MHCs, while type IIAC contains equal proportions of these two MHCs.

Although many animal muscles contain all four major MHCs to which we have referred, the standard ATPase histochemical techniques could not distinguish IIB from IIX fibers. Techniques have been developed in recent years, however, that involve formaldehyde fixation of some of the serial sections before the standard ATPase procedure, which appears to permit the distinction of IIB from IIX fibers (L. Martin et al. 1993; Lind and Kernell 1991).

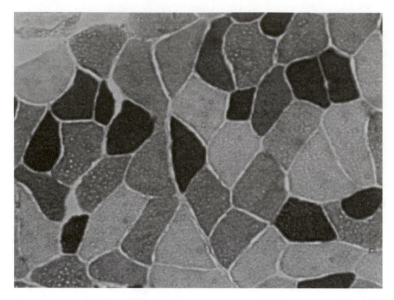

Figure 1.1 Photomicrograph of a cross section (10 μm thick) from human vastus medialis, treated for the demonstration of myofibrillar ATPase, following a preincubation treatment at pH 4.7. Darkest fibers are type I, lightest are IIA, and all others are IIX or IIAX. Width of entire image is 915 μm.

Myofibrillar ATPase Histochemistry vs. MHC Immunohistochemistry

The myofibrillar ATPase–based techniques require a subjective judgment of the color intensity of each fiber by the experimenter, who then makes corresponding judgments as to the MHCs present in the fiber. Several questions arise about the precision of these techniques. Fiber-type determination requires exact replication of all parameters during the reactions. For example, slight variations in preincubation times, pH, temperature during the reaction, washing procedures, and even the age of the reagents can alter color development. Another issue of major significance is how fibers containing proportions of several different MHCs will respond in the color reaction.

In fact, the myofibrillar ATPase–based histochemical techniques, with some modifications over the years, have been shown to be quite sensitive in detecting the types of myosin heavy chains present in the fiber, as verified in serial cross sections treated with antibodies against the specific myosin heavy chains or by gel electrophoresis of single fibers (Sant'ana Pereira, De Haan, et al. 1995; Sant'ana Pereira, Wessels, et al. 1995). However, myofibrillar ATPase–based

techniques are not as definitive in discerning fiber types as techniques using immunohistochemistry with antibodies specific for the various myosin heavy chains. Immunohistochemical techniques detect the presence of very small amounts of the protein and do not rely on the subjective judgment of color intensity. For example, fibers containing admixtures of type I and IIa myosin heavy chains frequently react as pure type I fibers on mATPase techniques (Klitgaard, Zhou, and Richter 1990). In addition, fibers containing types I + IIa + IIx cannot be distinguished using the mATPase procedure (Sant'ana Pereira, Wessels, et al. 1995). Even assuming that the previously designated type IIB fiber in humans actually expresses MHC IIx, fibers expressing IIx transcripts do not all behave the same on the acid preincubations used to distinguish the different type II subgroups (Smerdu et al. 1994).

A very good relationship has been shown between the optical staining intensity of the myofibrillar ATPase reaction and the combination of MHCs IIa and IIx in the fiber as determined using single-fiber electrophoresis (figure 1.2). This technique of determining the proportions of IIa and IIx MHC using measurement of color intensity requires, however, that one have assurance a priori that pure type IIA and IIX fibers are present in the section being analyzed to allow the range of optical density values to be set. Of course, this would not be necessary if the color reaction exhibited high reproducibility.

Finally, mATPase-based techniques are not sensitive to the appearance of novel isoforms that are not normally present but that might appear in cases such as inactivity, aging, nerve or muscle disease, or exercise training. These novel proteins could nonetheless be discerned using an antibody developed for this purpose.

Groups at University of London and Free University of Amsterdam have provided the most exhaustive information to date concerning the fiber types present in human muscle. Their approach (Ennion et al. 1995; Sant'ana Pereira, Wessels, et al. 1995; Sant'ana Pereira et al. 1997) has been to freeze-dry biopsy material and sub-

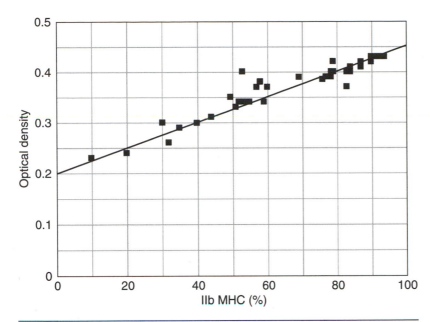

Figure 1.2 Measured optical density (OD) of human muscle fibers after preincubation at pH 4.6 and the proportion of MHC IIb (actually IIx) in mixed fibers.

Reprinted from Sant'ana Pereira et al. 1995.

tigators were able to resolve several issues relating to fiber types in human locomotor muscles.

First, they showed that the myofibrillar ATPase techniques for determination of fiber types, when performed in a rigorous and standardized fashion, can detect the myosin heavy-chain species in the fiber in the majority of cases (< 9 times out of 10; figure 1.3). In particular, they found a high correlation between the proportions of myosin heavy chains IIa and IIx in hybrid fibers and their optical density using quantitative image analysis techniques (see figure 1.2). Second, in the majority of cases, the presence of a particular myosin heavy chain was mirrored by the presence of the corresponding mRNA transcript. However, examples were found where the protein but not the corresponding mRNA was present, and vice versa. This phenomenon would be expected in fibers that were in the process of changing the expression of their heavy chains. Consistent with this idea, fibers in which the MHC mRNA does not match the corresponding protein are found in higher abundance in fibers of humans following endurance training and during the detraining period, and involve primarily IIA and IIX fibers (Andersen and Schiaffino 1997).

ject each of the subsequently dissected fibers to a variety of procedures, including myofibrillar ATPase histochemistry, immunohistochemistry with antibodies specific to the myosin heavy chains, single-fiber electrophoresis and Western blotting to quantify and identify MHCs, and determination of the mRNA species in the fiber at the time of sampling. In their experiments, isolation of the mRNA from small fiber segments allowed cloning of the gene corresponding to these mRNAs to yield many copies of the complementary DNA (cDNA), using reverse transcriptase polymerase chain reaction (RTPCR). This cDNA can then be sequenced. These inves-

Third, these investigators showed, using immunohistochemistry, that the IIB fiber in the human expresses a myosin heavy chain similar to the IIX of the rat and is therefore actually a

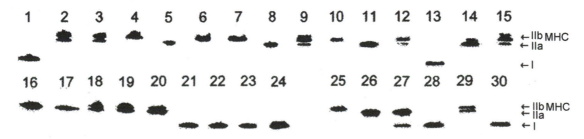

Figure 1.3 MHC in single fibers of human muscles (lane 1 is soleus, all others vastus lateralis). Mixtures of IIa + IIb MHC are frequent (lanes 2, 3, 9, 14, 15, 18–20, 29), while I + IIa mixtures are rare (lane 27). The I + IIb combination was not found.

Reprinted from Larsson and Salviati 1992.

IIX, not a IIB, fiber. This was confirmation of the findings of Smerdu and colleagues (1994), who reported that the nucleotide sequence in the untranslated region of the rat IIx and human IIb MHC genes showed 87% homology, while the rat and human (presumptive) IIb genes showed only 65% homology. Sant'ana Pereira and colleagues (Sant'ana Pereira, De Haan, et al. 1995; Sant'ana Pereira, Wessels, et al. 1995; Sant'ana Pereira et al. 1997) demonstrated specifically that the human IIB fiber responds like a rat IIX when subjected to incubation conditions designed to demonstrate the IIX fiber in the muscles of rats. This explains why humans appear to have only three fiber types, whereas mice and rats have four—humans are missing the IIB fiber.

Finally, these investigators were able to demonstrate nonhomogeneity of myosin heavy-chain expression along the length of some fibers. This finding, substantiating similar previous observations from their own laboratory and those of others (Staron 1991), raises the possibility that different portions of the same fiber may alter their proteins at different times, perhaps depending on the distance of the fiber segment from the nerve terminal, the adequacy of the available blood supply, or the vectors of stress and strain to which the fiber is subjected. To some extent, this is a disturbing finding for investigators who would like to relate the performance of a single fiber to its MHC composition and for those who would like to relate whole-muscle composition estimated from histochemical analysis of a small biopsy sample to the performance of that muscle. As we shall see at the end of this chapter, the relationships between muscle fiber proportions and measurements of muscle performance tend to be poor, perhaps partially due to this problem of nonuniformity of MHC type along the length of a variable proportion of fibers within muscles.

Functional Properties of Fibers Containing Different Myosin Heavy-Chain Profiles

In this section, I include properties that seem to vary according to, and are probably determined by, the principal myosin heavy-chain species in the fiber, including fibers that contain a mixture of these chains (hybrid fibers). I begin this section with a short discussion of technical approaches used in determining the functional properties of single fibers.

Techniques Used to Determine Single-Fiber Functional Properties

One of the principal techniques that has been used to generate information about the functional properties of single fibers is known as the permeabilized single-fiber preparation, which allows the determination of speed- and strength-related properties of single fibers. The technique involves isolation of a single muscle fiber and the subsequent permeabilization of its outer membrane using chemicals, mechanical peeling-off of the membrane, or freeze-drying. Chemical permeabilization is accomplished by immersing single fibers in a glycerol solution, sometimes containing a detergent; mechanical peeling is performed by the investigator virtually peeling off the membrane using microinstruments while visualizing the fiber under a microscope. The skinning solution can be modified to keep the sarcoplasmic reticulum intact. Both ends of the permeabilized fiber are then attached to a sensitive measuring device, where one end is fixed and the other end either measures or controls the force or velocity, or both, of the fiber's shortening or lengthening. The fiber is immersed in various physiological bathing solutions, whose composition can be varied (by changing calcium concentrations, for example), and the responses of the fiber are monitored.

The advantages of this experimental approach in determining the functional properties of muscle fibers are numerous. First, the procedures can be performed on very small muscle samples (from biopsies) and on segments of fibers as opposed to whole fibers. Second, since the investigator is examining the properties of the basic enzymatic and contractile machinery of the fiber, the fiber does not have to be kept "alive" in the same sense that one would have to do in an *in situ* preparation with an anesthetized animal; in fact, samples are routinely kept in glycerol for several days before these experiments or are freeze-dried and kept in a cold environment. Third, in the standard preparation in which the sarcoplasmic reticulum and the sarcolemma are rendered nonfunctional, it is possible to investigate the contractile properties of the myofibrils without having to pass through these components, which could prove limiting in their function under certain circumstances.

A fourth advantage of this technique is that the myofibrils can be exposed quickly to changes in their immediate environment simply by rapid change of the solution bathing the fiber. Contractile properties are unhindered in this preparation by the damping effect of neighboring noncontracting fibers and connective tissue. Finally, the fiber can be recovered after the experiment and subjected to techniques aimed at quantifying the proteins in the fiber (such as MHC isoforms by single-fiber electrophoresis).

There are, however, caveats to interpreting the results obtained using this technique. The main limitation is that the fiber contractile events *in vivo* involve the summation of individual twitches, which cannot be reproduced in this system due to the lack of the components (sarcolemma and sarcoplasmic reticulum) that promote these singular events. In addition, measurements of fiber cross-sectional area are more difficult to make than those of histological cross sections and are influenced by the method used in the skinning procedure (areas from freeze-dried fibers are different from those in which chemical skinning has been used). As we shall see further on, this causes some problems in attempting to derive specific tensions (i.e., fiber tension per unit of cross-sectional area).

Single-Fiber Unloaded Shortening Velocity (V_o)

CONCEPT

2 One parameter that is determined to a large extent by MHC type and that has a significant impact on whole-muscle performance is the maximal shortening velocity of the sarcomeres. Two techniques, the slack test (discussed here) and the force–velocity relationship (discussed in the next section), have been used to estimate maximal shortening velocity at the single-fiber level.

Using a permeabilized fiber preparation, investigators have estimated the maximal velocity of unloaded shortening using the slack test.

In this test, the individual muscle fiber is subjected to maximal tetanic contraction by immersing it in a solution of sufficient calcium concentration to maximally activate myofibrils. After maximal tension has been reached at an optimal sarcomere length, the fiber is subjected to a series of instantaneous, small decreases in length (quick releases of less than 20% of resting length). The isometric force decreases in the released fiber, and the time required for the fiber to begin to register a measurable force at the new shorter length is plotted as a function of time for releases of several magnitudes (figure 1.4). The slope of the resultant straight line is termed the unloaded shortening velocity, or V_o. After this experiment, the protein from the single fiber can be subjected to gel electrophoresis and Western blotting with MHC antibodies to identify its myosin heavy-chain content.

The principle of the slack test can be understood by imagining an experiment in which we wish to determine the maximal speed capability

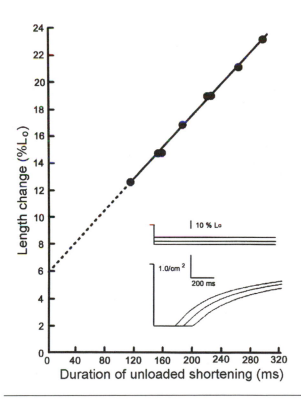

Figure 1.4 Determination of unloaded shortening velocity (V_o) using the slack test. The slope of the line is the maximal unloaded shortening velocity, in this case expressed as percentage of L_o per millisecond.

Reprinted from Reiser et al. 1985.

of a car that does not have a speedometer. One way to do this is to jack up the driving wheels of the car and tie one end of a rope 30 feet long to the bumper and the other end to a telephone pole. Now we start the car and, while driving the motor at its maximal capacity, kick out the jack and measure the time it takes for the slack to be removed from the rope (when it breaks). If we repeat this with several lengths of rope (60, 90, and 120 feet), we will eventually be able to draw a line by plotting time to rope breakage versus rope length. The slope of the line will be velocity (in feet per second). The results of several slack-test experiments are summarized in table 1.2.

Clearly, myosin heavy-chain composition has an impact on unloaded shortening velocity, in the direction I < IIa < IIx < IIb (figure 1.5). This is true for rats, rabbits, monkeys, and humans, even though the range of mean values among the fiber types increases as we progress from smaller to larger mammals. For example, in rats and rabbits, the mean shortening velocity of type IIX fibers is approximately three times faster than that of type I, whereas in humans it is 9 to 12 times faster (figure 1.6). This species difference reminds us again that the myosin heavy chains for a given fiber type, while reacting similarly to histochemical procedures and perhaps demonstrating reactivity to antibodies from another species, are indeed different proteins (species differences are discussed more in depth at the end of this chapter), often with quite different functional properties.

For fibers that contain mixtures of the myosin heavy chains, shortening velocities are in-

Table 1.2
Unloaded Shortening Velocity (V_o, in Fiber Lengths/s) of Single Fibers *in Vitro*

Species (source)	I	IIA	IIAX	IIX	IIXB	IIB
Rat (1)	0.6		1.1–1.6	1.7	1.9	1.9
Rat (2)	1.1	2.3		3.1		3.7
Rat (3)	1.1	2.5		3.3		3.6
Rat (4)	2.4			3.2	3.3–3.7	3.7
Rat (5)				1.6	2.1	2.4
Rabbit (6)	0.4	1.0				1.6–2.1
Monkey (7, 8)	0.7	– – – – – – – – – – – – –		4.3–5.1	– – – – – – – – – – – – –	
Human (9)	0.3	1.1	2.1	2.4		
Human (10)	0.3	1.1	1.9			
Human (11)	0.9	– – – – – – – – – – – – –		4.9	– – – – – – – – – – – – –	
Human (12)	0.4	1.9		5.6		
Human (13)	0.3	1.0	1.3	3.1		
Human (14)	0.3	1.2	1.9			

Horizontal dashed lines indicate no distinction among type II subgroups. Measurements taken at 12–15 °C. Notice the increasing difference between the slowest and fastest fiber types with increasing body mass. As explained in the text, human IIB is shown as IIX. References: (1) Galler, Schmitt, and Pette 1994; (2) Bottinelli, Betto, et al. 1994b; (3) Bottinelli, Canepari, et al. 1994; (4) Bottinelli, Betto, et al. 1994a; (5) Li and Larsson 1996; (6) Sweeney et al. 1988; (7) Widrick, Romatowski, Karhanek, et al. 1997; (8) Fitts et al. 1998; (9) Bottinelli et al. 1996; (10) Harridge et al. 1996; (11) Fitts, Costill, and Gardetto 1989; (12) Widrick, Trappe, Blaser, et al. 1996; (13) Larsson and Moss 1993; (14) Harridge et al. 1998.

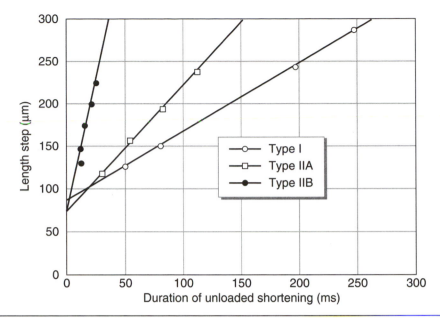

Figure 1.5   Slack-test results from three types of fibers in human quadriceps.

Reprinted from Larsson and Moss 1993.

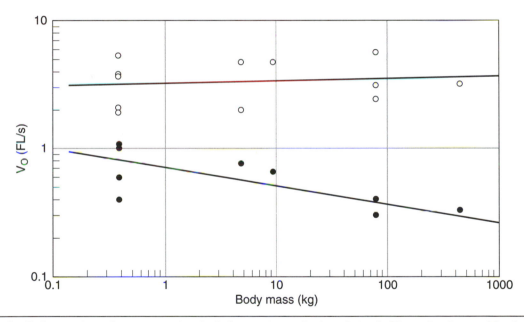

Figure 1.6   Relationships between unloaded shortening velocity of single fibers *in vitro* and body mass for type I fibers (filled circles) and the fastest fiber category (open circles), which is, for example, IIB in rats and IIX in humans. Notice how type I fibers, but not the fastest type II, get slower with increasing mass.

References for data can be found on page xi.

termediary, although it seems that the effect of mixtures on shortening velocity may not be a linear function of their proportions (Larsson and Moss 1993; figure 1.7).

Mean values do not give us the entire picture, however. Fibers containing the same MHC profile can differ substantially in V_o, and a signifi-

cant degree of overlap exists in V_o for fibers with different MHC profiles (figure 1.8). However, a report has shown no overlap in shortening velocities of fibers containing different MHCs in human gastrocnemius (figure 1.9). The wide range of values within each fiber type indicates either that other factors also contribute to

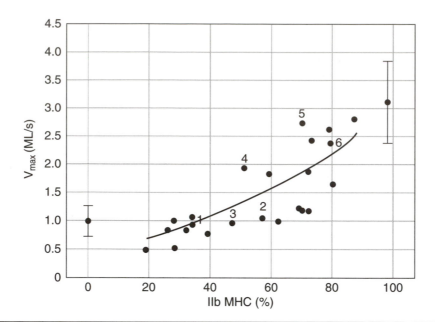

Figure 1.7 Maximal unloaded shortening velocity versus proportion of IIb MHC in human muscle fibers containing both IIa and IIb MHC.

Reprinted from Larsson and Moss 1993.

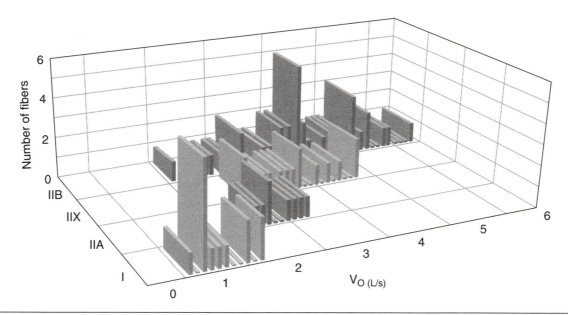

Figure 1.8 Distributions of V_o of fibers from rat soleus, plantaris, and tibialis anterior.

Adapted from Bottinelli et al. 1994.

determining V_o (such as myosin light-chain complement, discussed later) or that our measurement techniques are poor.

Single-Fiber Maximal Shortening Velocity (V_{max})

Another expression of maximal shortening velocity (V_{max}) is afforded by extrapolation of the

force–velocity curve to zero force. This procedure, like the measurement of unloaded shortening velocity, is done under maximum calcium-activating conditions. Muscle shortening is allowed such that the force is kept constant (clamped), and the velocity is measured early during the contraction. These points are used to generate a force–velocity curve, from which, among other things, one can estimate V_{max} as

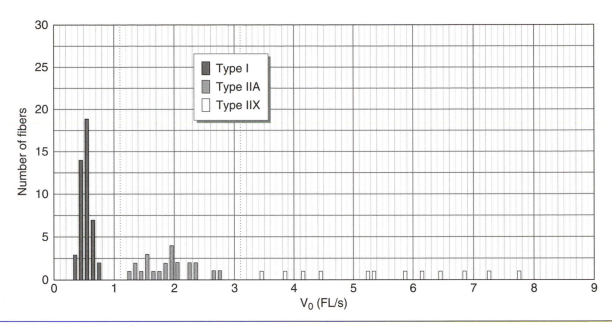

Figure 1.9   Distribution of V_o of fibers from human gastrocnemius.

Reprinted from Widrick et al. 1996.

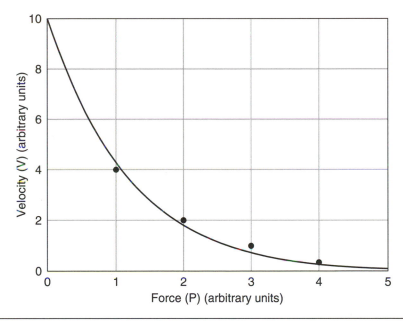

Figure 1.10   Force–velocity curve. The points fall along a rectangular hyperbola with the relationship $(P + a)(V + b) = C$.

Reprinted from Wohlfart and Edman 1994.

that point where the line extends to zero force. The force–velocity curve is described as part of a rectangular hyperbola, although it is probable that it is a composite of two hyperbolic functions located on either side of a break point near 80% of maximum isometric force (Wohlfart and Edman 1994). The formula is

$$(P + a)(V + b) = C \qquad (1.1)$$

where P is force, V is velocity, and a, b, and C are constants with the dimensions of force, velocity, and power (figure 1.10).

Determination of V_{max} using the force–velocity relationship results in values that are somewhat lower than values obtained using the slack test and vary less markedly among the fiber types. The reason for this discrepancy between the two methods is not clear. These two estimates of the velocity of sarcomere shortening, V_o and V_{max}, do not measure exactly the same thing, as witnessed by the facts that fibers do not have equal V_o and V_{max} values (V_o is usually higher) and that the V_o/V_{max} ratio is different among the different fiber types (Bottinelli et al. 1996; figure 1.11). It is thought that V_o may include a series elastic element not present during the measurement of V_{max}, while the latter, since it is

not a measured value but a value extrapolated using a series of points put into a mathematical relationship, may inherently be prone to error, especially because it is at the extreme edge of the curve. In addition, sarcomeres might be less uniform during loaded versus unloaded contractions (Widrick, Trappe, Costill, et al. 1996).

As with V_o, fibers with differing MHC complements show different V_{max} values, and a significant amount of variability is present within each fiber-type group (figure 1.12). There is, however, considerably less data available in the research literature pertaining to single-fiber V_{max} than there is on V_o.

It is probable that, for reasons given earlier, V_o is a more powerful index for determining the intrinsic shortening of sarcomeres than is V_{max}. However, the force–velocity curve used to generate V_{max} is of value for determining muscle fiber power at different shortening speeds, as discussed in the next section.

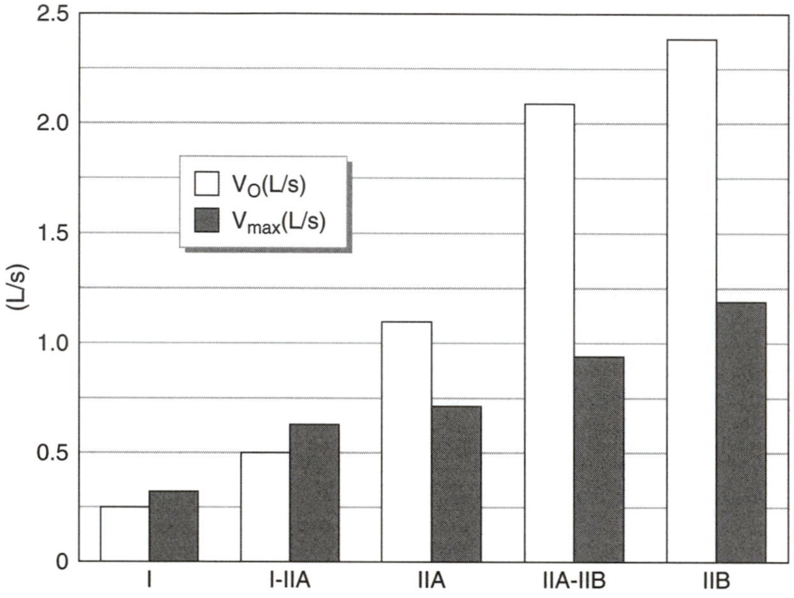

Muscle Fiber Power

Determination of the force–velocity relationship as outlined earlier also allows the determination of the

Figure 1.11 V_o versus V_{max} in human muscle fibers. The relationship is not the same across fiber types.

Adapted from Bottinelli et al. 1996.

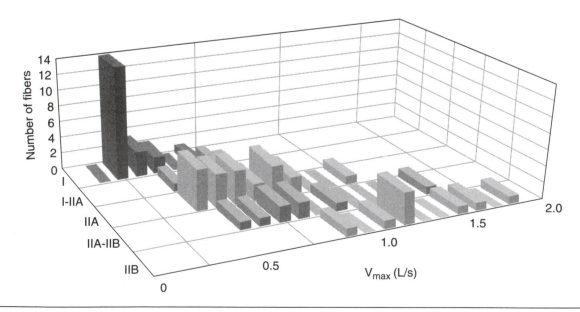

Figure 1.12 V_{max} of human muscle fibers.

Adapted from Bottinelli et al. 1996.

power–velocity, power–force, and peak power relationships among the different fiber types. By definition, maximum power will occur where, in figure 1.10, V/V_{max} is equal to P/P_o (i.e., where the product of force and velocity is highest), with P_o being maximum isometric tetanic force. For two muscle fibers with the same V_{max} and P_o, the one with the greater maximum power is the one with the straighter (less curved) force–velocity relationship. This degree of curvature of the force–velocity relationship is expressed as a/P_o; the higher this value, the less curved is the line describing this relationship and the higher the relative force at which peak power occurs.

Relative values for single-fiber peak power and a/P_o are different among the fiber types; type I fibers have the smallest a/P_o (and thus the largest curvature) and the lowest peak power, with the type IIX fibers situated at the other end of the continuum (figures 1.13 and 1.14).

A more appropriate and perhaps physiologically more pertinent measure of fiber shortening speed than V_{max} or V_o may be the speed of shortening under loaded conditions. In this situation, muscle fiber power can be determined. Rome and colleagues (Rome, Swank, and Corda 1993; Rome et al. 1988) have pointed out that fibers are generally recruited so as to generate contractions of peak power and thus they shorten at the velocity that accomplishes this end. More specifically, these authors estimated the speed of sarcomere shortening in muscles of carp during swimming at different speeds and compared these speeds to measurements of fiber V_{max} and V_o measured *in vitro*. They found that, during slow and fast swimming, slow and fast muscles respectively were recruited at or near those shortening speeds that yielded maximum mechanical power. An examination of optimal velocity (V_{opt}), or the velocity at which peak power occurs, shows a systematic variation among fiber types, which appears at least as pronounced as the differences in V_{max} or V_o. Perhaps more significantly, there was less overlap between the fiber types in V_{opt} than in V_{max} and V_o (figure 1.15). In addition, a systematic difference between type I and II fibers was found in the percentage of maximum velocity (V/V_{max}) at which peak power occurred.

Specific Tetanic Tension

Another property of individual fibers worth considering is the maximum tetanic tension per unit cross section, known as specific tetanic tension. The specific tetanic tension is of physiological significance for a number of reasons. First of all, a systematic difference in this property among fiber types would have to be

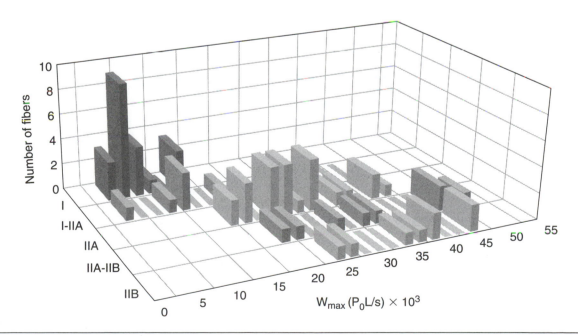

Figure 1.13   Peak power of human muscle fibers.

Adapted from Bottinelli et al. 1996.

considered in any attempt to relate the fiber composition of a muscle to its performance. Second, such a difference, if attributable to a difference in the density of myofibrillar proteins, would also have to be a consideration in interpreting the energy cost of contractions, as well as myofibrillar and myosin ATPase activities. However, the functional importance of this measurement may be limited: Fibers are not really perfect cylinders, after all.

The specific tetanic tension of a single fiber is not easy to determine. Some investigators

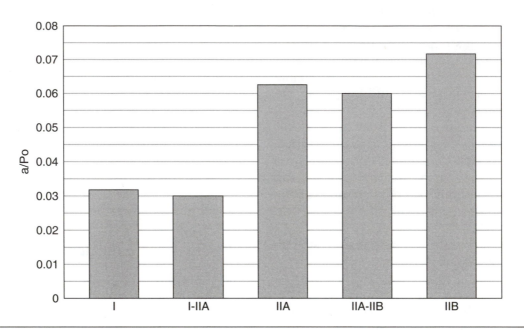

Figure 1.14 a/P_o in human muscle fibers. Higher values indicate less curvature to the force–velocity curve.

Adapted from Bottinelli et al. 1996.

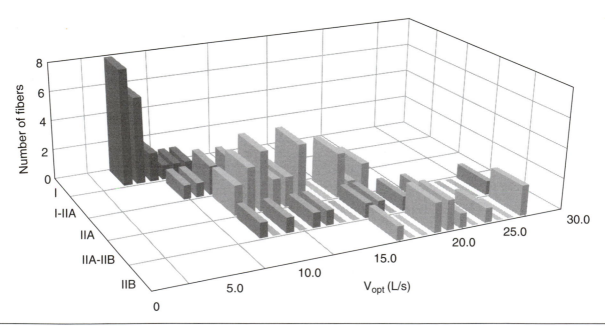

Figure 1.15 V_{opt}, or velocity at which peak power occurs, from single human fiber force–velocity measurements.

Adapted from Bottinelli et al. 1996.

have tried to determine this in *in situ* experiments by stimulating a single motor axon to generate a measurable tetanic force by its innervated muscle fibers, then depleting these fibers of glycogen to allow their identification in histological cross sections. The problem is to be sure that all stimulated fibers are identifiable in the sections, which is not an easy task, given that in most muscles that are pennate, all depleted fibers may not be present in a single cross section. These technical problems have produced differences in specific tetanic tension of more than fivefold between the slowest and fastest fiber types (this is discussed in more detail in chapter 2).

Fiber types appear to vary systematically in specific tetanic tension, in the order I < IIA < IIX. Specific tetanic tension values for human muscle fibers, expressed as a percentage increase over that of type I fibers, are summarized in figure 1.16. These values are taken from

studies where a permeabilized fiber preparation was used to measure functional properties and are therefore subject to the possible associated errors (as discussed earlier). Nonetheless, the results suggest that IIA fibers may be slightly stronger (20%) and type IIB fibers stronger still (50%) than type I fibers. These trends are similar to those found generally for mammalian species other than humans.

Myofibrillar ATPase Activity and Economy of Contraction

The energetic cost of contraction depends in part on the speed with which the cross-bridges consume ATP during contractions. Apart from the qualitative differences in acid and alkaline sensitivity of myofibrillar ATPase, which have been used to distinguish different fiber types, semiquantitative estimates of the maximal activity of the enzyme have also shown systematic

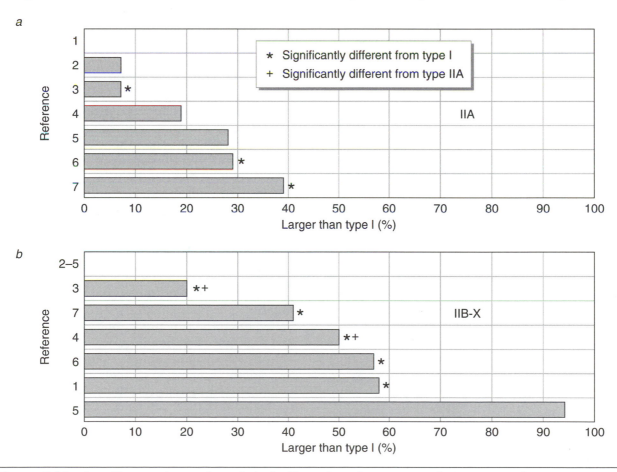

Figure 1.16 Specific tetanic tension of single fibers from human muscles, expressed as percentages above type I value. Asterisks indicate significantly different from type I, and crosses indicate significantly different from type IIA.

References for data can be found on page xi.

variation among the fiber types. This is to be expected, of course, since maximal shortening velocity reflects cross-bridge turnover rate, which in turn is greatly limited by the enzymatic activity of the ATPase.

Myofibrillar ATPase was measured quantitatively for the first time in single human muscle fibers (from vastus lateralis and rectus abdominis) by Stienen et al. (1996). They estimated the rate of hydrolysis of ATP during a maximum tetanic contraction in single fibers by coupling it to the oxidation of NADH, thus permitting them to follow this reaction spectrophotometrically. Their findings showed that the rate of ATP hydrolysis during a tetanic contraction in IIX fibers was slightly greater than four times that of type I fibers and 52% higher than IIA, with no significant difference found between type I and IIA.

Sant'ana Pereira et al. (1997) used a quantitative technique of calcium-activated myosin ATPase (ATPase at different concentrations, followed by a plot to determine the maximum velocity of the reaction) to determine the relative myosin ATPase of type I, IIA, and IIX fibers in the human vastus lateralis. They found type IIA and IIX fibers to have three times and four times, respectively, the estimated maximal mATPase of type I fibers; these results are thus similar to the results of Stienen et al. (1996). Values measured from human fibers were very similar to values in corresponding types in rat gastrocnemius, in which the maximal mATPase values for type IIB were almost five times that of type I fibers.

Although, as previously mentioned, we might expect a very good relationship between maximal ATPase activity and V_o or V_{max}, this does not appear to be the case within each fiber type (although a relationship still exists when mean ATPase for each fiber type is plotted against mean V_o; Bottinelli, Canepari, et al. 1994; figures 1.17 and 1.18). This lack of relationship within each fiber type may simply reflect the reproducibility of these rather sophisticated measurement techniques.

With knowledge of the ATPase hydrolysis rate during an isometric tetanic contraction and the force–time integral of the contraction, an investigator can estimate the energy cost of generating this type of contraction. Stienen et al. (1996) did this and found the energy cost for IIB fibers to be about four times that of type I fibers. Inter-

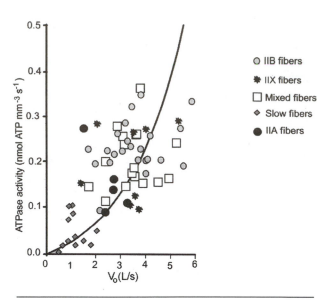

Figure 1.17   ATPase activity versus V_o in single rat fibers, using an *in vitro* NADH-linked assay.

Reprinted from Bottinelli et al. 1994.

estingly, the energy cost for isometric contractions is remarkably similar in rat and human fibers. For example, human-to-rat ratios for isometric tension cost are 0.85, 0.89, and 0.93 for types I, IIA, and IIX respectively (Stienen et al. 1996).

Thus, the type II fibers are stronger and faster contracting, but their contractions are more costly energetically than those of type I fibers. The order of recruitment of fibers as force increases is such that contractions become gradually less economical, since fibers are recruited gradually in the order I, IIA, IIAX, and finally IIX.

The Role of Myosin Light Chains

CONCEPT

3   Each myosin molecule possesses four light chains (MLC), two of which are called essential (for the stability of the myosin molecule) or alkaline (since they can be dissociated with alkaline solutions) and two of which are called regulatory, phosphorylatable, light chain 2, P-light chain, or DTNB (since they can be dissociated with 5,5'-dithio-2-nitrobenzoic acid). They are situated near the S1 subfragment of the heavy chain, with the essential light chains closer to the head region than the regulatory light chains (figure 1.19). Both MLCs are wrapped around the lever arm of the S1 subfragment, where they are in a position to trans-

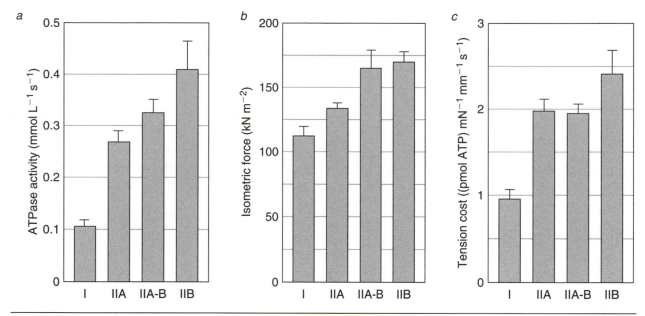

Figure 1.18 Influence of MHC composition on (*a*) myofibrillar ATPase activity, (*b*) isometric force, and (*c*) tension cost in rat muscle fibers.

Reprinted from Stienen et al. 1994.

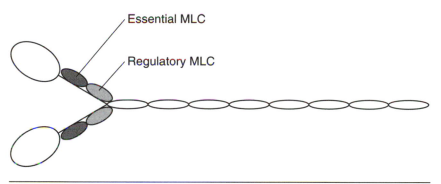

Figure 1.19 Location of myosin light chains on the myosin molecule.

Reprinted from Schiaffino and Reggiani 1996.

sin light chains for contractile function, using a unique motility assay in which myosin molecules are adsorbed by a nitrocellulose-coated coverslip and the rate of movement of fluorescently labeled actin filaments across them is measured via fluorescence microscopy. They showed that removal of light chains from the myosin molecules (both regulatory and essential) reduced to zero the velocity with which actin moved across myosin in their assay without influencing ATPase activity. When one or the other of the light chains was added, the shortening velocity was partially restored, and addition of both completely restored the original shortening velocity (figure 1.20). These investigators also demonstrated that the type of essential light chain added to the heavy-chain preparation appeared to influence shortening velocity. Two essential light-chain variants seen in fast muscles had different effects on the shortening velocity of actin filaments on myosin (their A2 variant moved actin about 1.5 times faster than the A1 variant). As we shall see in the following sections, the mechanisms by which

mit the movements of the S1 head to the filamentous backbone of the myosin molecule. Thus, they may play an important role in modulation of actin–myosin interactions and of the dynamics of the force stroke. There is some suggestion that their role might be one of stabilization of the myosin head during force development. The essential MLCs exist in several isoforms, two of which are found primarily in fast muscle fibers (MLC 1f, MLC 3f) and two in slow fibers (MLC 1sa, MLC 1sb). The regulatory light chains are present as fast (MLC 2f) or slow (MLC 2s) isoforms (Schiaffino and Reggiani 1996).

Lowey, Waller, and Trybus (1993a, 1993b) have demonstrated the necessity of the myo-

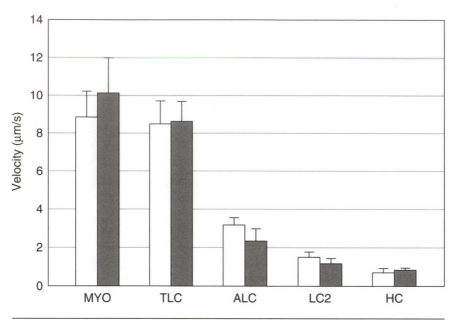

Figure 1.20   Velocity of movement of actin filaments along myosin molecules in an *in vitro* motility assay, in the presence of MHC only (far right, HC) and when light chains are added (LC2 = [regulatory] light chain 2; ALC = alkaline [essential] light chain; TLC = total light chains; MYO = native myosin). Myosin was either adsorbed directly by the nitrocellulose (dark bars) or was attached via an antibody (light bars).

Reprinted from Lowey, Waller, and Trybus 1993.

MLCs influence actin–myosin interactions during contractions are different for regulatory and essential light chains; the effect of essential light chains is modulated by the presence of several isoforms.

The Essential Light Chains

There is general consensus that the essential MLC isoform composition might have an influence *in vivo* on the maximum shortening velocity. Greaser, Moss, and Reiser (1988) showed that a reasonably good relationship ($r = 0.86$) existed between V_{max} and the proportion MLC 3f/(MLC 1s + MLC 1f) in single rabbit plantaris muscle fibers; however, their techniques did not permit simultaneous demonstration of the MHC composition of the same fibers. This point is significant because it has been demonstrated in single muscle fibers from the rat that the proportion MLC 3f/MLC 2f varies systematically as a function of MHC IIb/(IIb + IIx) (Li and Larsson 1996). Also, Sweeney and colleagues (1988) showed that the lower the relative content of the MLC 3f isoform, the lower the V_o measured *in vitro* in single pure IIB fibers (i.e., containing

only the IIb heavy chain) in rabbit tibialis anterior and psoas muscles. These investigators also reported differences among fiber types in their proportions of MLC 3f: IIB, IIA, and I fibers contained 38–48%, 26%, and 0%, respectively, of this MLC isoform. This trend has been confirmed for fast fibers in rat extensor digitorum longus (EDL), in which the ratio MLC 3f/(MLC 1f + MLC 3f) is 0.18, 0.23, and 0.32 for type IIA, IIX, and IIB fibers, respectively (Wada and Pette 1993). Similar data have been reported for fibers from rat plantaris, tibialis anterior (TA), and soleus (Bottinelli, Betto, et al. 1994b). However, these authors also indicated the variability around each of these means, which were from 0.05 to 0.35 for type IIA, from 0.05 to 0.55 for IIX, and from 0.105 to 0.65 for IIB. This variability is also present among human fibers; interestingly, means and ranges for this ratio are similar in human and rat fast fiber types (assuming as usual that previously reported human type IIB fibers are actually IIX).

The establishment of a cause–effect relationship between contraction velocity and essential MLC composition requires a reasonably large sampling of fibers containing the same MHC and thin filament proteins (since troponin and tropomyosin isoforms might covary with, and thus influence, the role of MLC 3) but widely varying essential and regulatory MLCs. This research has not yet appeared in the literature. Bottinelli and colleagues (Bottinelli, Betto, et al. 1994b; Bottinelli and Reggiani 1995a) have shown significant relationships ($r = 0.72$ to 0.99) between the ratio MLC 3f/MLC 2f and V_o in a modest sample of rat muscle fibers identified according to their MHC composition (figure 1.21). More recently, Bottinelli and Reggiani (1995b) were able to determine the effects of differing MLC 3f/MLC 2f ratios on *in vitro* contractile properties of a sample of single IIX fibers

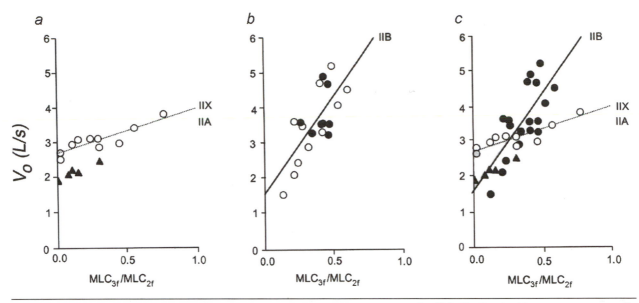

Figure 1.21 V_o versus MLC 3f/MLC 2f ratio in fibers from rat muscle: (*a*) plantaris fibers containing IIa (filled triangles) and IIx (open circles) MHC; (*b*) fibers containing IIb MHC from plantaris (open circles) and tibialis anterior (filled circles) with regression line; (*c*) all data points from *a* and *b* on the same plot.

Reprinted from Bottinelli et al. 1994.

from the rat plantaris. Their results indicated a high degree of relationship ($r = 0.9$) between V_{max} and this ratio; however, while the range of MLC 3f/MLC 2f was from less than 2 to more than 6, the corresponding range in V_{max} was from 1.4 to 2.0 fiber lengths per second (figure 1.22). No relationships were found between this ratio and the shortening velocity at a load of 5% of maximal tetanic force or with maximum power, both of which are physiologically more meaningful than V_{max}. These results tend to minimize the physiological influence of essential light-chain isoforms on muscle function.

Many fibers contain an admixture of fast and slow MLCs. Sugiura, Matoba, and Murakamis (1992) studied MLC combinations in single fibers of the rat lateral gastrocnemius and soleus. In their sample, almost all type II fibers of the high-oxidative region of the lateral gastrocnemius, as well as 26% of the type I fibers of the soleus, contained various admixtures of fast and slow MLCs. Many type II muscle fibers in humans, in addition to showing variation in essential MLC composition, also contain the regulatory MLC 2s in varying amounts. Larsson

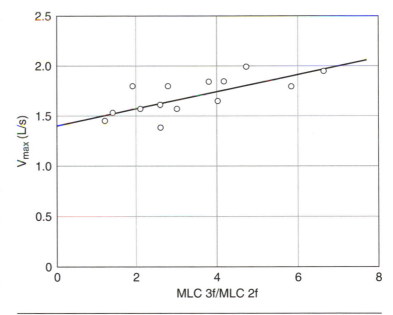

Figure 1.22 V_{max} versus MLC 3f/MLC 2f ratio in IIx fibers from rat plantaris.

Reprinted from Bottinelli and Reggiani 1995.

and Moss (1993) found that type II fibers coexpressing the MLC 2f and 2s isoforms had slower contractile speeds than those expressing only the 2f isoform. Perhaps, they contend, the regulatory light chains have a modulatory effect on myosin ATPase or on the influence that the essential MLC might exert on contractile dynamics. In addition, they supported previous findings (Reiser et al. 1985; Julian, Moss, and Waller 1981) of no clear relationship between MLC 3f content and shortening speed.

An interesting model with which to investigate the role of essential light-chain isoforms on V_{max} is one used by Moss and colleagues (1990), in which they were able to experimentally manipulate the content of MLC 3f and 1f in single rat fibers while keeping all other proteins unchanged. When they experimentally increased MLC 3f, at the expense of MLC 1f, V_{max} increased. Unfortunately, their technique allowed them to change MLC 3f content only by 10% to 15%.

In general, it is believed that the essential MLCs may significantly influence the maximal shortening velocity of the sarcomere, especially if the sarcomere possesses the fastest MHC (see figure 1.21). However, the research necessary to determine the exact mechanism by which essential light chains influence contractile properties and the significance of this influence during more common types of contractions has yet to be conducted.

The Regulatory Light Chains

The regulatory light chains are situated further from the S1 head of the MHC than the essential light chains, but they are still in a position to influence the dynamics of the actin–myosin cross-bridge (see figure 1.19). Their primary role is to alter cross-bridge function through their state of phosphorylation.

Phosphorylation of the regulatory light chain at a specific serine residue appears to increase the rate by which myosin cross-bridges move into the force-producing state. This phosphorylation is brought about via the action of an MLC kinase, which is in turn activated via an increase in myoplasmic calcium and calmodulin (Sweeney, Bowman, and Stull 1993; figure 1.23). Some examples of regulatory MLC phosphorylation, from the work of Moore and Stull (1984), are shown in figure 1.24. Since the rate of inactivation of the kinase is relatively slow, phosphorylation of MLC can occur at fairly low frequencies of stimulation. The increased twitch force seen with phosphorylation of the regulatory light chain is probably due to an increased rate of cross-bridge cycling.

MLC phosphorylation appears to increase the sensitivity of the contractile elements to calcium, so that force generation is enhanced at low, but not tetanic, frequencies of stimulation. Stull and his colleagues (Moore and Stull 1984; Sweeney et al. 1993) have proposed that MLC phosphorylation decreases the association of the S1 myosin heads from the thick filament, thereby increasing the rate at which attachment to actin can take place.

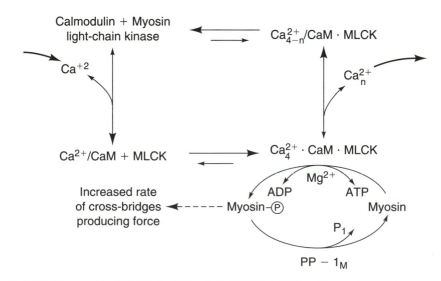

Figure 1.23 Scheme for regulation of MLC phosphorylation in muscle fibers. Myoplasmic calcium binds to calmodulin (CaM), which subsequently activates myosin light-chain kinase (MLCK). Phosphorylation of MLC increases the rate of transition of cross-bridges to a force-generating state.

Reprinted from Sweeney, Bowman, and Stull 1993.

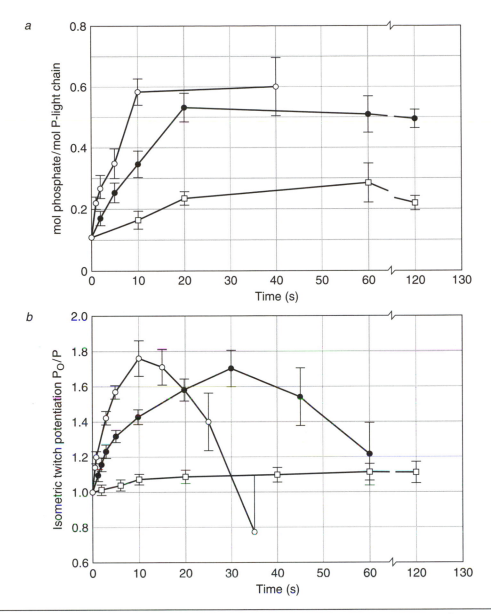

Figure 1.24   Effects of stimulation frequency (open squares = 0.5 Hz, filled circles = 5 Hz, open circles = 10 Hz) on (*a*) myosin light-chain phosphorylation and (*b*) twitch potentiation in rat gastrocnemius *in situ*. Reprinted from Moore and Stull 1984.

MLC phosphorylation as a significant physiological event is confined primarily to fast fibers, since they contain higher levels of the required kinase and fewer phosphatase enzymes (Moore and Stull 1984).

Hofmann and colleagues (1990) found that partial removal of the regulatory MLC 2 from skinned rabbit muscle fibers reduced V_o. These authors suggested that the rate of dissociation of cross-bridges decreased with extraction of MLC 2. It also resulted in increased active tension in calcium concentrations ranging from $10^{-6.6}$–$10^{-5.7}$ M (i.e., pCa 6.6–5.7).

The Role of Troponin-Tropomyosin

The components of the troponin–tropomyosin complex include troponin C (TnC), which binds with calcium during activation; troponin I (TnI), the inhibitory component that prevents actin–myosin interaction under low-calcium conditions; troponin T (TnT), which binds to tropomyosin and interacts with TnI and TnC; and tropomyosin (Tm), which influences seven actin monomers along the actin filament (figure 1.25). All of these components are present as isoforms and covary significantly with the

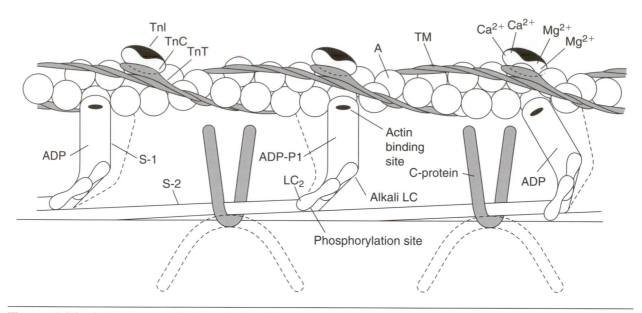

Figure 1.25 Schematic of thick and thin filaments. A = actin, TM = tropomyosin, Tn = troponin, LC = myosin light chain.

Reprinted from Moss, Diffee, and Greaser 1995.

myosin heavy chain found in the fiber (see table 1.1).

Using the *in vitro* approach, researchers have determined in single fibers the relationship of calcium concentration versus extent of myofibril activation as a function of the composition of the troponin–tropomyosin complex. Slow fibers are different from fast fibers in the following characteristics relating to calcium activation of myofibrils: (1) they require less calcium to activate than fast fibers, (2) the curve of the relationship of calcium concentration to force is less steep than in fast fibers, and (3) the curve is to the left of that for fast fibers (figure 1.26). These differences are not due to the properties of the myosin heavy chains, since substitution of TnT isoforms alters these properties *in vitro* (Schachat, Diamond, and Brandt 1987; figure 1.27). In addition, it appears that filament proteins other than TnC are involved, since replacement of endogenous TnC in rabbit muscles with that from heart had no effect on the steepness of the calcium–tension curve (Moss, Diffee, and Greaser 1995). These differences probably reflect differences attributable to the different

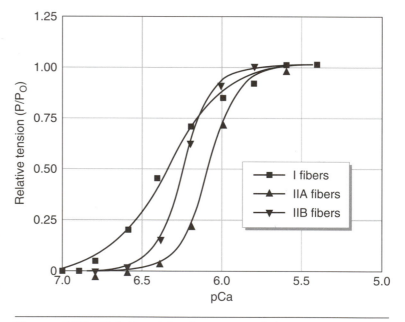

Figure 1.26 Relationships between calcium concentration (pCa) and force in rat muscle fibers *in vitro*.

Reprinted from Schiaffino and Reggiani 1996.

isoforms of Tm and TnT in cooperativity among the regulatory strands during calcium activation. The current hypothesis is that differences exist among the various TnT isoforms in the extent to which they interact, via physical overlapping chemical groups, with neighboring Tm filaments in a cooperative manner. Thus, the ex-

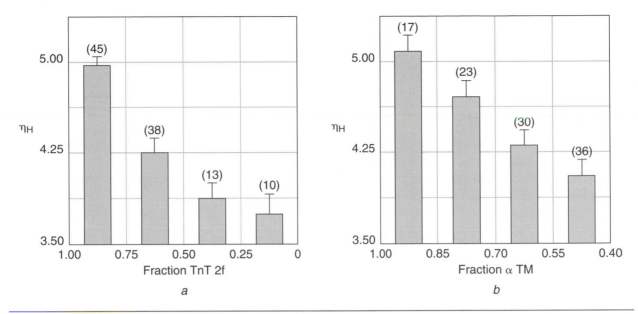

Figure 1.27 Effects of differences in (*a*) troponin T (TnT) and (*b*) tropomyosin (Tm) isoform composition on the steepness of the calcium concentration–tension curve in rabbit fast fibers. Steepness is expressed as the apparent Hill coefficient (η_H).

Reprinted from Schachat, Diamond, and Brandt 1987.

tent to which events occurring on one Tm filament influence events on a neighboring filament (in a cooperative manner) may be determined by the TnT isoform.

Even among type II fibers, calcium activation curves vary as a function of the Tm-TnT composition. All fibers appear to contain at least low levels of TnT 2f, with other forms more associated with specific fiber types: TnT 4f with IIB fibers, TnT 1f with IIX fibers, and TnT 3f with IIA fibers (Galler, Schmitt, et al. 1997). Slow fibers contain the slow TnT isoforms TnT 1s and TnT 2s.

Fibers containing the highest proportions of TnT 2f and Tm α_{fast} showed the lowest calcium activation thresholds and steepest calcium–force curves (figure 1.28). However, in this study (Schachat et al. 1987), fibers containing significant amounts of TnT 4f and 1f were not sampled. Interestingly, we now know that the TnT isoform that occurs in highest proportions in the fastest muscle fibers (IIB) is TNT 4f (Galler, Schmitt, et al. 1997).

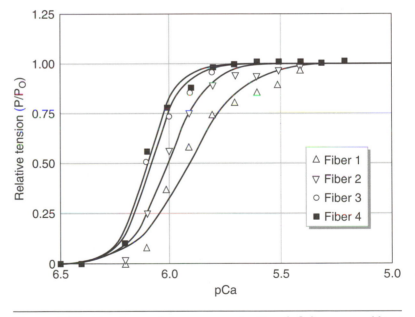

Figure 1.28 Tension–calcium concentration (pCa) curves of four different rabbit type II fibers (each shown by a different symbol) containing variable proportions of TnT 2f and Tm α_{fast}. TnT 2f and Tm α_{fast} increase in proportion from "triangle" fiber to "square-circle" fibers.

Reprinted from Schiaffino and Reggiani 1996.

Thus, some additional physiological properties, other than maximal or optimal shortening speed, are modulated by isoforms of contractile proteins. Such is the case with the troponin-tropomyosin system, which, via several isoforms in different combinations within the sarcomere, can alter the calcium sensitivity and degree of cooperativity of calcium binding by contractile proteins. Such modulation has an impact on the tension that a muscle fiber generates in response to a physiological train of nerve impulses.

Isomyosins

CONCEPT

4 Up to this point, we have considered the differences among fibers in myosin heavy and light chains, which are measurable when these proteins are separated from one another under denaturing conditions. Pyrophosphate-polyacrylamide gel electrophoresis (PP-PAGE) is a technique that allows the separation of the various isomyosins, each of which represents the myosin molecule as a constitutive element of the myosin filament *in vivo* (two myosin heavy chains with four associated myosin light chains). Since pure type I fibers contain type I MHC and three possible types of myosin light chain (MLC 1sa, MLC 1sb, and MLC 2s), one might expect type I fibers to exist within a muscle with one of three possible combinations of light chains: two regulatory light chains MLC 2s and either two MLC 1sa, two MLC 1sb, or one each of MLC 1sa and 1sb. Similarly, for each type II fiber, one would expect, in addition to the specific MHC, two MLC 2f and either two MLC 1f, two MLC 3f, or one of each of the latter two. In fact, PP-PAGE analysis of muscle homogenates of samples containing only one fiber type demonstrates the existence of these three isomyosins in rabbit and rat muscles (Wada and Pette 1993; Wada, Hämäläinen, and Pette 1995; Termin and Pette 1991; figure 1.29). Interestingly enough, isomyosins are also found when analyzing extracts from single fibers, indicating that in fibers containing only one type of MHC, different molecules within the same fiber can have different combinations of light chains (Wada and Pette 1993; Wada, Hämäläinen, and Pette 1995). This has been shown to be the case in human fibers as well as in muscles of rabbits and rats (Wada et al. 1996).

The powerful PP-PAGE technique can provide us with information not available when we ex-

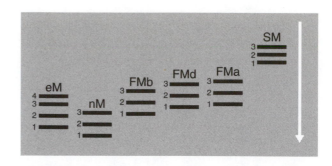

Figure 1.29   Schematic representation of the electrophoretic mobilities of IIB-based (FMb), IIX-based (FMd), IIA-based (FMa), and I-based (SM) isomyosins. Embryonic (eM) and neonatal (nM) isomyosins are also shown.

Reprinted from Termin and Pette 1991.

amine tissue homogenates under denaturing conditions. To demonstrate the difference in information that results from these two approaches, imagine a muscle fiber containing two different type II myosin heavy chains and three different light chains, according to standard polyacrylamide gel electrophoresis (PAGE) under denaturing conditions. Under PP-PAGE, we would be able to determine whether this represented one isomyosin (molecules consisting of two different heavy chains with the same three light chains in equal amounts) or several (in which case the number of possible combinations of heavy and light chains is fairly impressive). The presence of heterodimers (myosin molecules consisting of two different MHCs) in human muscle fibers has not been demonstrated. The demonstrated combinations of heavy and light chains to form different isomyosins, even in the absence of MHC heterodimers, now number at least 31 in human muscle (Wada et al. 1996; table 1.3).

Although the predominant isomyosins in human muscle fibers of different heavy-chain complements have not been determined as of this writing, it has been for several rabbit muscles for type IIA, IIX, and IIB fibers that are pure (not hybrid) in myosin heavy-chain content (see table 1.4). Among fast fibers, FM2 (an MLC 1f/MLC 3f heterodimer) is the predominant isomyosin, with the relative proportion of the fast isomyosins FM1 to FM3 increasing from IIA to IIB fibers. All three isomyosins for any given myosin heavy chain are present in single fibers, lending support to the proposal that light-chain

Table 1.3
Possible Combinations of Myosin Heavy Chains and Light Chains in Human Myosin Molecules

No.	HEAVY CHAINS	LIGHT CHAINS				
		1s	2s	1f	2f	3f
1	$(MHC\ I)_2$	X X	X X			
2	$(MHC\ I)_2$	X X	X		X	
3	$(MHC\ I)_2$	X X			X X	
4	$(MHC\ I)_2$	X	X X	X		
5	$(MHC\ I)_2$	X	X	X	X	
6	$(MHC\ I)_2$	X		X	X X	
7	$(MHC\ I)_2$		X X	X X		
8	$(MHC\ I)_2$		X	X X	X	
9	$(MHC\ I)_2$			X X	X X	
10	$(MHC\ IIa)_2$	X X	X X			
11	$(MHC\ IIa)_2$	X X	X		X	
12	$(MHC\ IIa)_2$	X X			X X	
13	$(MHC\ IIa)_2$	X	X X	X		
14	$(MHC\ IIa)_2$	X	X	X	X	
15	$(MHC\ IIa)_2$	X		X	X X	
16	$(MHC\ IIa)_2$		X X	X X		
17	$(MHC\ IIa)_2$		X	X X	X	
18	$(MHC\ IIa)_2$			X X	X X	
19	$(MHC\ IIa)_2$			X	X X	X
20	$(MHC\ IIa)_2$				X X	X X
21	$(MHC\ IIb)_2$	X X	X X			
22	$(MHC\ IIb)_2$	X X	X		X	
23	$(MHC\ IIb)_2$	X X			X X	
24	$(MHC\ IIb)_2$	X	X X	X		
25	$(MHC\ IIb)_2$	X	X	X	X	
26	$(MHC\ IIb)_2$	X		X	X X	
27	$(MHC\ IIb)_2$		X X		X X	
28	$(MHC\ IIb)_2$		X	X X	X	
29	$(MHC\ IIb)_2$			X X	X X	
30	$(MHC\ IIb)_2$			X	X X	X
31	$(MHC\ IIb)_2$				X X	X X

Possible combinations are based on data from Billeter et al. (1981) and Wada et al. (1996) and do not include possible MHC heterodimers or slow-twitch alkali MLC heterodimers.

Adapted from Wada et al. 1996.

Table 1.4
Distribution of Isomyosins FM1, FM2, and FM3 in Fiber Types of Single Rabbit Limb Muscles

Fiber type	FM3 %	FM2 %	FM1 %	FM1/FM3 %
IIA	36.4 ± 2.9	42.1 ± 6.1	21.5 ± 4.0	0.58 ± 0.10
IID/X	29.8 ± 4.1[a]	46.8 ± 4.6	23.4 ± 2.9	0.80 ± 0.18
IIB	30.5 ± 4.4[d]	40.2 ± 1.1[c]	29.3 ± 4.4[c,e]	1.00 ± 0.28[b,e]

[a] IIA vs. IID, $p < 0.01$

[b] IIB vs. IID, $p < 0.05$

[c] IIB vs. IID, $p < 0.01$

[d] IIA vs. IIB, $p < 0.05$

[e] IIA vs. IIB, $p < 0.01$

Reprinted from Wada et al. 1995.

complement may in some way serve to fine-tune the contractile speed as it is crudely fixed by the myosin heavy-chain type. Such a situation would explain to some degree the wide variability in expressions of maximum sarcomere shortening speed that exists among fibers with the same myosin heavy-chain composition and would allow a smoother continuum of contractile speeds across muscle fibers, perhaps in the order of their recruitment. The question of the organization of the different isomyosins in the same fiber (random? different myofibrils? different sarcomeres?) has yet to be answered.

Regulation of Myoplasmic Calcium Levels

A major difference exists between fast and slow fibers in the amount of calcium released per activation, which can be ascribed to a corresponding difference in the quantity of ryanodine receptors per muscle fiber (two- to threefold higher in fast than in slow fibers; Y. S. Lee et al. 1991; Damiani and Margreth 1994). This is consistent with the estimated volume of sarcoplasmic reticulum and transverse tubules as approximately double in type II compared with type I fibers in human muscles (Alway et al. 1988). This rate difference has implications for the maximal rate at which myofibrils can be activated (Salviati and Volpe 1988).

The initial step in the calcium release mechanism in skeletal muscle is the activation of the

ryanodine calcium release channels by the dihydropyridine receptors (DHPRs) of the T-tubules. DHPRs are more abundant in fast than in slow fibers (Péréon et al. 1998). In addition, some slow fibers express the cardiac alpha-1 subunit of the five-subunit DHPR complex. The physiological significance of this difference has yet to be determined but may indicate that skeletal muscle can exhibit some heartlike properties of excitation-contraction, in which extracellular calcium is important (Péréon et al. 1998).

Slow fibers are also more sensitive to caffeine-induced potentiation of sarcoplasmic reticulum calcium release than fast fibers. The calcium-imaging experiments of Pagala and Taylor (1998) suggest that caffeine-sensitive mechanisms of calcium-induced calcium release are more important in slow than in fast fibers.

Fast and slow fibers also differ in the rates at which their sarcoplasmic reticula take up calcium. The four- to sevenfold difference in the transport system, which is an ATPase system, is due to differences in both protein content (more than double, per unit of fiber volume) and isoform of the ATPase (Briggs, Poland, and Solaro 1977; Heiner et al. 1984; Dux 1993). In type I fibers, the sarcoplasmic reticulum ATPase is partially inhibited due to its interaction with nonphosphorylated phospholamban, which is present only in this fiber type (Damiani and Margreth 1994).

Calsequestrin is a calcium-binding protein that is found in the lumina of the terminal cisternae of the sarcoplasmic reticulum and probably plays a role in the storage and transfer of calcium. Its concentration is lower in slow than in fast muscle fibers (Dux 1993), which probably reflects the differences in percentage volume of protein of the terminal cisternae.

Parvalbumin is an 11-kilodalton cytosolic calcium-binding protein that is thought to serve as a reversible calcium buffer that receives calcium from troponin C until it is sequestered by the sarcoplasmic reticulum, thus facilitating relaxation (J.A. Rall 1996). Its concentration varies among the fiber types such that IIB > IIA > I, with type I demonstrating a virtual lack of the protein (Schmitt and Pette 1991). In fast fibers, parvalbumin is estimated to be the major contributor (50–73%) to the calcium decay constant after a single action potential, whereas in slow muscles that possess negligible amounts of parvalbumin, this rate constant is determined primarily by the calcium ATPase (Carroll, Klein, and Schneider 1997). In genetically-altered mice in which the parvalbumin gene is not expressed, twitch contractions of fast muscles are larger and of slower time course, attesting to parvalbumin's influence on the duration of the active state (i.e., the time that calcium is raised in myoplasm; Schwaller et al. 1999). In addition, in soleus muscles of rats transfected with parvalbumin cDNA, the rate of twitch relaxation is higher (Müntener et al. 1995).

Interestingly, when the buffer ethylenediamine tetraacetate (EDTA) is injected into soleus as an artificial parvalbumin, semifused contractions at subtetanic frequencies show the sag phenomenon typical of fast muscles (L.D. Johnson, Jiang, and Rall 1999). This property, which is used traditionally to distinguish fast from slow motor unit types, is discussed in chapter 2.

Enzymes of Energy Metabolism

Table 1.5 summarizes the results of several studies in which the activity of enzymes of energy metabolism has been measured in single human fibers, using either quantitative histochemical techniques or techniques in which single fibers are dissected out, homogenized, and the enzyme activities measured biochemically. The literature in which enzyme activities have been classified by subclass of type II fibers in human muscles is sparse, and thus most of

the data show enzyme activities grouped for types I and II only. In general, type II fibers show lower activities (approximately 60%) for oxidative enzymes (succinate dehydrogenase [SDH], citrate synthase, enzymes of beta-oxidation of fats) and higher activities (approximately double) for glycolytic enzymes (glycerol-3-phosphate dehydrogenase, lactic dehydrogenase, phosphorylase, phosphoglucoisomerase, pyruvate kinase) and for enzymes involved in phosphagen metabolism (myokinase, creatine kinase). The magnitude of difference between type I and II fibers is relatively higher for adenylate kinase and lactic dehydrogenase, the activities of which have been reported to be approximately three and four times higher, respectively, in type II than in type I fibers. A major difference across species in these relationships is the relative mean activity of oxidative enzymes. For purposes of comparison, data on single-fiber enzyme levels in muscles of the rat, which is the nonhuman species for which published data are most abundant, are also included in table 1.5. In rat muscle fibers, IIA fibers have higher activities for oxidative enzymes, as opposed to human fibers, where IIA levels are lower than in type I. In addition, IIB fibers in rats have higher overall activities for glycolytic enzymes than their counterparts in humans, the IIX fibers. Finally, the enzymes lactic dehydrogenase and adenylate kinase appear to show greater differences between type I and II fibers in humans than in rats (see table 1.5).

The range of enzyme activities within each fiber type attests to the significant degree of variation present at this level and may in fact contraindicate the calculation of mathematical averages of enzyme activities by fiber type. Even within the same animal, the relationships between type I and II fibers in enzyme activities probably depend on the composition of the parent muscle (i.e., primarily slow or fast fibers). For example, in the rat, oxidative enzyme activities are higher and glycolytic enzyme activities lower in type I and IIA fibers of soleus than in the corresponding fibers of extensor digitorum longus (Takekura and Yoshioka 1990).

Figure 1.30, from the work of Reichmann and Pette (1982), shows the relationships in SDH activity among cat tibialis anterior fiber types (the relationships among the cat fiber types are more similar to those of humans than those of rats). In this example, fiber SDH activity, measured

Table 1.5

Relative Enzyme Profiles in Single Fibers From Rat and Human Muscles

	I	II combined	IIA	IIX	IIB
			FIBER TYPE		
			HUMAN MUSCLE		
Oxidative enzymes	1	0.64 (0.44–0.92)	0.74 (0.62–0.85)	0.36 (0.21–0.51)	
Glycolytic enzymes, CK, MK	1	2.0 (1.6–2.4)	3.0 (1.9–4.0)	3.8 (2.4–5.0)	
LDH	1	3.8 (1.7–5.3)	3.5 (3.3–3.8)	4.5 (4.0–5.0)	
AK	1	2.8 (2.6–3.7)	3.1 (2.7–3.4)	5.3 (4.7–6.3)	
			RAT MUSCLE		
Oxidative enzymes	1	0.6 (0.5–0.7)	1.2 (0.7–1.5)	0.67 (0.63–0.70)	0.5 (0.3–0.7)
Glycolytic enzymes, CK, MK, AK	1	1.7 (1.6–1.8)	3.2 (1.2–5.0)	2.1	6.0 (1.7–10.0)

CK = creatine kinase, MK = myokinase, LDH = lactate dehydrogenase, AK = adenylate kinase. Values are expressed as a proportion of corresponding levels in type I fibers. Means and ranges were calculated from normalized data from several sources, including, in the case of oxidative enzymes, mitochondrial volume percentage from ultrastructural studies. Data concerning type II combined and type II subgroups are not always consistent (for example, human and rat glycolytic enzymes, where II combined is smaller than IIA and IIX), due to different enzymes and techniques used in the various studies surveyed. References for human fibers: Essen et al. 1975; Chi et al. 1983, 1987; Hintz, Chi, and Lowry 1984; Howald et al. 1985; Tesch, Thorsson, and Essen-Gustavsson 1989; T.P. Martin et al. 1992; Kent-Braun et al. 1997; Chilibeck, Syrotiuk, and Bell 1999; Evertsen et al. 1999. References for rat fibers: Hintz et al. 1980; Nemeth and Pette 1981; Hintz, Chi, and Lowry 1984; Roy et al. 1987; Takekura and Yoshioka 1987, 1989; Fitts, Brimmer, et al. 1989; Dunn and Michel 1997; Rivero, Talmadge, and Edgerton 1998.

using a quantitative histochemical assay, varies across fiber types such that I > IIA > IIB, although the wide variation in values for a given fiber type is apparent. Acknowledgment of this variability has led researchers to express single-fiber enzyme activities not as means, but in scatter plots. Visualized in this way, data from different fiber types and the effects of various interventions, such as altered activity, can be more easily scrutinized (figure 1.31).

As alluded to previously, variation among fiber types in enzyme activities is, to a certain extent, species specific. Apart from differences in absolute enzyme activities, the relative enzyme activities among the fiber types vary considerably across mammalian species. For example, in rat, guinea pig, rabbit, and cat hindlimb muscles, IIA fibers have higher mean SDH activities than type IIB fibers, while in human leg muscles (vastus lateralis and tibialis anterior), no such apparent difference exists. In the mouse, on the other hand, IIB fibers have higher SDH activities than IIA fibers (Reichmann and Pette 1982). Type I fibers have a higher mean SDH activity than type IIA fibers in human and cat but not in rat hindlimb muscles (Nemeth and Pette 1981).

Do Fiber Properties Vary Along Their Length?

Of course, in discussing the properties of fibers as we have been doing, we must make the assumption that these properties are at least reasonably constant along the muscle length. But are they? In 1979, Pool, Moll, and Diegenbach, using their newly developed semiquantitative histochemical technique, showed that there was

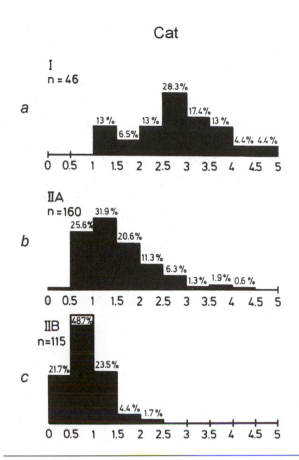

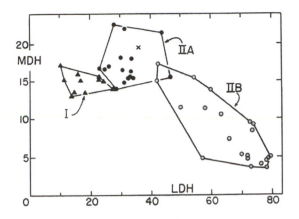

Figure 1.31 Malate dehydrogenase (MDH) activity plotted against lactic dehydrogenase (LDH) activity in 48 typed fibers from rat plantaris. Activities are in moles per kilogram of dry weight per hour.

Reprinted from Hintz et al. 1984.

Figure 1.30 Distribution of succinate dehydrogenase (SDH) activities in type (*a*) I, (*b*) IIA, and (*c*) IIB fibers of cat tibialis anterior.

Reprinted from Reichmann and Pette 1982.

considerable variation in SDH staining at different levels in mouse muscle fibers. Pette, Wimmer, and Nemeth (1980) measured the same enzyme in rabbit muscle fibers, in sections taken at 32-micrometer intervals, for distances up to 840 micrometers along the fiber. Their data suggested no appreciable variation in enzyme activity, or therefore content, along the fiber length. Similar results were reported more recently by Sant'ana Pereira, Wessels, and colleagues (1995), using the semiquantitative histochemical technique to measure myofibrillar ATPase along the fiber length.

Hintz and colleagues (Hintz, Chi, and Lowry, 1984) reported some variation along the length of rat plantaris fibers when they microdissected freeze-dried fibers and analyzed pieces biochemically. Using a coefficient of variation of 5% as an indicator of a variation just above analytical error, they found a coefficient of variation of 13% for fumarase, 9% for malate dehydrogenase, and a rather large 34% for glycogen phos-

phorylase along a 3.2-millimeter length. Thus, variation along the length of the fiber may depend on the protein; in this case, the authors suggested that phosphorylase varied the most among the enzymes measured because of corresponding length variation in its substrate, glycogen.

Several investigators have shown this possibility more recently. Staron published evidence in 1991 that a few examples of the same muscle fiber could show variability in the acid and alkaline stability of its myofibrillar ATPase in serial cuts taken at 1-millimeter intervals along its length. More recently, Ennion and colleagues (1995) found evidence of this in a few single human muscle fibers, in cuts taken within a few millimeters of one another. It is becoming more evident that these phenomena are not isolated events or technical errors. Peuker and Pette (1997) gave ample proof, using measurements of both protein and mRNA, that considerable nonuniformity can exist over several millimeters in a number of fibers. In their study of rabbit muscle fibers, they found variations in MHC mRNA isoform expression along the length of the fiber that were unambiguous in 17 cases.

The functional importance of such variation, if it is present in a significant proportion of fibers, is not known at present. It certainly is not difficult to envisage, given that muscle fibers have nuclei extending along their lengths, with each potentially receiving slightly different signals for gene expression. The general consensus at present, I believe, is that the variation along fiber length is generally minor and may explain an important proportion of the variation of properties among fibers within each fiber type.

Fiber Types and Performance

CONCEPT 5 The fiber proportions of muscles portrayed in figure 1.32 are actually means of several means found in the literature and may represent muscles of a typical, healthy, adult man. However, it is not unusual to find marked variability among subjects in fiber-type composition for the same muscle. The most information in this regard comes from the vastus lateralis muscle, which is probably the most frequently biopsied muscle of the body. In a study of 203 adult women and 215 men, Simoneau and Bouchard (1989) reported a considerable degree of variability in the percentages of fiber types in this muscle. The range of type I fiber percentages for men was from 15% to 79%, and that for women from 18% to 85%. Simoneau and Bou-

chard (1995), as a result of analysis of similar data in the literature from identical and fraternal twins and from siblings, have estimated that about 45% of this variability is hereditary, about 40% is the result of the environment, and the remainder is due to error. What these estimates mean in terms of potential for fiber-type changes is not fully understood as yet. We know, for example, that one important environmental component is activity level and that some fiber-type conversions are possible as a result of increased chronic activity level (discussed in chapter 4). The percentages proposed by Simoneau and Bouchard might suggest, for example, that two individuals who have 20% and 80% type I fibers in their vastus lateralis must differ in more than their genes. The contributions of various factors to fiber-type composition are crude estimates, however, and we do not know whether genes place a ceiling on our capacity to change fiber types with changing environment.

Such a wide degree of variation in fiber-type composition of a locomotor muscle such as vastus lateralis allows us to consider the effects

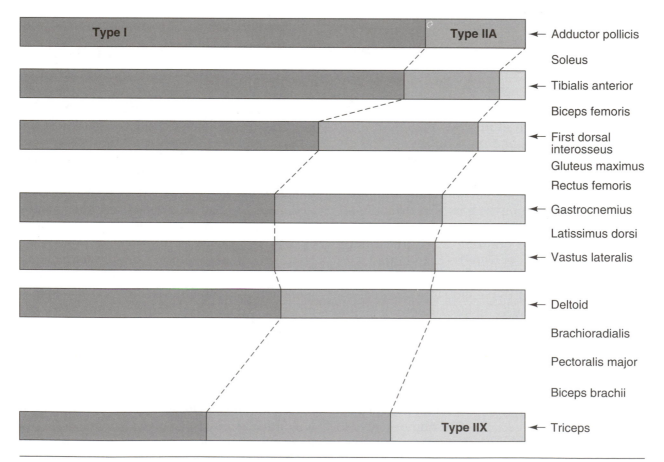

Figure 1.32   Fiber-type proportions in various human limb muscles.

of fiber-type composition on physical performance. There is some evidence from studies where muscles have been stimulated electrically and their fiber-type or myosin heavy-chain composition subsequently analyzed that fiber composition differences produce the expected differences in muscle performance. Rice and colleagues (1988) found modest but significant correlation coefficients between lateral gastrocnemius histochemical properties (percentage of fast fibers) and the time to peak tension of the evoked twitch ($r = 0.49$) and the 10 Hz/50 Hz tetanic force ratio ($r = 0.55$).

Harridge and colleagues (1996) more recently compared the isometric contractile properties of triceps surae, quadriceps, and triceps brachii. As expected from these muscles' fiber-type compositions, all indices of contractile speed (twitch time course, maximum rates of rise and fall of tetanic force) showed that triceps surae was slowest, triceps brachii was the fastest, and quadriceps was between the two. This relationship paralleled the percentages of type I fibers and the proportions of MHC I in muscles of these muscle groups: Soleus had the highest, triceps brachii the lowest, and vastus lateralis was between

these two. Correlations were excellent between indices of contractile speed and of type II fiber composition when all muscles were included but were poor and nonsignificant when these relationships were examined within each muscle.

Such poor relationships as these between fiber type–related and standardized performance variables when muscle is electrically stimulated do not bode well for finding good relationships using voluntary performance as the performance variable. However, we must consider the representativeness of the biopsy vis à vis the muscle or muscle group being tested and whether the correct measurements are being taken. For example, in the only study of its kind, Lexell and colleagues (1983) demonstrated a fairly systematic superficial-to-deep variation in fiber composition in whole cross sections of human vastus lateralis muscles taken from cadavers. Whether this variability can be accommodated by always sampling muscle biopsy specimens from the same muscle depth is not known.

Table 1.6 is a summary of studies that have found significant correlations between an expression of type II fiber composition (e.g., percentage of fibers or percentage of area constituting

Table 1.6
Correlations Between Percentage of Type II Fibers and Performance Measures

Performance measure	Correlation	Subjects[a]	Reference
ABSOLUTE FORCE			
Absolute isometric force of quadriceps	0.55	1	(1)
Maximum one-legged isometric strength	0.55 to 0.58*	1	(2)
EXPLOSIVE SPEED–POWER			
Squat jump (cm) with 40-kg load	0.45	2	(3)
Vertical jump, countermovement jump (cm)	0.49	2	(4)
Peak acceleration of unloaded knee extension	0.4	1	(5)
Time from 10% to 30% maximum voluntary contraction during maximum leg extension	0.48	2	(4)
Maximum rates of isometric force development	0.3 to 0.5	1	(6)
Maximum stair-climbing velocity	0.37	1, 2	(7)
40-m maximum running velocity from a flying start	0.73	2	(8)

(continued)

Table 1.6 (continued)

Performance measure	Correlation	Subjects[a]	Reference
ISOKINETIC TORQUE/VELOCITY, POWER			
% maximum isokinetic torque at 180°/s	0.50	1	(9)
Maximum isokinetic torque at 180°/s	0.52	2	(8)
Maximum isokinetic torque at 180°/s	0.91**	2	(10)
Maximum isokinetic torque at 180°/s	0.48*	2	(11)
% maximum isokinetic torque at 115°/s to 400°/s of leg (hip and knee) extension	0.44 to 0.75	1	(12)
Maximum isokinetic torque at 92°/s to 288°/s	0.55 to 0.7*	2	(13)
Maximum isokinetic torque at 30°/s to 180°/s	0.87 to 0.90***	1	(14)
Knee extension peak power, optimal velocity estimated from isokinetic torque–velocity relationship	0.5 to 0.55	1	(15)
Work, peak power, and rate of power production during isokinetic knee extensions at 6°/s to 300°/s	0.5 to 0.7	1	(16)
Maximum isometric torque, maximum power during dynamic knee extensions	0.6 to 0.69	2	(17)
Electromechanical delay during isokinetic knee extensions	−0.58	1	(18)
ANAEROBIC POWER/CAPACITY			
Torque decline during repeated maximal knee extensions (50/min) at 189°/s	0.86	1	(19)
Torque decline during repeated maximal knee extensions (50/min) at 189°/s	0.57****	1	(20)
% increase in EMG/torque ratio during repeated maximal knee extensions (50/min) at 189°/s	0.84	1	(21)
Power decrease during Wingate test	0.68 to 0.76*	1	(22)
Power decrease during Wingate test	0.67	2	(8)
Anaerobic power, Wingate test	0.59	1	(23)
Anaerobic capacity, Wingate test	0.81	1	(23)
Average power, Wingate test	0.59 to 0.72	2	(8)
Wingate peak 5-s power output, total work	0.76 to 0.89	1	(22)
Wingate peak power	0.54	1	(24)

Performance measure	Correlation	Subjects[a]	Reference
ENDURANCE RELATED			
Isometric endurance (leg extension) at 50% maximum voluntary contraction	−0.51	1	(25)
Isometric endurance (leg extension) at 50% maximum voluntary contraction	−0.51	1	(26)
Workload corresponding to the onset of blood lactate accumulation (WOBLA, watts)	−0.57	1	(27)
Velocity, 2000-m run	−0.60	2	(8)
Running performance time, 1 to 6 miles (gastrocnemius)	0.52 to 0.55	2	(28)
9-min run distance	−0.58****	1	(1)
Gross efficiency (ergometer work/O_2 consumption) during an all-out 1-h ergometer ride	−0.75	2	(29)
Gross efficiency during cycling at 52% to 71% of $\dot{V}O_2$max	−0.75	2	(30)
Initial $\dot{V}O_2$ response to exercise ($\Delta\dot{V}O_2/\Delta$work rate)	−0.84	1	(34)
Energy cost in $kJ \cdot kg^{-1} \cdot km^{-1}$ of a 45-min run at 3.33 m/s	0.6	2	(31)
Time to fatigue at 88% $\dot{V}O_2$max	−0.62	2	(32)
Marathon running velocity	−0.64	2	(33)

[a] 1 = nonathletes, 2 = athletes

* Percentage area of type II fibers

** Absolute total area of type II fibers

*** Percentage area of type IIA fibers

**** Percentage of type IIB fibers

References: (1) Jansson and Hedberg 1991; (2) Tesch and Karlsson 1978; (3) Häkkinen, Newton et al. 1998; (4) Viitasalo, Häkkinen, and Komi 1981; (5) Houston, Norman, and Froese 1988; (6) Viitasalo and Komi 1978; (7) Komi et al. 1977; (8) Inbar, Kaiser, and Tesch 1981; (9) Thorstensson, Grimby, and Karlsson 1976; (10) Johansson et al. 1987; (11) Thorstensson, Larsson, and Karlsson 1977; (12) Coyle, Costill, and Lesmes 1979; (13) Gregor et al. 1979; (14) Ryushi and Fukunaga 1986; (15) MacIntosh et al. 1993; (16) Ivy et al. 1981; (17) Tihanyi, Apor, and Fekete 1982; (18) Viitasalo and Komi 1981; (19) Thorstensson and Karlsson 1976; (20) Tesch 1980; (21) Nilsson, Tesch, and Thorstensson 1977; (22) Froese and Houston 1987; (23) Kaczkowski et al. 1982; (24) Bar-Or et al. 1980; (25) Hulten and Karlsson 1974; (26) Häkkinen and Komi 1983b; (27) Karlsson and Jacobs 1982; (28) Foster et al. 1978; (29) Horowitz, Sidossis, and Coyle 1994; (30) Coyle et al. 1992; (31) Bosco et al. 1987; (32) Coyle et al. 1988; (33) Sjodin, Jacobs, and Karlsson 1981; (34) Barstow et al. 2000.

type II fibers) and physical performance in humans. All studies are of the vastus lateralis, with one exception, which is noted in the table. I have not included studies presenting correlations between fiber composition and metabolic variables (such as enzyme activities, lactate, and $\dot{V}O_2$max) or any nonsignificant correlations. In most studies shown in table 1.6, the range of fiber compositions among the subjects was also judged to be sufficiently large to warrant the correlational approach used. In general, the results show rather convincingly that fiber composition has an impact on maximal performances in a rather predictable way. The "benefits" of having more type II fibers probably include the combination of their intrinsic properties that were discussed earlier in this chapter: contractile speed, power, perhaps specific tetanic tension (which might explain the positive relationships with maximal isometric force), and the relationships between myofibrillar ATPase and metabolic enzyme activities (which contribute to endurance-related correlations).

Summary

Although the classic methods for determining fiber types include techniques that indicate the presence of specific myosin heavy chains, it is now clear that fiber function depends on more than merely myosin heavy-chain composition. Fibers show a remarkable degree of variability in sarcomeric protein composition and, as a consequence, in functional properties. This variability may indicate the transitory nature of fibers, which adjust continually to their changing environment and might vary in their degree of activation. When one considers the different combinations of protein isoforms, enzymes of energy metabolism, and calcium-regulating proteins that are known to exist among muscle fibers, it becomes evident that the original concept of fiber typing has limitations. Rather, fibers appear to exist on a continuum from low (pure type I) to high (pure IIX) in contractile speed, maximal force per unit area, power, and energy cost of contractile activity. This continuum idea has implications for the orderly recruitment of fibers during voluntary contractions, which is considered in the next chapter.

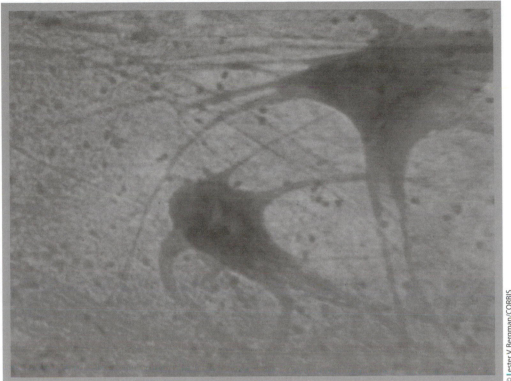

Motoneurons and the Muscle Units They Innervate

> **"A** spinal cord will do the acts, simple or complex, which it has inherited from ancestors or acquired by education the faculties to do so, but if required to do new and strange acts, to associate in action muscles which have not acted together before, it will manifest an utter stupidity and impotence."
>
> H. Maudsley, *The Physiology of Mind,* 1899

Most Important Concepts From This Chapter

1 Each muscle unit (the muscle fibers innervated by a motoneuron), like the single fibers it comprises, exhibits properties of contractile speed, fatigue resistance, and tetanic strength. The latter is a function not only of fiber size and specific tetanic tension, but also of fiber number.

2 Muscle units can be classified according to these properties into at least three and perhaps four groups according to their relative tetanic force, the presence or absence of "sag" during a semifused contraction, and their fatigue resistance. The groups are type S (weak, no sag, high fatigue resistance), type FF (strongest, sag present, highly fatigable), type FR (intermediate in strength and fatigue resistance, sag present), and FI (between FF and FR).

3 The motor unit (muscle unit and its innervating motoneuron) exhibits a variation of properties that allows a logical recruitment blueprint for the performance of simple isometric contractions. For the motoneuron, properties that vary systematically among the motor units include rheobase current, input resistance, afterhyperpolarization (AHP) duration, and propensity for late adaptation and, perhaps, for bistability.

4 Motoneuron and muscle unit data are applied to a recruitment model. This approach can be used to demonstrate how a model system with fixed parameters will generate isometric forces of different levels and how changes in the parameters, representing either different muscles or the same muscle adapted to increased or decreased usage, alter recruitment.

5 Differences in the patterns of recruitment of motor units during isometric contractions occur when contractions involve increasing or decreasing force levels, concentric or eccentric contractions, and unilateral or bilateral contractions.

6 Recruitment patterns during prolonged, complex activities, such as running and cycling, are less well known at present.

7 Most individuals appear to be able to recruit their muscles maximally, although considerable intra- and interindividual variation occurs in this capability.

The motor unit (consisting of a single alpha-motoneuron and the muscle fibers it innervates), and not the muscle fiber, is the force unit used by the nervous system to construct voluntary movements. The patterns with which muscle fibers are grouped together into muscle units (the muscle fiber component of the motor unit) and the patterns with which the intrinsic properties of the motoneuron vary with muscle unit properties constitute important background information for understanding how motor units are recruited and thus how movements are generated. I describe our current ideas about the properties of the muscle unit, followed by those of the motoneuron that innervates it. This information sets the stage for the discussion of the recruitment of motor units during different types of tasks.

The Muscle Unit and Muscle Unit Types

CONCEPT

1 We have already seen in chapter 1 the variability that is present in protein composition and functional properties among fibers within a muscle. Since muscle units are groups of muscle fibers innervated by a common motoneuron, it is not surprising to find that muscle units within a given muscle are not all the same. Perhaps more interestingly, several muscle unit properties—contractile speed, strength, specific tension, and fatigability—covary significantly. The seminal work of R. Burke and associates during the late 1960s and early 1970s (Burke et al. 1973; Burke and Tsairis 1974; Burke 1967) on the muscle unit composition of the cat medial gastrocnemius resulted in the generation of the three-category and later the four-category scheme of muscle unit classification that is still widely used. This system has since been shown to be generally applicable to muscle unit populations in several other cat limb muscles and to limb muscles of various other species. There is some question, however, as to the applicability of this general scheme to human motor units (Stephens and Usherwood 1977; Bigland-Ritchie, Fuglevand, and Thomas 1998).

2 This information came from experiments on anesthetized cats, in which the isometric contractile properties of single muscle units were measured either by stimulation of single axons or by injection of suprathreshold current pulses into the somata of their innervating motoneurons. The relative fatigue resistance of motor units was estimated using a standardized stimulation protocol, now used widely and known as the *Burke fatigue protocol,* which consists of a train of impulses lasting 330 milliseconds at a frequency of 40 hertz, once per second. Muscle units were grouped according to their drop in peak force generation during two minutes of stimulation: Type S (slow) units showed no decrease, type FR (fast, fatigue resistant) declined less than 25%, and FF (fast, fatigable) declined 75% or more. A fourth category, FI, designated units in which the force decline was between 25% and 75% of initial force. This criterion, as well as the presence

(F types) or absence (S) of "sag" in force during subtetanic stimulation (figure 2.1), distinguished F from S units. Other properties seemed to vary systematically with sag and fatigue resistance, such as isometric twitch contraction time and maximal muscle unit force as measured by tetanic stimulation (the frequency above which muscle force does not increase; figure 2.2).

It is intuitively satisfying that this four-part classification system, based on physiological properties, was conceived before we knew of the heterogeneity in myosin heavy-chain composition that could explain these motor unit groupings. It is highly probable, from the limited data that we have at present, that the type S, FR, and FF units are composed of fibers containing primarily type I, IIA, and IIB (or, in humans, IIX) fibers, respectively (Larsson, Ansved, et al. 1991; Larsson, Edström, et al. 1991). To appreciate the elegance with which muscle unit properties appear to be organized, it is a good idea at this time to examine each of the properties that are used to separate units into their categories.

Muscle Unit Fatigability

As indicated in the previous chapter, we do not have information on fatigability of single mammalian fibers of various types due to the technical difficulties associated with maintaining their viability *in vitro.* In addition, we are not

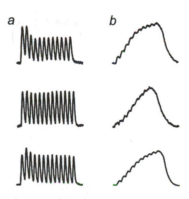

Figure 2.1 Various examples of sag in force during unfused contractions (25 Hz, 1 s) in (*a*) three fast muscle units of the rat gastrocnemius. Compare with (*b*) three slow units at the same frequency and duration of stimulation. Forces have been normalized for ease of comparison.

Reprinted from Gardiner 1993.

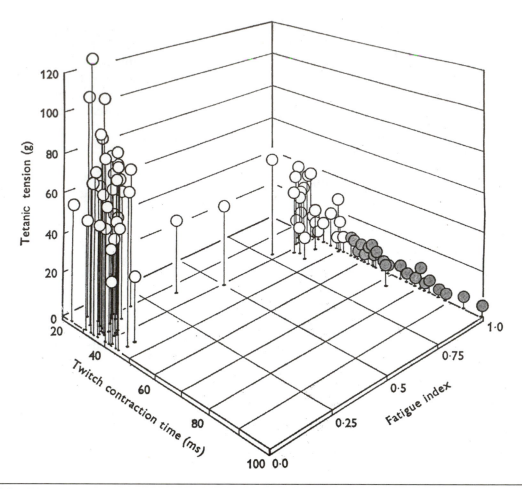

Figure 2.2 Physiological profiles of a sample of cat medial gastrocnemius motor units. Shaded units demonstrated no sag of force during unfused tetanic contraction.

Reprinted from Burke et al. 1973.

sure that this measurement made *in vitro* at the single-fiber level would extrapolate well to the properties at the motor unit level, where blood flow, compliance of the surrounding tissues, and number of fibers are factors that influence fatigue resistance. We do, however, have fatigue resistance information from motor unit stimulation experiments.

Generally, motor unit size (determined primarily by innervation ratio) varies systematically with fatigability: Larger motor units are less fatigue resistant than smaller units (see figure 2.2). It should be recognized that the Burke fatigue test of two minutes' duration, which was very useful in distinguishing motor unit types, may not be a true test of the long-term fatigability of the unit during locomotor tasks *in vivo*. Force decline as measured by the test, however, is not attributable to events such as neuromuscular transmission or sarcolemma propagation fail-

ure and is thus localized in the muscle fibers (at least in the cat; in the rat, some propagation failure occurs during this test; see Gardiner and Olha 1987). Jami and colleagues (1983) demonstrated quite clearly that muscle units fatigued by using the Burke protocol were not as incapable of generating force as one would like to believe. They showed that cat peroneus tertius motor units, when fatigued using this protocol, showed a gradual increase in force when the stimulation was continued for a longer period of time than the 330-millisecond train duration used during the fatigue regimen (figure 2.3). This suggests the possible involvement of the calcium release mechanism in the fatigue resulting from the Burke protocol; this mechanism possibly became less sensitive to the stimulus but could respond if the stimulus intensity was increased, in this case by increasing the time of stimulation. Obviously, this type of fatigue may

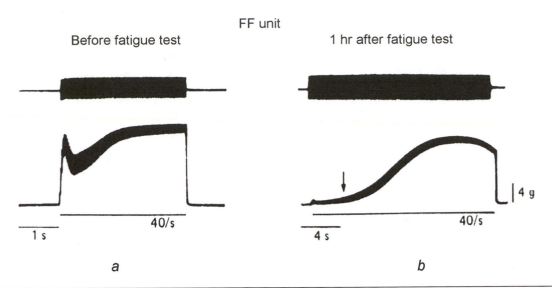

Figure 2.3 Fatigability of an FF unit of cat peroneus longus muscle. (*a*) Before fatigue resulting from the Burke procedure, the unit reached peak isometric force in response to stimulation at 40 Hz within about 0.5 s. (*b*) One hour after the fatigue protocol, peak force was not attained until after more than 10 s of stimulation. The arrow in *b* shows where force plateaued in prefatigue contraction. This demonstrates that fatigue induced by this protocol involves more than a failure of the contractile components.

Reprinted from Jami 1983.

or may not be pertinent to fatigue during longer, or different, types of fatiguing tasks.

When higher frequencies are used to avoid this problem, neuromuscular propagation failure becomes a contributing factor to fatigue (Kugelberg and Lindegren 1979; Larsson, Ansved, et al. 1991; figure 2.4).

While an inverse relationship between muscle unit strength and fatigue resistance appears evident from figure 2.2, the results from the two-minute fatigue test leave very little room for distinction among individual units within each type group. A very clear relationship between fatigue resistance and muscle unit tetanic force is evident from the experiments of Botterman and Cope (1988a, 1988b; Cope et al. 1991) and is worth discussing here (figures 2.5 and 2.6). Their strategy was to stimulate single motor units in cat hindlimb muscles to contract using a stimulation frequency that would evoke a fixed percentage of their maximal isometric tetanic tension (25% for F units, 85% for S units). As stimulation continued and force decreased due to fatigue, they gradually increased the frequency until the unit could no longer maintain the target force. Using this approach, they found a wide variation in fatigue times among fast units (from 11 to 2,000 seconds) and a significant inverse rela-

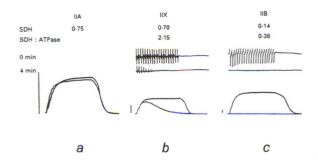

Figure 2.4 Responses of rat tibialis anterior motor units to trains of 20 pulses at 100 Hz, twice per second, for four minutes. Note the electromyographic (EMG) decrement in (*b*) IIX and (*c*) IIB motor units.

Reprinted from Larsson et al. 1991.

tionship between fatigue time and tetanic tension. This approach was all the more interesting when applied to type S motor units, which all had the same fatigue index (1.0) according to the Burke fatigue test. By force-clamping these S motor units at 85% of initial maximal tetanic force using the same procedure explained earlier, they found endurance times to vary 50-fold (up to 9,000 seconds). No relationship, however, was found for S units between endurance time and tetanic force, as had been found for F units.

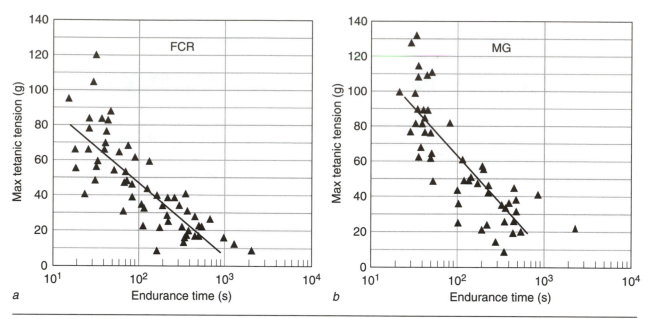

Figure 2.5   Semilogarithmic plot of maximum tetanic tension versus endurance time for (*a*) cat flexor carpi radialis (FCR) and (*b*) medial gastrocnemius (MG) motor units. Units were force-clamped at 25% of maximal tetanic tension.

Reprinted from Botterman and Cope 1988.

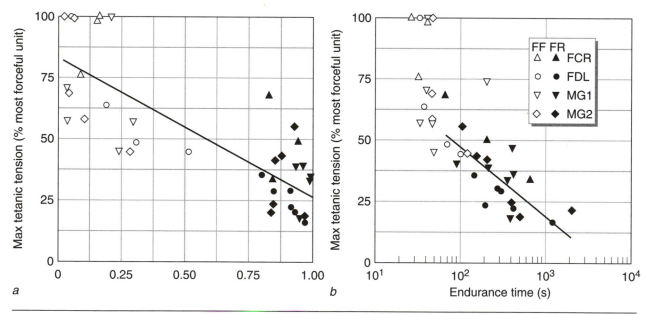

Figure 2.6   Comparison of fatigue index using (*a*) the Burke criterion and (*b*) endurance time when force-clamped at 25% of maximal tetanic force, for motor units from four cat muscles.

Reprinted from Botterman and Cope 1988.

In the previous chapter we discussed the differences among fiber types in their profiles of metabolic enzymes. Is there any relationship with fatigability? Kugelberg and Lindegren in 1979 measured fatigue resistance of rat tibialis motor units and subsequently measured the intensity of the stain for the mitochondrial enzyme succinate dehydrogenase (SDH) in fibers that had been identified as belonging to the unit following glycogen depletion. Their results showed a significant linear relationship. This was subsequently done in motor units from rat extensor digitorum longus using fatiguing stimulation for up to 60 minutes (Nemeth, Pette, and Vrbova 1981) and cat tibialis posterior using the Burke protocol of 2 minutes (Hamm et al. 1988); investigators measured fatigue resistance and then measured metabolic enzymes quantitatively in fibers identified after glycogen depletion. The relationships in these rather limited samples are also linear and significant, suggesting that a cause–effect relationship may exist between mitochondrial enzyme activity and fatigue resistance (figure 2.7).

More recently, Larsson and colleagues (Larsson, Ansved, et al. 1991; Larsson, Edström, et al. 1991) performed this type of experiment on rat tibialis anterior motor units; SDH staining intensity was measured photometrically in glycogen-depleted fibers. Their findings showed that the staining intensity for calcium-stimulated ATPase increased and that for SDH decreased, in the order type I, IIA, IIX, IIB. A significant correlation ($r = 0.88$) was observed between resistance to fatigue and SDH staining intensity. However, these experiments and those of Lindegren and colleagues mentioned earlier must be interpreted with the awareness that a significant degree of neuromuscular transmission failure (variable among motor unit types) occurred in their experiments due to the high frequency (100 hertz) of stimulation used (see figure 2.4).

A stimulation protocol more closely resembling the pattern one might see during voluntary recruitment might yield additional information about the fatigue resistance of motor units. This was studied by Bevan and colleagues (1992), who examined stimulation protocols in which frequency was higher at the beginning of the burst (figure 2.8). Their findings confirmed that muscle can be made more fatigue resistant by optimizing the frequency of stimulation.

In summary, fatigue resistance varies systematically among the muscle unit types, but the extent of this variability depends greatly on the technical approach used to investigate it. Experiments involving more physiologically relevant stimulation protocols, like those of Botterman and colleagues (Botterman and Cope

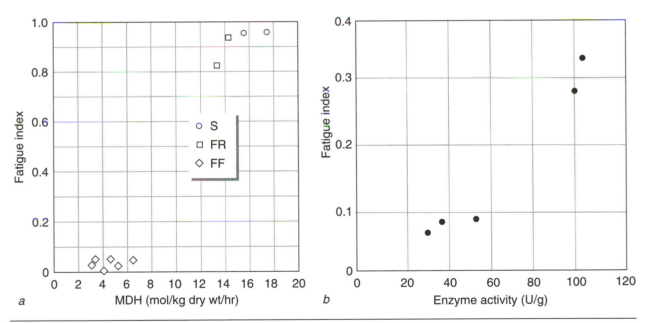

Figure 2.7 Enzyme properties versus fatigue resistance in single motor units. (*a*) Fatigue index using the Burke protocol versus malate dehydrogenase (MDH) activity measured biochemically in single cat tibialis posterior fibers. (*b*) MDH activities in single fibers of rat extensor digitorum longus (EDL) motor units versus fatigue resistance measured as a function of initial tetanic tension remaining after 30 minutes of stimulation (10 pulses at 40 Hz, once per second).

a Reprinted, by permission of Wiley-Liss, Inc., a subsidiary of John Wiley & Sons, Inc., from Hamm et al., 1988, "Association between biochemical and physiological properties in single motor units," *Muscle & Nerve 11*: 248. © 1988 Wiley-Liss, Inc.
b Reprinted from Nemeth, Pette, and Vrbová 1981.

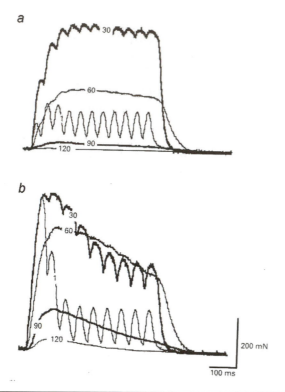

Figure 2.8 Fatigue responses can depend highly on the stimulation pattern used. Shown here are the responses of a cat tibialis posterior motor unit to two stimulation patterns. (*a*) Stimulation was with trains of 500 ms, once per second, at constant interpulse interval (1.8 times the twitch contraction time). (*b*) The same total number of pulses per train were delivered, except that the first three had short interpulse intervals (10 ms). Note how muscle appears more fatigue resistant in *b*.

Reprinted from Bevan et al. 1992.

1988a, 1988b; Cope et al. 1991) and Bevan and colleagues (1992) described earlier, will be necessary in order to determine the factors that affect this important property.

Muscle Unit Contractile Speed

Twitch time course (time to peak tension and half-relaxation time, in milliseconds) has been traditionally used as the index of contractile speed of motor units. The separation of motor units into S versus FR and FF is reasonably clear (see figure 2.2), perhaps as clear as using V_o or V_{max} to determine myosin heavy-chain composition at the single-fiber level (see chapter 1). In fact, the wide variation seen in muscle unit twitch contraction times within a muscle may be related to the variability seen in velocity of

shortening measured *in vitro* at the single-fiber level, but this is not known for certain. The twitch is, theoretically speaking, not the best index of contractile speed in the strict sense of the term, since it can be influenced by many factors, such as muscle length, series elastic component, and previous history of fiber activation.

A more accurate, but technically more difficult, property with which to estimate contractile speed would be shortening velocity, as has been done at the single-fiber level *in vitro*. This has been done for muscle units in the rat soleus *in situ* (Devasahayam and Sandercock 1992). The findings from this study showed no significant relationship between measures of shortening velocity and twitch contraction time (figure 2.9). The maximum rate of rise of the tetanus, another index of muscle unit contractile speed, is also poorly related to twitch contraction time (Goslow, Cameron, and Stuart 1977). In spite of these problems, however, twitch contraction time does show a reasonably good relationship to motor unit type, the latter determined using the fatigability and sag criteria. S twitches are the longest and FF twitches the shortest, with considerable overlap among type F groups and slight overlap between type S and F groups (figure 2.10). The distinction of fast and slow muscle units on the basis of twitch contraction time is not completely inappropriate, since contraction time influences the degree of fusion of individual twitches when stimulated at subtetanic frequencies. In fact, the effect that twitch contraction time has on the percentage of maximum tetanic tension when stimulation is at a subtetanic frequency may point to this percentage as a more appropriate index for separating fast from slow units.

The utility of using the sag criterion to separate F from S units has been questioned, since sag may not be evident in some F units and may be measurable in some S units (Burke 1990). This phenomenon will take on more significance when we know what causes sag and whether it plays a role in tension excursions during voluntary movements. There is some evidence, for example, that the presence of parvalbumin within the muscle fiber is essential for the expression of the sag phenomenon, suggesting that fluctuating calcium levels during excitation of the contractile apparatus might be involved (Müntener et al. 1995; J.A. Rall 1996; Schwaller et al. 1999).

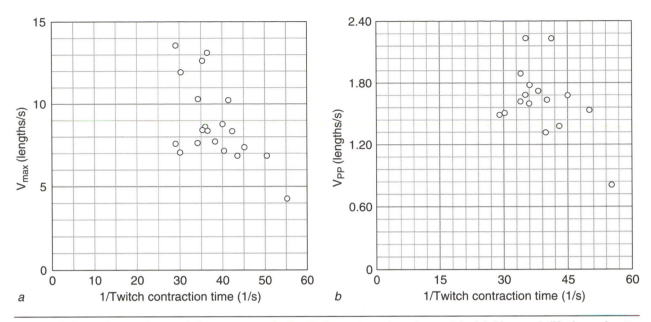

Figure 2.9   Relationships of the reciprocal of twitch contraction time with (*a*) V_{max} and (*b*) the velocity at which peak power occurs (V_{pp}) in rat soleus motor units *in situ*. Correlation coefficients in both cases were less than –0.5, and were not statistically different from zero.

Reprinted from Devasahayam and Sandercock 1992.

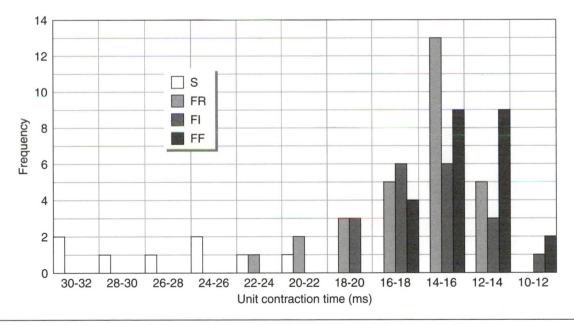

Figure 2.10   Distribution of muscle unit time to peak tension in rat gastrocnemius.

Muscle Unit Tetanic Tension

As suggested by figure 2.1, average muscle unit twitch and tetanic force increase in the direction S < FR < FI < FF. This seems to be consistent across limb muscles and species (Garnett et al. 1979; Zardini and Parry 1998; Burke et al. 1973; Gardiner and Olha 1987). An issue that has not been resolved at this writing is the precise reason for this variation, in which the difference between the smallest and largest unit can exceed 100-fold. Major contributors clearly are fiber number, or innervation ratio, and average fiber cross-sectional area. Another factor may be tension generation per unit of cross section, also known as specific tension. As we saw for measurement of single fibers *in vitro,* there is some basis for believing that specific tetanic tensions of type II fibers may be 20% to 50% higher than those of type I fibers. Information from *in situ* motor unit studies is not totally consistent with this. Some estimates of muscle unit specific tension indicate as high as a sixfold difference between type S and type FF units (J.C. McDonagh et al. 1980). Because all depleted fibers in a motor unit may not be visible in a single muscle cross section when using the glycogen depletion method, other means of estimating muscle unit specific tension have been used. These involve measuring percentages of fiber types and motor unit types in the muscle to arrive at the relative innervation ratio and calculating differences in average fiber cross section among the fiber types. It is possible that measurements of specific tension *in vitro,* involving activation of individual fibers at supramaximal calcium concentrations, are underestimates of this property in a muscle fiber or motor unit that is contracting *in situ* within the structural confines of a muscle, under physiological conditions where excitation–contraction coupling mechanisms are intact. Whether or not different muscle units have different specific tensions and the magnitude of the difference, if indeed a difference exists, remain unknown.

Are Muscle Units Uniform?

The question of whether or not all fibers within a given motor unit are the same has been debated for several years now and unfortunately has not yet been resolved. The experimental approach has been to examine the variability in either histochemical or biochemical properties among muscle fibers that have been identified by glycogen depletion as belonging to the same motor unit. The report from Nemeth, Pette, and Vrbova in 1981 showed quite convincing evidence that enzymes in fibers belonging to the same unit (as identified by glycogen depletion) were virtually identical (figure 2.11). Of course, this conclusion is based on the premise that all fibers in the unit have used sufficient glycogen to be equally visible on the subsequent glycogen stain. It is also based on the assumption that those fibers of the unit not iden-

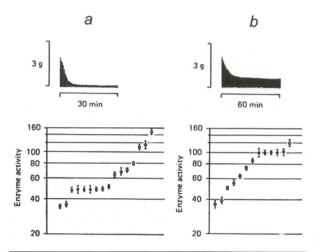

Figure 2.11 Malate dehydrogenase content of fibers and fatigue characteristics of the corresponding motor units in rat extensor digitorum longus. Single fibers from motor units (filled circles) are similar, compared with the dispersion of values from randomly selected fibers (open circles). Fibers are arranged in order of increasing enzymatic activity.

Reprinted from Nemeth, Pette, and Vrbová 1981.

tified as glycogen depleted (either because they were not in the sample examined or were not sufficiently depleted) would not be different in enzyme activity from those that were, so that the sampled fibers represent the entire unit. We can see that these assumptions might be problematic.

More recently, evidence has been presented that muscle units may not be uniform (Larsson 1992). In this experiment, it was found that muscle fibers within a motor unit in rat tibialis anterior were spread out into both high-oxidative, deep and low-oxidative, superficial regions of the muscle. Examination of the oxidative enzyme activity in single glycogen-depleted fibers (using quantitative histochemical techniques) revealed that, within a given motor unit, fibers that were within the deep muscle region were more oxidative than fibers in the superficial region.

It thus appears that, at least in rat muscles, motor units are not uniform in mitochondrial enzyme activity and perhaps in other properties. Since it is easier to see all muscle unit fibers in a cross section of rat muscle than it is in larger animals (or humans), this evidence suggests that our previous notions of homogeneity of fibers within the muscle unit should be revisited. Clearly, studies similar to those of Larsson must be continued.

Determining Human Muscle Unit Types

Generally speaking, it may be that human muscle units are organized in a way that is different from those of cats, rats, and mice—and most other mammalian species, for that matter (Bigland-Ritchie, Fuglevand, and Thomas 1998). The main problem in answering this question unequivocally has been the technical constraints in measuring motor unit properties in humans. Three techniques have been used to do this: intramuscular microstimulation, intraneural microstimulation, and spike-triggered averaging.

Intramuscular microstimulation involves the insertion of a fine-wire electrode into the muscle and application of electrical pulses of varying voltages until an all-or-none electromyogram (EMG) is recorded from the muscle of interest. Intraneural microstimulation is similar, except that it involves the insertion of a tungsten micro-electrode, insulated except for the tip, into a peripheral nerve and adjusting its position until stimulation evokes a measurable twitch contraction and EMG in the muscle of interest.

Spike-triggered averaging is a technique whereby the properties of muscle unit isometric twitches are extracted from the whole-muscle force record during a voluntary contraction. This is done by recording the compound action potential of any single motor unit during a sustained contraction and triggering the computer to average the force for a fixed period of time after the appearance of this distinct, reproducible spike. Since all units other than the unit of interest contribute their forces randomly around this one, the twitch response, which is synchronized with its action potential, emerges from the averaged force signal after several hundred or even several thousand after-spike force episodes are averaged.

These techniques have both advantages and disadvantages. The stimulation techniques allow us to study muscle units in a manner similar to that used in the anesthetized animals from which our initial ideas about muscle units have come. However, as with the animal experiments, these techniques give us little information about how the units are used during movement. The spike-triggered averaging technique gives us more information in this arena, since motor units are characterized according to the whole-muscle force at which they are recruited (because subjects must maintain a sustained voluntary contraction during this procedure). Thus, we can obtain information regarding the strength of each recorded muscle unit as a function of the whole-muscle force at which it is recruited. However, this technique requires that the same action potential be recognizable throughout the experiment, which allows for little tolerance in electrode movement. In addition, the frequency of firing of the unit of interest must be low enough to ensure that summation of twitches, which would distort the true shape of the twitch, does not occur. This means that voluntary contractions are usually of low force, and thus the motor units studied are generally of low threshold. Some differences in twitch properties have been found when comparing spike-triggered averaging and intramuscular stimulation (Thomas, Bigland-Ritchie, et al. 1990). In any case, whatever techniques are used, there is a dearth of information regarding muscle unit properties in humans, especially for the large limb muscles, because of the problems associated with recording small forces from these large muscles. Most of our information regarding human muscle unit properties is consequently from hand and arm muscles. In figure 2.12, I include a comparison of motor unit profiles from animal and human experiments to give some idea of the similarities and differences. As one might expect, human motor units are stronger and have slower twitches than those of cats. Despite these quantitative differences, for the purposes of further discussion, we assume that motor unit organization is similar between humans and cats.

The Motoneuron Component of the Motor Unit

CONCEPT

3 The properties of motoneurons covary with the properties of their innervated muscle fibers in ways that determine how easily and in which order the motor units are initially recruited during voluntary activation, the

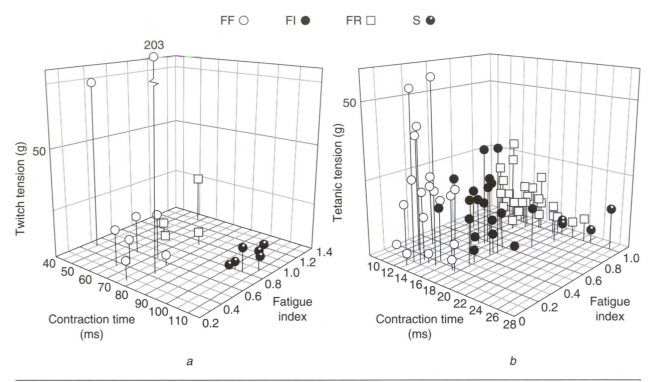

Figure 2.12   Distribution of motor units in (*a*) human and (*b*) rat gastrocnemius according to twitch contraction time, twitch tension, and fatigue index.

a Adapted from Garnett et al. 1979.

extent of additional recruitment as voluntary drive increases, and their recruitment behavior during sustained activation. In this section, we consider this coordination of nerve–muscle properties. In later chapters, we consider how these properties are implicated in fatigue and in adaptation during exercise training. A summary of the properties to be discussed and their relative differences among the motor unit types is presented in table 2.1.

Motoneuron Size

Decerebrate cats were initially used to demonstrate the orderly recruitment of motor units according to the size of their motoneurons (measured by the amplitude of the action potentials in the ventral root). In these ground-breaking experiments (Henneman 1981; Henneman, Somjen, and Carpenter 1965), recruitment of hindlimb motoneurons was elicited by stretching the muscle. This is possible in this situation since decerebration releases inhibition of the gamma motoneurons, thus resulting in increased extensor tone. Henneman and his colleagues showed that motoneurons' responsiveness to stretching of their innervated muscle was a function of the

amplitude of the action potential recorded in the ventral root and thus of motoneuron size. This has become known as the *Henneman size principle* of motor unit recruitment. Furthermore, it appeared that differences in motoneuron size, and not in quality of synaptic input to motoneurons, were the determining factor in this stretch-induced recruitment order, since recruitment order was the same with a variety of excitatory and inhibitory stimuli, arising from both ipsilateral and contralateral sources, via physiological or electrical stimulation, mono- and polysynaptically, alone and in combination.

Motoneuron Size vs. Unit Type

Since estimated motoneuron size appears to co-vary with motor unit type (see table 2.1), one might wonder which is more important in fixing recruitment order: motoneuron size or type of fibers innervated? In fact, size appears to be a more important consideration than the type of muscle fibers innervated, since recruitment according to motoneuron size is still evident in the slow soleus, which contains only S-type motor units (Binder et al. 1983). Although stimulation of cutaneous afferent nerves might

Table 2.1
Properties of Cat Motoneurons Innervating Different Muscle Unit Types

	S	FR	FF	Source
Soma diameter (micrometers)	49	53	53	4
Total membrane area (micrometers)	249	323	369	4
Stem dendrite number	12	12.6	10	4
Input resistance (megaohms)	1.6–2.6	0.9–1	0.6	1, 4
Rheobase (nanoamperes)	5.0	12.0	21.3	1
Threshold depolarization (millivolts)	14.4	18.5	20.1	3
Afterhyperpolarization duration (milliseconds)	161	78	65	1
Minimum firing frequency (impulses per second)*	10		22	8
Maximum firing frequency (impulses per second)*	20		70	8
Current/frequency slope (impulses per second per nanoampere)	1.4	1.4	1.4	5, 6, 7, 8
Late adaptation	+	+++	+++	2
Membrane bistability	++		+	9

* Primary range of firing

References: (1) Zengel et al. 1985; (2) Spielmann et al. 1993; (3) Gustafsson and Pinter 1984a; (4) Burke et al. 1982; (5) Kernell 1965a; (6) Kernell 1983; (7) Kernell 1992; (8) Kernell 1979; (9) R.H. Lee and Heckman 1998a, 1998b.

alter the strict size-related recruitment during stretch, attesting to the possibility of qualitative differences in synaptic inputs onto small versus large motoneurons (Kanda, Burke, and Walmsley 1977), this phenomenon appears to be more the exception than the rule (B.D. Clark, Dacko, and Cope 1993) and may be different in soleus than in medial gastrocnemius (Sokoloff and Cope 1996). Consistent with the relationships apparent among motoneuron size estimates and muscle unit properties, it is not surprising, then, that motor units in the triceps surae of decerebrate cats are recruited during stretch in the order of increasing axon conduction velocity and tetanic force and decreasing twitch contraction time and fatigue index, in a very reproducible manner, with very few exceptions (Cope et al. 1997).

This orderly recruitment based on motoneuron size appears to be the same during the stretch reflex in awake human subjects (Calancie and Bawa 1985). In addition, the response of recruited motor units in human wrist extensor muscles to tendon taps (to excite primary spindle afferents) was shown to be consistent with the orderly recruitment of motor units based on muscle unit size, motoneuron size, and Ia afferent efficacy (Schmied et al. 1997; figure 2.13). Finally, both percutaneous electrical stimulation (Rothwell et al. 1991; Gandevia and Rothwell 1987) and transcranial magnetic stimulation (Bawa and Lemon 1993; Rothwell et al. 1991) of the cortex result in recruitment of motor units in hand and forearm muscles that is consistent with the size principle.

How Is Motoneuron Size Determined?

Motoneuron size is not an easy thing to determine. Consider that the cell body constitutes only about 5% of the total cell volume. Thus, while one can easily see the cell body using simple histological techniques, the possibility for error in determining cell size is high. While it is true that the generation of action potentials occurs at the cell body, the electrophysiological events leading up to the action potential are membrane-associated events. For this reason, estimates of cell size must take into account the

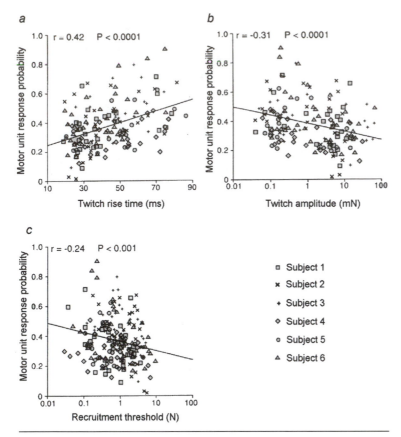

Figure 2.13 Discharge of motor units in wrist extensors was recorded while stimulating the homonymous muscle spindles by means of tendon taps. Changes in firing rate probabilities were more pronounced in (*a*) slower, (*b*) weaker, and (*c*) lower-threshold units, supporting the idea that recruitment via spindle afferent excitation occurs according to the size principle.

Reprinted from Schmied et al. 1997.

surface area of membrane taking part in these events, including proximal dendrites.

There are several ways that cell size can be expressed, taking into account the total cell membrane area. One way to do this is to deduce the relative cell size from the caliber of its axon. Figure 2.14 shows that this seems to be fairly accurate; this is the basis for Henneman's original proposal for the size principle, since he showed that motor axons were recruited in decerebrate cats in order of increasing amplitude of the axonal spike and thus inferred that larger axons, and thus larger motoneurons, were recruited after smaller ones.

A second technique has been to attempt to measure total surface membrane area histologically. This involves estimating the combined vol-

umes of the cell body and the trunk dendrites near the soma and making some assumptions about the length of dendrites from these measurements. An example of the results of this technique from a study of a sample of cat hindlimb motoneurons is shown in figure 2.15. (Unfortunately, we have no data like this from human motoneurons yet.) An important observation from this figure is that estimated total membrane area showed a greater difference among motor unit types than did soma diameter: The former were different from each other, but not the latter. This is due to the probability that the number and size of dendrites differ systematically among the various muscle unit types (see table 2.1).

A third estimate of motoneuron size is provided by measurement of cell capacitance, using electrophysiological techniques. Motoneurons behave much like a resistor and capacitor in parallel, such that a sustained injection of subthreshold current produces a voltage change that is a combination of the response of these two elements (figure 2.16). By measuring the time constant of the voltage change and assuming that the motoneuron's soma and dendrites can be modeled as a cylinder, one can determine the total cell capacitance. Since it is assumed that motoneuronal membranes possess a standard capacitance per unit of membrane area, one can estimate the total cell membrane area. Representing the motoneuron and its dendrites as a cylinder and making assumptions about the value and constancy in different parts of the motoneuron of the specific membrane capacitance limit the precision of this technique (Kernell and Zwaagstra 1989). For all intents and purposes, currently none of the techniques discussed is superior to the others in the determination of cell size: They are all estimates.

Motoneuron Input Resistance

Motoneurons are integrative transducers in that they integrate the voltage responses of the soma and dendrites to thousands of synaptic

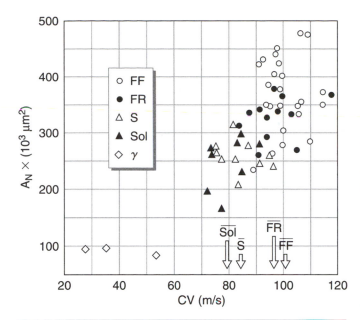

Figure 2.14 Axon conduction velocity (abscissa) versus estimated total membrane surface area (ordinate) in cat motoneurons innervating soleus and medial gastrocnemius. Means for conduction velocity of motor unit types are indicated by arrows.

Reprinted, by permission of Wiley-Liss, Inc., a subsidiary of John Wiley & Sons, Inc., from R. Burke et al., 1982, "An HRP study of the relation between cell size and motor unit type in cat ankle extensor motoneurons," *Journal of Comparative Neurology* 209: 23.

currents and transduce this integrated signal into a voltage change at the initial segment of the axon, which is the most excitable portion of the motoneuron. Here, if threshold conditions are met, a single action potential or a series is generated and propagated down the axon to the neuromuscular junctions of the muscle fibers. Thus, the intrinsic excitability of a motoneuron plays a major role in determining the probability of its being recruited during excitation of a motor pool. Motoneuron intrinsic excitability has been investigated using intracellular electrophysiological techniques.

Motoneuron excitability is determined by two factors: input resistance (R_{in}) and threshold depolarization (V_{th}). The latter, which is the amount of depolarization from resting membrane potential that is necessary for an action potential to be generated, is discussed in the next section. Input resistance is simply Ohm's law applied to the motoneuron, which, because of the composition of its membrane, presents a resistance to the outflow of current that flows through it:

$$R_{in} = V/I \qquad (2.1)$$

where V is potential in volts and I is current in amperes.

Measuring Input Resistance

Neuron input resistance is measured by injecting small, subthreshold depolarizing or hyperpolarizing currents into a motoneuron at or near its resting membrane potential and measuring the resultant voltage change. A larger change in membrane potential in response to a small standard current injection indicates a more excitable cell. Small current amplitudes are used to avoid relatively large deviations from resting membrane potential that might bring into play voltage- and time-dependent nonlinearities in membrane response.

Cell input resistance varies considerably within a motor pool (figure 2.17) and among motor unit types (table 2.1). Zengel and colleagues (1985) suggest that in cat medial gastrocnemius, cell input resistance may be one of the best passive membrane properties of motoneurons by which to predict the properties (speed, fatigue resistance) of the innervated muscle unit and thus that this property and the contractile characteristics (as well as the number) of fibers in the muscle unit are somehow linked.

Specific Membrane Resistivity

What are the prime determinants of cell input resistance? The original thought was that cell volume was the prime determinant and that specific membrane resistivity (expressed in ohms times centimeters squared) was constant across all motoneurons. Thus, larger cells with more membrane and thus more resistances in parallel would have lower resistances as a consequence (because more current can flow out across the more numerous resistances in parallel). It appears, however, that specific membrane resistivity may vary systematically among motoneurons, with smaller cells having disproportionately more-resistive membranes per unit of membrane area. This has been difficult to verify experimentally, since it involves detailed morphological and electrophysiological measurements on the same cells. These measurements have been made, however, and suggest

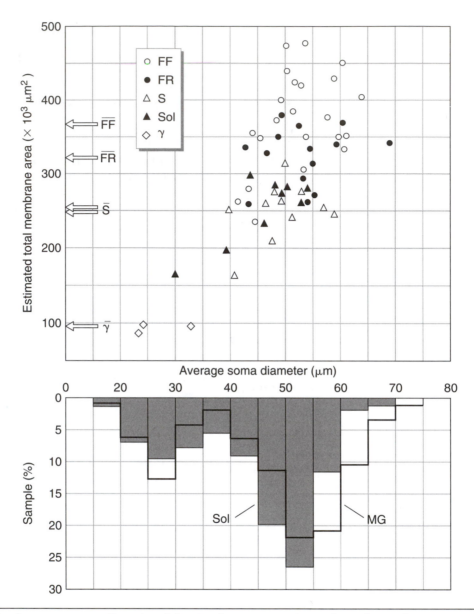

Figure 2.15 Average soma diameter (abscissa) versus estimated total membrane area (ordinate) in motoneurons innervating cat medial gastrocnemius and soleus. Means of total membrane area for the motor unit types are indicated by arrows.

Reprinted, by permission of Wiley-Liss, Inc., a subsidiary of John Wiley & Sons, Inc., from R. Burke et al., 1982, "An HRP study of the relation between cell size and motor unit type in cat ankle extensor motoneurons," *Journal of Comparative Neurology* 209: 21.

that smaller cells have disproportionately higher membrane resistivity than larger cells (Burke et al. 1982; Kernell and Zwaagstra 1981; figure 2.18).

Perhaps the strongest support for systematic variation in specific membrane resistivity among motoneurons is simply the difference between the range of motoneuron size and the range of cell input resistance. In fact, while we have seen that input resistance varies signifi-

cantly and systematically among motoneurons of different muscle unit types, measurements of cell size using light microscopy indicate that the motoneurons innervating different muscle unit types are very similar in size (compare the range of values for input resistance of cat hindlimb motoneurons in figure 2.17 with the range of motoneuron areas in figure 2.14).

There are two major caveats for these conclusions about specific membrane resistivity and

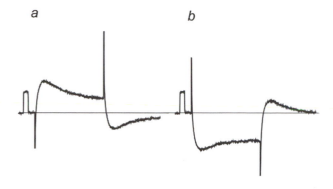

a   *b*

Figure 2.16 Intracellular membrane voltage response of the same rat tibial motoneuron to (*a*) depolarizing and (*b*) hyperpolarizing current pulses of 1 nA lasting 150 ms. Notice the voltage "overshoot" followed by sag that occurred after initiation and termination of the current pulses (which are indicated by large on and off voltage artifacts). Calibration pulses at the beginning of the traces are 1 mV for 10 ms.

motoneuron size. First, when estimating total cell membrane area either using microscopic images or electrophysiologically, several assumptions are necessary. Morphological analyses use the trunk diameter of each of the stem dendrites to estimate the total surface area of the membrane covering the dendrites; this is an important consideration, since together the dendrites constitute over 90% of the neuron's surface membrane. Electrophysiological estimates rely on the use of a constant value for membrane capacitance, around 1 microfarad per square centimeter, in order to arrive at whole-cell capacitance and thus membrane area. A second caveat has to do with the assumption that membrane resistivity is constant throughout the cell. Indications are now that this assumption is probably false and that specific resistivity of dendritic membranes is higher than that of somal membrane (W. Rall et al. 1992). This issue raises the questions of whether measurements of whole-cell input resistance are meaningful at all and whether the location of the recording electrode within the cell influences this measurement for any given cell. The interesting facts remain, however, that whole-cell input resistance varies somewhat systematically across motor unit types and that the membranes of motoneurons that innervate different muscle unit types are probably qualitatively different in that membrane specific resistivity also seems to vary systematically.

Rheobase Current and Threshold Depolarization

Of perhaps more practical significance in discussing excitability of motor units within a recruitment scheme is the amount of transmembrane current required to generate an action potential in the innervating motoneuron. Like input resistance, this is determined using intracellular electrophysiological techniques.

The magnitude of rheobase current (I_{rh}) depends on a number of factors, including the resting membrane potential (RMP), the amount of voltage change required to reach the threshold for spike generation (known as threshold depolarization, V_{th}), and the cell input resistance. Cell size influences I_{rh} to the extent that it influences R_{in}: Small cells have disproportionately higher R_{in} (as previously discussed). It appears that, at least in anesthetized animals *in situ,* there is no systematic difference among motoneurons in resting membrane potential, so we can forget this as a factor for the time being (although this may sometimes become significant *in vivo*). Let us also assume that the membrane behaves linearly between resting membrane potential and the voltage at which action-potential generation occurs. Under these circumstances, the amount of current needed to generate an action potential is dictated to a large extent by R_{in}: the larger the R_{in}, the smaller the current required for generating an action potential:

$$I_{rh} = V_{th}/R_{in} \qquad (2.2)$$

If variation among motoneurons in I_{rh} could be completely explained by corresponding variations in R_{in}, one would expect a linear and proportional relationship between input conductance (C_{in}, the inverse of R_{in}) and I_{rh}, with the slope of this relationship corresponding to the average value of V_{th} (Gustafsson and Pinter 1984a). In fact, these two independent measurements are very highly correlated, but the range of I_{rh} exceeds that of C_{in} by a factor of around 2. In figure 2.19, it is apparent that this linear relationship is in fact not so linear; low-rheobase cells are mostly below the line of best fit, while high-rheobase cells are above the line. Since the slope of the line gives the average amount of depolarization from the resting membrane potential that is necessary for generation of an action potential (V_{th}, which ranges from 10 to

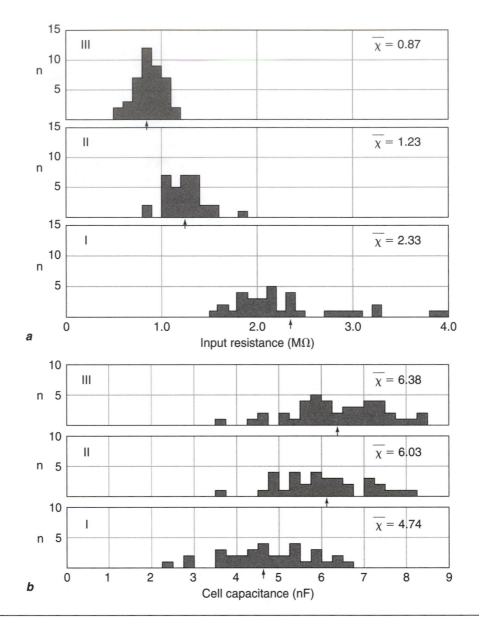

Figure 2.17   Distribution of (*a*) input resistance and (*b*) cell capacitance in a sample of cat lumbar motoneurons *in situ*. Categories I, II, and III correspond to motoneurons with low, medium, and high rheobase currents, respectively. Note the apparent lack of correspondence between input resistance and capacitance.

Reprinted from Gustafsson and Pinter 1984.

25 millivolts in cat motoneurons), the suggestion is that low-rheobase cells require less depolarization to generate an action potential than high-rheobase cells. This difference cannot be attributed to variation in resting membrane potential or spike height.

Another factor that contributes to motoneuronal excitability is nonlinearity in membrane response between RMP and the voltage at which the action potential is generated. For example, the product of I_{rh} and R_{in} (known as rheobase voltage) should give V_{th} under linear conditions, but in fact, the actually measured V_{th} is somewhat greater than the product of I_{rh} and R_{in}. This is due to a persistent inward current, most likely calcium mediated, that occurs at subthreshold voltages between RMP and threshold (Gustafsson and Pinter 1984a; Binder, Heckman, and Powers 1996). This inward rectification, also known as anomalous rectification,

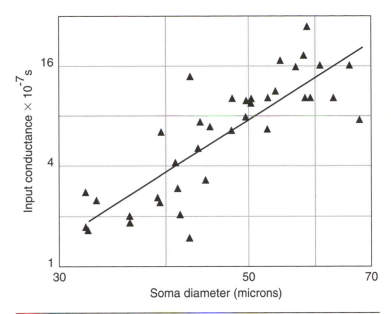

Figure 2.18 Double logarithmic plot of the relationship between input conductance and soma diameter in cat hindlimb motoneurons. The slope of the regression line is 3.3, which is twice as great as the value of 1.6 that one would expect if smaller and larger motoneurons had similar specific membrane resistivities.

Reprinted from Kernell and Zwaagstra 1981.

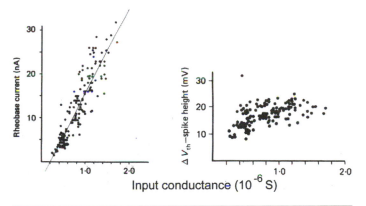

Figure 2.19 Left, relationship between rheobase current and input conductance in cat hindlimb motoneurons. Notice the departure from linearity at low and high rheobase groups (the former are more to the right of the line, the latter to the left). Right, threshold depolarization versus input conductance for the same study. Threshold depolarization tends to be smaller for smaller cells.

Reprinted from Gustafsson and Pinter 1984.

appears to be slightly more significant in smaller than in larger cells, although this variability among cells is most likely secondary to variations in passive properties (Gustafsson and Pinter 1984a).

Thus, small cells may have an additional "advantage" over larger cells with regard to their susceptibility to be recruited. Their passive properties may accentuate the effect of inward currents that operate at depolarized voltages, and at the same time they may require less depolarization (i.e., they have lower V_{th}) to generate an action potential. This latter consideration, assuming that no systematic differences exist among cells in resting membrane potential, would suggest that the voltage at which action potentials occur (the firing level) is slightly (around 6 millivolts) more negative in small type S motoneurons.

Minimal Firing Rates and Duration of Afterhyperpolarization

Immediately following the somaldendritic spike of the motoneuron, the repolarization continues at a level below resting membrane potential for a short period of time before returning to resting membrane potential. The temporary, prolonged hyperpolarization that follows the action potential is termed the slow afterhyperpolarization (sAHP, or simply AHP), and its duration varies systematically among motoneurons of different types (figure 2.20). The AHP is attributed to the activation of a calcium-activated potassium conductance (G_{KCa}), which gradually declines after the spike (Binder, Heckman, and Powers 1996; Kudina and Alexeeva 1992). Injection of a calcium chelator into the cell or exposure of the cell to the bee venom apamin (a specific G_{KCa} channel blocker) or to cobalt or manganese significantly reduces the AHP, indicating that it is controlled by calcium influx. It appears to be the only electrophysiological variable that can unequivocally distinguish fast from slow motor units, in both the cat and the rat (Zengel et al. 1985; Bakels and Kernell 1993; Gardiner 1993). The reason for the difference in AHP time course between slow and fast motoneurons is unknown, but may involve differences in the density, location, or activation and deactivation kinetics of the G_{KCa} channels or in the regulation of the intracellular calcium levels that activate them. Its importance for our discussion is in the rhythmic firing patterns of motoneurons in response to sustained excitation.

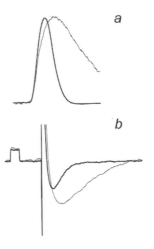

Figure 2.20   (*a*) Muscle unit isometric twitch and (*b*) corresponding AHP response of the innervating motoneuron from rat gastrocnemius motor units. The fast unit is designated by a thick line, the slow unit by a thin line. Calibration at the beginning of AHP is 10 ms (for both twitch and AHP) and 1 mV (AHP only). Twitch forces have been normalized for ease of comparison.

Reprinted from Gardiner 1993.

Figure 2.21, from Kernell's seminal work (1965c), shows the minimal firing frequency of cat hindlimb motoneurons in response to sustained currents injected through a microelectrode as a function of AHP duration (shown as the reciprocal of the time to afterdepolarization). Significant correlations were also found between AHP time course and maximal firing frequencies, although they were not as impressive (r = 0.58 to 0.76). In this paper, the author also demonstrated the relationships between limits of motoneuron firing rates and the corresponding expected muscle fiber responses at those frequencies. Minimal firing frequency was shown to correspond to a frequency that would result in slightly fused twitch responses of the innervated muscle fibers, while the maximal firing rates in the primary and secondary ranges (explained later in this chapter) correspond to approximately 80% and 95%, respectively, of maximal tetanic force. This appears to be the same for fast and slow muscles; thus, the firing frequencies of slow and fast motoneurons, dictated to some extent by their AHP, appear to correspond quite conveniently to the contractile speed of the innervated muscle fibers. The fact that minimal firing frequencies of motoneurons produce not individual twitches but slightly fused contractions most likely has an implication for the smoothness of contractions

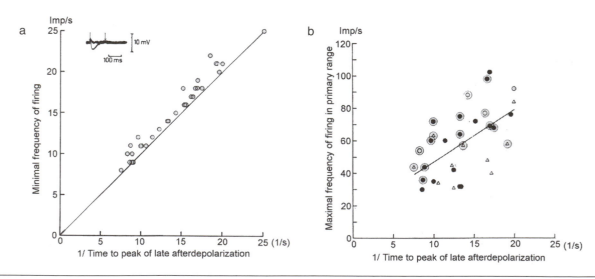

Figure 2.21   Relationships between AHP time course (the reciprocal of time to peak of afterdepolarization) and (*a*) minimal and (*b*) maximal firing frequencies. Different symbols in *b* denote cells with different capabilities of sustained firing in the secondary range. (Primary and secondary ranges are explained later in text.)

Reprinted from Kernell 1965c.

at low force requirements, when few motor units are recruited at their lowest frequencies.

There are, however, factors other than AHP duration that control minimal firing rates of human motoneurons during voluntary movement. Kudina and Alexeeva (1992) estimated the characteristics of soleus and flexor carpi ulnaris motoneuron afterpotentials in human subjects from the recovery curve of motoneuron excitability after a single discharge evoked by afferent stimulation or gentle voluntary muscle contraction. They found that the minimal rates of motoneuron firing were not correlated with the estimated AHP duration (figure 2.22), suggesting that other spinal or supraspinal mechanisms controlling motoneuron firing are masked in *in situ* experiments with animals in which anesthesia or spinalization are involved. They also found that some fast motoneurons demonstrated early recovery of excitability, which they attributed to the delayed depolarization that follows the spike, before the downstroke of the AHP. These motoneurons were capable of firing double discharges ("doublets") with a relatively rapid interspike interval (5–15 milliseconds).

Additional evidence that motoneuron firing rates may not be controlled only by AHP time course comes from decerebrate cats in which fictive locomotion has been induced via stimulation of the mesencephalic locomotor region. Under these conditions, AHP amplitudes in motoneurons during trains of impulses are significantly decreased compared with amplitudes seen during comparable frequencies of firing induced by intracellular current injections. In addition, motoneurons showed an increased gain (increase in firing frequency per unit current injected) during locomotion compared with the resting condition (Brownstone et al. 1992). Thus, the AHP–firing frequency relationship may constitute a blueprint, evident in anesthetized animals, upon which influences from afferent sources can induce modifications. We will see examples of this (e.g., the effect of serotonin) further on in this chapter.

The AHP is not an absolute refractory period in that a second action potential can be generated during the AHP with strong enough currents. Figure 2.23 shows the evolution of the AHP in a motoneuron firing at low and high

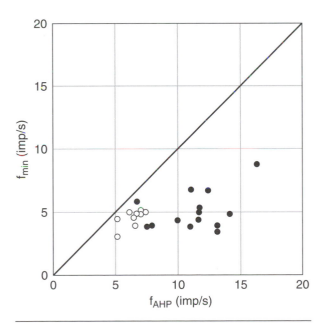

Figure 2.22   Minimum motor unit firing rates (ordinate) versus estimated AHP duration (abscissa) for human soleus (open symbols) and flexor carpi ulnaris (filled symbols) motor units. Firing rates are lower than one would predict from the AHP duration (line of identity).

Reprinted from Kudina and Alexeeva 1992.

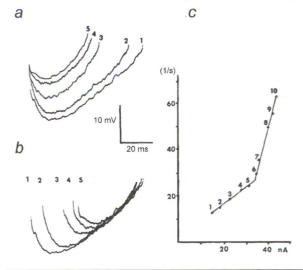

Figure 2.23   Membrane potential trajectories in a cat motoneuron firing at different rates in response to current injection. The frequency–current relationship is shown in *c*, where numbers correspond to the recordings in *a* and *b*. Alignment of AHPs in *a* is according to original DC levels, and in *b* the same recordings are aligned to a common firing level (i.e., the voltage at which the following spike occurs). Note how AHP gets shorter and smaller in amplitude as firing rate increases.

Reprinted from Schwindt and Calvin 1972.

frequencies. What happens to the AHP when changing from a single spike to a series of spikes during rhythmic firing is dealt with in the next section.

The Motoneuron Current–Frequency Relationship

Up to this point, we have considered recruitment during a relatively short excitation of the motoneuron pool. Let us consider now the additional factors that determine the behaviors of the system when more prolonged, sustained excitations are instituted.

In moving from generation of single action potentials to trains of impulses, we begin to see the influence of time-dependent and voltage-dependent membrane nonlinearities in which motoneuron excitability can change as a function of the time and intensity of sustained current injection. For example, when rheobase current is sustained, trains of impulses do not occur until the current intensity is increased approximately a further 50% (called the current threshold for rhythmic firing; Kernell and Monster 1981). A principal contributing factor is the sag property of the membrane in response to prolonged current injection (see figure 2.16). This

is a time-dependent, mixed-cation conductance active around resting membrane potential (RMP), which gradually alters membrane voltage in the hyperpolarizing direction during depolarizing pulses and in the depolarizing direction during hyperpolarizing pulses. Interestingly, this sag property of membranes appears to be more significant in motoneurons that innervate fast muscle units. Contribution of G_{KCa} to this sag response becomes more significant with larger depolarizations. It has been proposed that this membrane sag may explain part of the difference between fast and slow motoneurons in the time course of their afterpotentials (Gustafsson and Pinter 1985).

The current–frequency relationship for a typical motoneuron is shown in figure 2.24. With increasing current intensity above the current threshold for rhythmic firing, frequency of firing increases, as one might expect. What is less expected is that the current–frequency relationship has two linear portions, termed the primary and secondary ranges of firing (Granit, Kernell, and Shortess 1963a; Kernell 1965b). This phenomenon has been observed in anesthetized cat motoneurons and in preparations of several species of motoneurons *in vitro*. Whether or not this is a property of all moto-

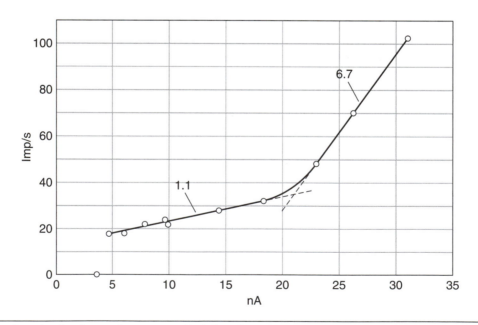

Figure 2.24   Steady discharge frequency (1–1.5 s after onset of current injection, ordinate) versus current strength (in nanoamperes, abscissa) for a cat motoneuron. Primary and secondary range slopes (in impulses per second per nanoampere) are shown.

Reprinted from Kernell 1965b.

neurons is not known; it may be that this property is lost in some motoneurons as a result of electrode impalement. The mechanism by which moto-neurons switch from primary to secondary range of firing with increasing current injection is unknown. One hypothesis is that the increase in the thresh-old voltage for spike generation results in a membrane trajec-tory between spikes that is more depolarized due to accom-modation at the initial segment (figure 2.25). This allows activa-tion of the persistent inward current, which increases firing frequency at the same level of injected current and thus in-creases gain.

Maximum sustainable firing frequencies of motoneurons are difficult to measure. Although firing frequencies at the begin-ning of an intense injection of current can be very high and can exceed the frequencies necessary for tetanization of their innervated muscle fibers, these high frequen-cies fall off very rapidly due to adaptation.

One would suspect that the most efficient system for gradation of whole-muscle force up to the maximum would be one in which the first recruited motoneuron is still increasing in firing frequency at the time when the last unit is re-cruited. The best model for this is one in which the slopes of increased firing frequency as a function of injected current are approximately equal for all motoneurons in the pool (Kernell 1984). There is some evidence that this is in-deed the case (Kernell 1979).

Late Adaptation

Adaptation is the process by which firing fre-quency decreases with sustained constant-intensity excitation. Late adaptation is adapta-tion that occurs after tens of seconds and even after minutes, as opposed to the very rapid adap-tation that occurs at the beginning of a burst. Figure 2.26 shows examples of early and late adaptation (Sawczuk, Powers, and Binder 1995).

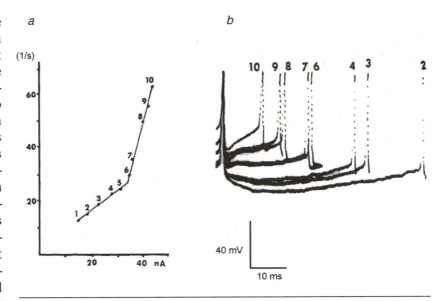

Figure 2.25 Change in firing level (voltage at which spike is gener-ated) during repetitive firing in a cat motoneuron. Numbers in *b* corre-spond to their frequency in *a*. In *b*, recordings have been aligned to the peak of the preceding spike. Note how the firing level increases as rate increases, especially in the secondary range of firing (from spike 6 to 10).

Reprinted from Schwindt and Calvin 1972.

Late adaptation varies among the motor unit types; it is more developed in motoneurons that innervate fast muscle units (Spielmann et al. 1993). The mechanism most likely involves an increased outward current, mediated by a calcium-activated potassium conductance (G_{KCa}). By this mechanism, calcium entering the cell during rhythmic action potential generation activates an outward potassium conductance, which increases the amplitude and duration of the outward potassium current following the spike. As calcium accumulates in the cell, the excitability for subsequent spike production during continued current injection becomes lower. The apparent systematic difference be-tween fast and slow motoneurons (see figure 2.26) in late adaptation to sustained excitation may indicate different rates of calcium entry and accumulation during rhythmic firing, differing capacities to buffer accumulating calcium, or different spatial relationships between calcium accumulation and channel sites in the moto-neuron's membrane. Of particular interest is that late adaptation is evident not only during sustained excitation, but also during inter-mittent excitation, such as during rhythmic

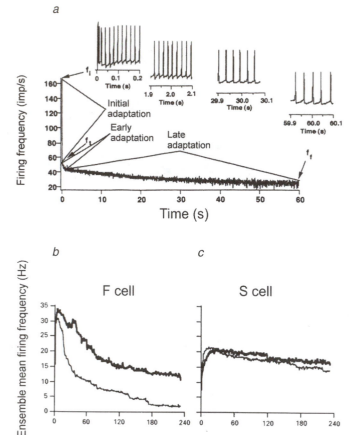

Figure 2.26 Motoneuron adaptation. (*a*) Adaptation during repetitive discharge of rat hypoglossal motoneurons *in vitro,* in response to sustained intracellular current injection. Three phases of frequency decrease have been identified, as shown. (*b* and *c*) Frequency responses of cat hindlimb motoneurons to sustained (thin lines) and intermittent (600 ms at 1 Hz, thick lines) excitation via extracellular current application. Current intensity was 1.25 times the threshold for repetitive firing. Note that fast motor units show more late adaptation.

a Reprinted from Sawczuk, Powers, and Binder 1995.
b-c Reprinted from Spielmann et al. 1993.

exercise (Spielmann et al. 1993). The phenomenon of late adaptation is particularly interesting because it relates to the types of contractions that are involved in exercise. As we shall see during our consideration of neuromuscular fatigue (chapter 3), the process of motoneuronal adaptation may be an important component, not necessarily in limiting the force produced, but certainly in increasing the excitatory "effort" required to sustain a specific level of neuromuscular effort.

Motoneuron Membrane Bistability

Membrane bistability concerns a shift between two states of excitability (hence bistability) via the development of long-lasting plateau potentials that outlast the brief periods of excitation that generate them. An example of membrane bistability is shown in figure 2.27. This change from one level of excitability to another is caused at least in part by a voltage-sensitive, non-inactivating, inward, calcium conductance, which is maximally activated near threshold voltage, but it may also involve a calcium-mediated, non-specific-cation current (R.H. Lee and Heckman 1998b). It can be generated by direct intracellular or synaptic excitation and lasts for up to several seconds after the excitation has been terminated, due probably to the slow inactivation kinetics of the channels involved. It manifests itself as a sustained decrease (i.e., shift toward threshold) in the membrane potential or, if the cell is already firing, as an increase in firing rate, which can outlast the original stimulation by several seconds. It can be reversed by an inhibitory influence, such as a hyperpolarizing pulse. It can be considered as a type of warm-up phenomenon for motoneurons (Bennett et al. 1998). Nonvolatile anesthetics, such as barbiturates, used during *in situ* experiments inhibit the generation of plateau potentials, and thus they are seen only in experiments where these are not used, such as in decerebrate animals or with tissue slices *in vitro* (Guertin and Hounsgaard 1999).

Several substances operate to uncover plateau potentials in motoneurons, such as serotonin (5-HT), norepinephrine, thyrotropin-releasing hormone (TRH), angiotensin II, substance P, oxytocin, cholecystokinin, and somatostatin (Hultborn and Kiehn 1992). The most intensely studied modulator of plateau potentials in motoneurons is serotonin. This modulator appears to allow the expression of plateau potentials in motoneurons by enhancing the inward rectifier (sag) current and by reducing the G_{KCa}, which is the outward current responsible for the AHP (figure 2.28). The various modulators listed earlier probably function by several different mechanisms, all of which result in uncovering of the bistability phenomenon.

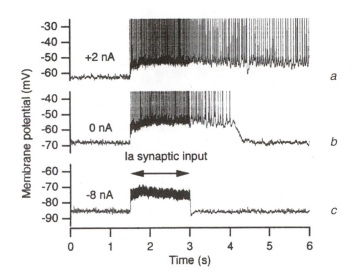

Figure 2.27 Example of bistability in a motoneuron in a decerebrate cat. (*a* and *b*) High-frequency, low-amplitude tendon vibration of the triceps surae for 1.5 s evoked firing that continued after the stimulation ceased. (*c*) At hyperpolarized membrane potential, bistability was not present.

Reprinted from Lee and Heckman 1998.

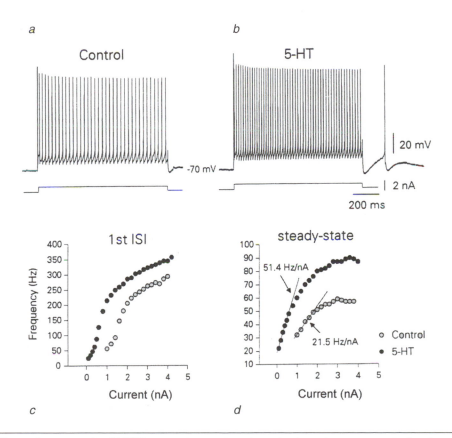

Figure 2.28 Effect of serotonin (5-HT) on guinea pig trigeminal motoneurons in brain stem slices *in vitro*. Serotonin increased firing frequency for a given current (*b* vs. *a*) and increased the frequency–current slope (*c* and *d*).

Reprinted from Hsiao et al. 1997.

Recent work suggests that low-threshold motoneurons (those that innervate type S and FR muscle units) demonstrate more marked bistability (show self-sustaining firing for longer periods) than higher-threshold motoneurons (R.H. Lee and Heckman 1998b). This suggests a functional importance for this property in postural and longer-lasting tonic and rhythmic locomotor activities. Indeed, when monoamines in the spinal cords of rats were depleted using specific neurotoxins, the EMG pattern of the postural soleus muscle changed from a tonic to a more phasic one (Kiehn et al. 1996).

As can be imagined, detecting the presence of motoneuron bistability in an awake, voluntarily moving animal or human is difficult, since the intensity of excitation of motoneurons is not known or controllable. Eken and Kiehn (1989) examined EMG records of single motor units in rat soleus and found behavior resembling bistability in the presence of excitatory and inhibitory stimulation (of afferents and skin, respectively). Gorassini, Bennett, and Yang (1998) provided evidence that this phenomenon may be of physiological significance in humans performing voluntary contractions (figure 2.29). In their experiments, they asked their subjects to perform low-force, constant-effort, isometric contractions of the tibialis anterior muscle and monitored the firing rate of a low-threshold control unit. When the firing rate of this control unit stabilized at a constant frequency, vibration of the muscle tendon resulted in the recruitment of a second test unit, which maintained its firing after the vibration was terminated (the vibration lasted 0.5 to 1.5 seconds). This occurred in spite of no change in the firing rate of the control unit after vibration, thus suggesting that descending drive to the motoneuronal pool was the same after as before the vibration. Sustained activity of the second unit occurred in about 50% of the units that were evoked by the vibration and lasted from a few seconds to a few minutes after cessation of the vibration. These researchers also found that repeated vibrations resulted in an increased number of spikes of the newly

recruited unit. In most cases, the sustained firing remained after vibration, even when the subjects were instructed to decrease their effort slightly, during which time the firing rate of the control unit actually decreased. Thus, even during effort requiring constant-level drive of motoneuron pools, the same facilitated firing behavior may be at work in humans as has been demonstrated in reduced animal preparations. The implications for a possible reversal of the order of recruitment of motor units during voluntary effort due to this phenomenon are not clear yet. However, if bistability is organized in a manner similar to input resistance, such that bistability decreases in the order S > FR > FF, the order of recruitment might remain unchanged if all motoneurons in the pool were given the same plateau-promoting stimulus. More information than this is lacking at present.

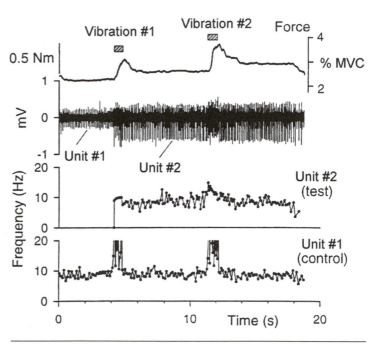

Figure 2.29 Evidence of motoneuron bistability in humans during sustained firing of a tibialis anterior motor unit recruited by brief vibration. The subject was instructed to maintain constant dorsiflexion force while firing of unit 1 was monitored. Brief vibration of the tendon recruited a second unit (unit 2) that continued to fire after the vibration, without an increase in firing rate of the control unit.

Reprinted from Gorassini, Bennett, and Yang 1998.

The Heckman–Binder Model of Motor Unit Recruitment

CONCEPT

4 Now that we have laid the groundwork for a system whereby recruitment according to specific rules is possible, we are ready to apply these to a model of recruitment. Following is a simple model, published by Heckman and Binder in 1991, by which a designated set of motor units making up a pool is recruited to produce a brief, isometric force at several force levels up to maximum. It is instructive because it demonstrates how recruitment and rate coding, in conjunction with the known relationships among motoneuron excitability, motor unit size, and contractile speed, interact to generate simple contractions. Given that the ranges and distributions of motoneuron excitabilities and motor unit sizes vary among

muscles, as does the role of recruitment versus rate coding as a consequence, the patterns change accordingly from pool to pool. Nonetheless, the general ideas are the same as in this model, and the possible influence of these factors on recruitment strategies can be determined by changing the data set. Although the originally published model used data from experiments with cats, in this application of the model, data from rats are used. Data for humans are not available. Data are now available for the rat and the cat, and this model can also be used to determine the functional impact of several models of increased and decreased usage, which were investigated using the rat model, on motor unit recruitment.

The data set, representing a pool of 100 motor units in rat gastrocnemius, is presented in figure 2.30. The first three columns represent the motor unit number, the maximum tetanic force

Motor unit	F_{max}	i threshold	f threshold	G1	f transition	G2	Firing freq	P	Tf
Number	grams	nA	Hz	Hz/nA	Hz	Hz/nA	Hz		s
1	3.046	0.923	8.000	8.0	45.000	12.0	0.000	1.382	28.313
5	3.394	1.020	8.380	8.0	47.000	12.0	0.000	1.384	29.594
10	3.884	1.156	8.860	8.0	49.000	12.0	0.000	1.396	30.970
15	4.445	1.309	9.340	8.0	51.000	12.0	0.000	1.406	32.345
20	5.088	1.484	9.820	8.0	53.000	12.0	0.000	1.416	33.718
25	5.823	1.681	10.300	8.0	55.000	12.0	0.000	1.425	35.091
30	6.665	1.905	10.780	8.0	57.000	12.0	0.000	1.433	36.463
35	7.628	2.159	11.260	8.0	59.000	12.0	0.000	1.441	37.835
40	8.731	2.446	11.740	8.0	61.000	12.0	0.000	1.449	39.205
45	9.993	2.772	12.220	8.0	63.000	12.0	0.000	1.455	40.576
50	11.437	3.141	12.700	8.0	65.000	12.0	0.000	1.462	41.946
55	13.090	3.560	13.180	8.0	67.000	12.0	0.000	1.468	43.315
60	14.982	4.034	13.660	8.0	69.000	12.0	0.000	1.474	44.684
65	17.148	4.571	14.140	8.0	71.000	12.0	0.000	1.479	46.053
70	19.626	5.179	14.620	8.0	73.000	12.0	0.000	1.484	47.422
75	22.463	5.869	15.100	8.0	75.000	12.0	0.000	1.489	48.790
80	25.710	6.650	15.580	8.0	77.000	12.0	0.000	1.494	50.158
85	29.426	7.536	16.060	8.0	79.000	12.0	0.000	1.498	51.526
90	33.679	8.539	16.540	8.0	81.000	12.0	0.000	1.503	52.893
95	38.547	9.676	17.020	8.0	83.000	12.0	0.000	1.507	54.261
100	44.118	10.964	17.500	8.0	85.000	12.0	0.000	1.510	55.628

Input current **0.0** nA   Total force **0.0** grams

Figure 2.30 Database for an application of the Heckman–Binder motor unit recruitment model. Motor units in rat gastrocnemius are presented, from weakest to strongest and from most excitable to least excitable.

Adapted from Heckman and Binder 1991.

of the unit (F_{max}), and the current threshold necessary for minimal firing (i threshold). The units are listed from lowest to highest for both F_{max} and i threshold, based on actual data (Bakels and Kernell 1993; Gardiner 1993). Data for i threshold were derived by multiplying rheobase by 1.5 (Heckman and Binder 1991; Kernell and Monster 1981; Kernell 1965c, 1984). When we activate our model pool, we will inject the same effective current (representing the amount of current generated at the soma) into all motoneurons and see which units are and are not recruited, the firing frequencies of those that are, the resultant force of each of the constituent muscle units, and the total muscle force.

For each motoneuron, besides its i threshold, we must know the minimum frequency at which it begins firing (the f threshold); the gain of the motoneuron, or the rate of change in frequency per unit change in current, above i threshold and in the primary range of firing (G1); the frequency at which the primary range turns into the secondary range (f transition); and the motoneuron gain in the secondary range (G2). Minimum frequency was estimated based on the assumption that twitches begin to summate in force at the lowest frequency (Kernell 1984); therefore, minimal frequencies are approximated by the interpulse intervals equivalent to contraction time plus half-relaxation times of the corresponding muscle units and increase gradually from unit 1 to 100. G1 and G2 are based on unpublished data for rat motoneurons, and just as appears to be the case in cat, I made these constants not differ systematically among motoneurons.

The transition from primary to secondary range of firing (f transition) was assumed to occur at approximately 80% of the maximal tetanic force of the unit (Heckman and Binder 1991; Kernell 1984). The force at the lowest frequency of steady firing was assumed to be 15% of maximal tetanic force and to be the same for all units (twitch/tetanic ratio is about 12% in rat motor units, and the twitches are slightly summated at their motoneuron's lowest firing frequency). Like cat motor units, rat motor units have force–frequency curves that are sigmoidal in shape. With this information, we can generate force–frequency curves for each motor unit, using the formula

$$F = F_{max} \times (1 - e^{(-\text{frequency}/Tf) \times P}) \qquad (2.3)$$

where F is force, F_{max} is the maximal tetanic force of the unit, Tf is a constant relating to the speed of the unit (and thus the left–right position of the curve), and P determines the degree of sigmoidism of the curve. Using the slowest unit (unit 1), frequencies corresponding to 15% and 80% of maximal tetanic force were used to generate a force–frequency curve, varying P and Tf until the curve resembled a physiologically accurate one. This was also done with motor unit 100, and values for P and Tf were subsequently varied proportionally throughout the entire pool. Figure 2.31 shows the force–frequency curves generated using this technique.

Figure 2.31 also shows the generation of total muscle isometric force as effective current injected into all motoneurons increases. The force at each current level was arrived at by the summation of the forces of each of the individual motor units recruited. The maximal force derived from this model is comparable to a rat lateral gastrocnemius containing 100 motor units (Seburn and Gardiner 1996). We can now consider how the system generates different submaximal forces through recruitment and rate coding. For example, what mechanisms are involved in generating a contraction equivalent to 50% of maximal voluntary contraction (MVC)? Similarly, what force is generated when we activate 50% of our motor unit population?

Figure 2.32 shows an example in which the same effective current of 3 nanoamperes has been delivered into the somata of all motoneurons in the pool. The magnitude of the current is sufficient to recruit 50% of the motor units. Note that the total isometric force generated is only about 102 grams, which amounts to 8% of the maximal force-generating capacity of the muscle. This is due to the relatively small tension-generating capacity of the small units and the fact that most of the recruited units are not rate-coded to generate near-maximal tensions. Also depicted in this figure is the contribution of each of the recruited motor units and the percentage of its specific maximal tetanic tension at this recruitment level.

Figures 2.33 and 2.34 show examples of recruitment increased to the point where larger numbers of motor units are recruited. In figure 2.33, the current is sufficient to recruit almost all of the motor units. The total force is still only around 50% of the final force capability of the muscle. In figure 2.34, the current is now

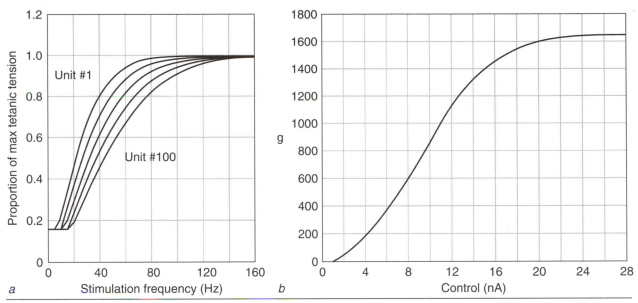

Figure 2.31 (*a*) Force–frequency relationships of muscle units 1, 25, 50, 75, and 100 (from left to right), created according to the Heckman–Binder model. (*b*) Whole-muscle force versus current injected into all motoneurons in the pool.

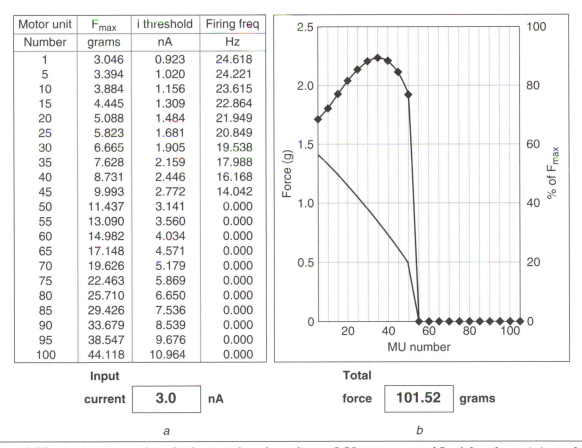

Motor unit	F_max	i threshold	Firing freq
Number	grams	nA	Hz
1	3.046	0.923	24.618
5	3.394	1.020	24.221
10	3.884	1.156	23.615
15	4.445	1.309	22.864
20	5.088	1.484	21.949
25	5.823	1.681	20.849
30	6.665	1.905	19.538
35	7.628	2.159	17.988
40	8.731	2.446	16.168
45	9.993	2.772	14.042
50	11.437	3.141	0.000
55	13.090	3.560	0.000
60	14.982	4.034	0.000
65	17.148	4.571	0.000
70	19.626	5.179	0.000
75	22.463	5.869	0.000
80	25.710	6.650	0.000
85	29.426	7.536	0.000
90	33.679	8.539	0.000
95	38.547	9.676	0.000
100	44.118	10.964	0.000

Input

current **3.0** nA

Total

force **101.52** grams

a　　　　　　　　　　　　*b*

Figure 2.32 In this example, which uses data from figure 2.30, a current of 3 nA has been injected into all motoneurons. Note the recruitment of motoneurons up to 45, at firing rates corresponding to the magnitude of the injected current, compared to the i threshold. Shown in the graph at right are the absolute force of each individual recruited muscle unit (points joined by line, scale at left) and the force of each unit as a percentage of its own maximal tetanic force (solid line, scale at right). Total muscle force is shown at bottom right.

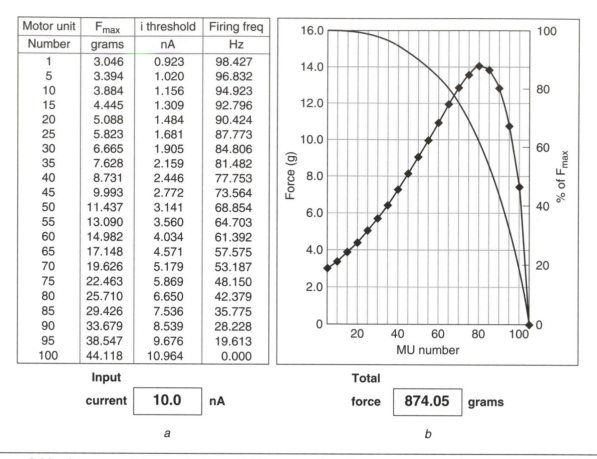

Motor unit	F_{max}	i threshold	Firing freq
Number	grams	nA	Hz
1	3.046	0.923	98.427
5	3.394	1.020	96.832
10	3.884	1.156	94.923
15	4.445	1.309	92.796
20	5.088	1.484	90.424
25	5.823	1.681	87.773
30	6.665	1.905	84.806
35	7.628	2.159	81.482
40	8.731	2.446	77.753
45	9.993	2.772	73.564
50	11.437	3.141	68.854
55	13.090	3.560	64.703
60	14.982	4.034	61.392
65	17.148	4.571	57.575
70	19.626	5.179	53.187
75	22.463	5.869	48.150
80	25.710	6.650	42.379
85	29.426	7.536	35.775
90	33.679	8.539	28.228
95	38.547	9.676	19.613
100	44.118	10.964	0.000

Input
current **10.0** nA

a

Total
force **874.05** grams

b

Figure 2.33   Same conditions as in figure 2.32, except that the current injected is 10 nA. Now almost all units are recruited, with the first 20 units generating near tetanic forces. Whole-muscle force is still only about 50% of maximal force capacity.

sufficient to recruit all units to near their maximal forces (note that the last 20% of the units are not fully recruited).

The model is limited by the lack of available data in some cases and by assumptions regarding linearity of response. For example, the model does not put limitations on maximal firing frequency and allows the early-recruited motor units to reach supratetanic frequencies before the recruitment of the last motor units, an inefficient and therefore undesirable situation (see page 58, "The Motoneuron Current–Frequency Relationship"). In fact, there is evidence that the frequencies of low-threshold units *in vivo* begin to plateau at high, submaximal frequencies to prevent this phenomenon (Monster and Chan 1977; figure 2.35). Heckman and Binder (1993) proposed that this could occur via less-than-linear summation of effective currents, which would influence low-threshold units before high-threshold units, or could arise from synaptic

organization that differs qualitatively between low- and high-threshold units.

In addition, the model assumes that one can linearly summate individual motor unit forces. In fact, the interaction of forces among motor units is complex (Troiani, Filippi, and Bassi 1999).

Motor Unit Recruitment During Different Types of Voluntary Contractions

CONCEPT

5   There are several techniques used currently to determine the recruitment of muscle fibers and motor units during voluntary movement, and each has its advantages and limitations.

Most of our knowledge in this area comes via the use of fine-wire electrodes that are inserted

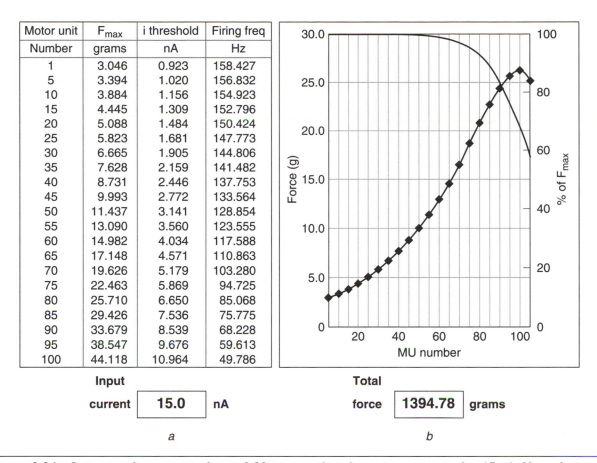

| Motor unit | F_{max} | i threshold | Firing freq |
Number	grams	nA	Hz
1	3.046	0.923	158.427
5	3.394	1.020	156.832
10	3.884	1.156	154.923
15	4.445	1.309	152.796
20	5.088	1.484	150.424
25	5.823	1.681	147.773
30	6.665	1.905	144.806
35	7.628	2.159	141.482
40	8.731	2.446	137.753
45	9.993	2.772	133.564
50	11.437	3.141	128.854
55	13.090	3.560	123.555
60	14.982	4.034	117.588
65	17.148	4.571	110.863
70	19.626	5.179	103.280
75	22.463	5.869	94.725
80	25.710	6.650	85.068
85	29.426	7.536	75.775
90	33.679	8.539	68.228
95	38.547	9.676	59.613
100	44.118	10.964	49.786

Input

current **15.0** nA

a

Total

force **1394.78** grams

b

Figure 2.34 Same conditions as in figure 2.32, except that the current injected is 15 nA. Now almost all units are recruited to their maximal force-generating capacity.

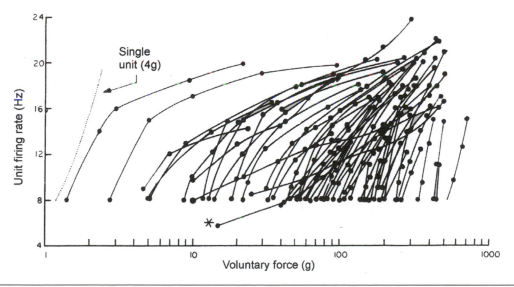

Figure 2.35 Firing-rate behavior of 60 motor units in extensor digitorum communis as a function of voluntary isometric contraction strength. The asterisk indicates a unit that is behaving unusually.

Reprinted from Monster and Chan 1977.

into the muscle in question to record the activity of muscle fibers in the electrode's immediate vicinity. In some cases, an attempt is made to reduce the movement of the electrode by hooking it into the muscle or under the skin and by reducing the movement of the contraction. Identification of the motor unit is possible by its characteristic waveform, which can be scrutinized throughout the experimental procedures to ensure that it has not changed and thus neither has the position of the electrode. Such studies examine the force threshold and firing-frequency properties of motor units under various conditions (such as recruitment of motor units when muscle is used in abduction vs. flexion or during concentric vs. eccentric contractions). Techniques for implanting these electrodes have improved over the years, so that reasonable high-force contractions can be examined. However, the type of motor unit is not known. An extension of this technique, which is obviously less invasive, is to derive mathematically the waveforms of the individual motor unit potentials of which a surface EMG is composed (Mambrito and De Luca 1983).

A permutation of this technique involves passing a fine tungsten electrode through the muscle during the contraction and recording the firing frequencies of muscle fibers near which the electrode passes (Woods, Furbush, and Bigland-Ritchie 1987; Bigland-Ritchie et al. 1983). In this technique, the primary concern is the sampling of as many firing fibers as possible. The result is a sample of firing motor units (unfortunately not including units that were firing but have ceased to do so) for each contraction condition, allowing the investigator to calculate the mean and variation in firing frequency for the activated motor units.

The technique of spike-triggered averaging has been used in conjunction with intramuscular motor unit recording to generate information about the order with which units are recruited during simple contractions. In this technique (Milner-Brown, Stein, and Yemm 1973a, 1973b), single-motor unit EMG impulses are used to trigger a signal averager, which sums the whole-muscle force immediately following the spike. With enough spikes, the characteristics of the time-locked isometric twitch that follows the spike in question eventually emerges from the force signal. This has been our primary source of identification of the contractile properties (twitch speed and strength) of human muscle units as a function of their force threshold during low-force, isometric contractions. Initial experiments suggested, as expected, that motor units are recruited from weakest to strongest and from slowest to fastest (figure 2.36).

Task-Related Partitioning of Motor Unit Recruitment

A discussion of the orderly recruitment of motor units in a muscle during voluntary movement must take into consideration that many muscles take part in various movements. Many muscles have been shown to consist of distinct anatomical compartments, which can be identified by (1) separate intramuscular motor and sensory nerve branches that subserve a distinct, regionalized subvolume of muscle fibers; (2) Ia excitatory postsynaptic potential (EPSP) partitioning, in which the strength of connections to motoneurons from Ia afferents from the same neuromuscular compartment is 1.6 to 2.3 times stronger than that from other compartments in the same muscle (Stuart, Hamm, and Vanden Noven 1988); and (3) EMG evidence during various movements indicating that different compartments of a muscle are used differently, depending on the task (Wickham and Brown 1998; Hensbergen and Kernell 1992).

Consistent with the idea of muscle compartmentalization is the concept of task groupings of motor units, whereby motor units in a muscle are recruited differently, depending on the task that they are called on to perform. The concept of task groups was originally proposed by Loeb (1987) as a functional compartmentalization of the motor apparatus, which does not necessarily correspond to the anatomical segregation at the muscle, nerve-branch, or motor-nucleus level. The idea of task groups of motor units is consistent with, but not dependent on, the presence of muscle compartmentalization.

Following is a discussion of the recruitment of motor units during several different types of voluntary movements, with emphasis on the degree to which orderly and reproducible recruitment occurs across different movement types. Keep in mind that the contributions of recruitment and rate coding to force increase can vary considerably among muscles. For

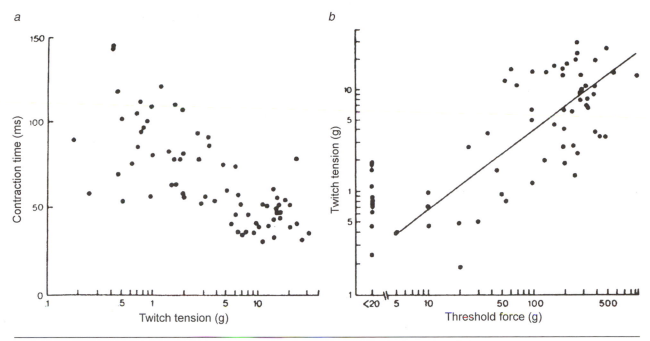

Figure 2.36   Relationships (*a*) between twitch contraction time and twitch force and (*b*) between twitch force and recruitment threshold in motor units of the human first dorsal interosseus (FDI), using spike-triggered averaging.

Reprinted from Stephens and Usherwood 1977.

example, rate coding is more important in adductor pollicis muscle, where no significant recruitment occurs after 30% of MVC, while recruitment of motor units in biceps brachii is still evident at 88% of MVC (Kukulka and Clamann 1981). In some examples, we will also see how the performance of different tasks by the same muscle changes or fails to change, as the case may be, the pattern of motor unit recruitment.

Maintained Isometric Contractions

Constant-force isometric contractions, when maintained for more than a few seconds, involve interesting changes in motor unit behaviors. De Luca, Foley, and Erim (1996) showed that motor unit firing rates in human muscles (first dorsal interosseus and tibialis anterior) appear to decrease gradually during the maintenance (up to 15 seconds) of constant force output, without evidence of the recruitment of additional motor units (figure 2.37). How can muscle force be maintained if motor unit frequencies are declining and no new units are being recruited? De Luca, Foley, and Erim (1996) proposed that the declines in firing frequency partially compensate for the tension potentiation

that occurs in fast motor units at low to moderate stimulation frequencies. In fact, according to these authors, several things occur during this type of contraction that would all tend to influence the voluntary generation of muscle force in various ways: twitch potentiation and staircase (the gradual increase in force at a constant stimulation frequency), afferent signals from muscle (which could either increase or decrease motoneuronal firing frequency), and intrinsic membrane properties such as adaptation (whereby a constant transsynaptic activation of the motoneuron results in a gradually decreasing firing frequency).

Slow-Ramp Isometric Contractions

Among the first results using spike-triggered averaging as a technique to extract the twitch characteristics of units as they are recruited was the finding that units are recruited generally from smallest to largest twitch force and from slowest to fastest twitch contraction time (see figure 2.36). This recruitment order has been shown using this technique in a variety of muscles, primarily muscles of the hand and wrist (Riek and Bawa 1992; Calancie and Bawa

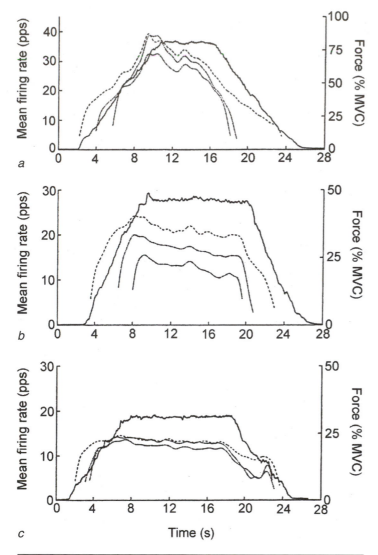

Figure 2.37 Examples of firing rates of motor units in tibialis anterior during isometric dorsiflexion of the ankle at (*a*) 80%, (*b*) 50%, and (*c*) 30% of MVC. Force scale is at right; firing rate scale is at left; force record (showing plateau) is the thick line. The other three lines in each of *a, b,* and *c* show mean firing rates of detected motor units. Note how firing rates decrease throughout the constant-force interval at all force levels.

Reprinted from De Luca, Foley, and Erim 1996.

1985; Thomas, Ross, and Calancie 1987; Dengler, Stein, and Thomas 1988; Romaiguère et al. 1989) and the tibialis anterior (Van Cutsem et al. 1997).

Experiments in which motor unit firing rates have been examined as a function of isometric contraction up to MVC offer us some insights into the complex way in which recruitment and rate coding are used to augment muscle force. In any given muscle containing a number of

motor units of differing strengths and excitabilities, force increase during a slow ramp up to MVC begins with the recruitment of low-threshold units, followed by their subsequent rate coding while additional units are recruited. This dual process of recruitment and rate coding proceeds until all units are recruited at MVC at or very near the firing frequencies that will result in their maximal tetanic force. An example of how units behave during increasing isometric muscle force generation has already been considered (see figure 2.35).

The system is most likely designed to avoid the situation in which the first-recruited units reach maximal firing rates before the highest-threshold units are maximally recruited. Such a situation would result in the inefficient generation of surplus action potentials on the part of the low-threshold units and perhaps even inactivation of their spike-generating mechanisms. Avoiding this inefficiency is designed into the system in a number of ways. First, thresholds of motoneurons are more closely spaced in low-threshold units, and this spacing increases as the threshold of the motor units increases (Bakels and Kernell 1994). In this way, recruitment plays a larger role than rate coding for the early-recruited units, with rate coding becoming more important as force increases. Second, there may be a synaptic bias in the motor system by which one system facilitates rate coding during smaller forces and another system becomes more significant with higher-threshold units as force increases (Heckman and Binder 1993). An example of this is shown in figure 2.38. The systems that might produce this leveling off of rate coding of low-threshold units as force increases are not known.

De Luca, Foley, and Erim (1996), using their technical approach of extracting single-motor unit signals from the whole-muscle EMG signal, suggested that during ramp-and-hold isometric contractions, motor unit firing rates demonstrate an "onion skin" phenomenon (see figure 2.37). According to this idea, at any particular submaximal force, later-recruited motor units

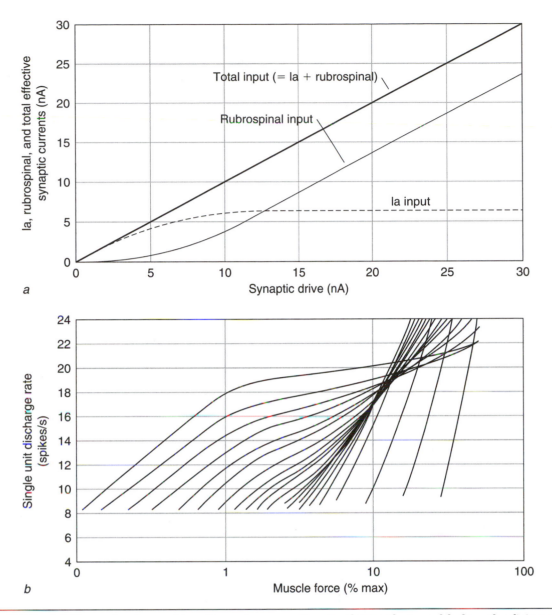

Figure 2.38 (*a*) Example of a systematic motoneuron input scheme that could slow the firing rate increase of low-threshold units as force increases, and (*b*) the resultant firing rates in a simulated motoneuron pool. The model in *a* shows two inputs to the motoneuron pool: one that has more influence on low-threshold units (Ia input) and the other on high-threshold units (rubrospinal input). Compare *b* with actual recorded motor unit firing rates presented in figure 2.35.

Reprinted from Heckman and Binder 1993.

begin firing at higher initial frequencies but always fire at lower frequencies than previously recruited units. This is consistent with the Heckman–Binder model presented previously (see figures 2.30 to 2.34). The results emanating from the laboratory of De Luca (De Luca and Erim, 1994) indicate that motor unit firing rates finally converge to similar values near MVC. In several ways, their results are similar to those of Monster and Chan (1977; see figure 2.35) in that low-threshold units demonstrate a plateau phenomenon after their initial rate-coding slope. They differ, however, in that De Luca and colleagues find very little evidence of crossover of motor unit firing rates near MVC. However, we cannot say with any certainty what really happens to firing rates at extremely high forces, due to technical difficulties. But there seems to be consensus that they probably become more similar than at submaximal force levels.

Interestingly, derecruitment of motor units during a slow decrease in isometric force may involve a different strategy than their recruitment during the preceding slow force increase. Romaiguère, Vedel, and Pagni (1993) had subjects perform slow-ramp isometric contractions and relaxations of the wrist extensor muscles. They reported that derecruitment thresholds of motor units were, on average, about 25% lower than recruitment thresholds, thus implying that more motor units were active during the relaxation ramp than at the same force level during the increasing-force ramp. They also reported that the pattern of change in firing frequency differed between the ascending and the descending ramp. Thus, more motor units firing at lower frequency were used to generate the decreasing force than were used for the same increasing force (figure 2.39).

Isometric Contractions of the Same Muscle in Various Directions

Thomas, Ross, and Calancie (1987) examined the recruitment of motor units in abductor pollicis brevis muscle during isometric contractions in two directions: abduction and opposition of the thumb. Using spike-triggered averaging, they found that the threshold force and twitch amplitudes were positively correlated in both movements, as one would expect if the size principle of orderly recruitment were obeyed.

In addition, thresholds of motor units for both movements were very similar, and the same units were recruited in both tasks, suggesting that the recruitment of motor units was virtually identical for these two isometric movements (figure 2.40). Similar results have been presented for motor unit recruitment in the extensor carpi radialis during wrist extension and radial deviation, two movements in which this muscle is extensively recruited (Riek and Bawa 1992). However, extensor digitorum communis, which is primarily a finger extensor but also takes part in wrist extension, is a muscle that demonstrates different task groups of motor units. This complex muscle functions as an extensor for all four fingers, via separate tendons, and as a wrist extensor and therefore exemplifies a muscle within which motor units are recruited as task groups, depending on the desired task. However, as with simpler muscle–joint systems, recruitment within each task seems to proceed according to the size principle. Likewise, in flexor carpi ulnaris, a wrist flexor, recruitment of motor units was similar for isometric wrist extension, anisometric wrist extension, ulnar deviation, and cocontraction of extensors and flexors (K.E. Jones, Bawa, and McMillan 1993).

K.E. Jones and colleagues (1994) recorded the recruitment of first dorsal interosseus units during three tasks that were near isometric: abduction, a rotation task (loosening a threaded knob), and a pincer task (pressing between thumb and index finger). The same units were

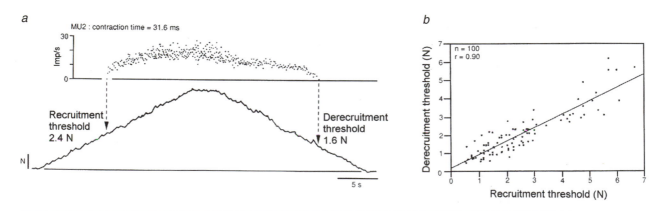

Figure 2.39 (a) Recruitment and derecruitment of a motor unit in extensor carpi radialis during isometric imposed-ramp contraction and relaxation. Derecruitment threshold is lower than recruitment threshold. (b) Relationship between recruitment and derecruitment thresholds for 20 extensor carpi radialis motor units. Note that derecruitment threshold is systematically lower.

Reprinted from Romaiguère, Vedel, and Pagni 1993.

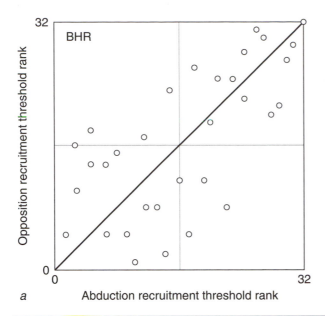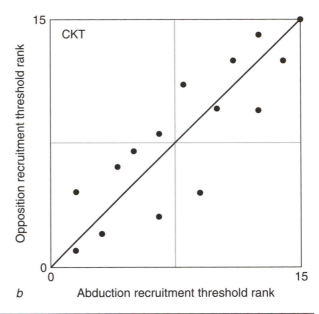

Figure 2.40   Recruitment threshold rank of motor units of abductor pollicis brevis during isometric contractions in two directions: abduction and opposition. Data from two subjects are shown. Recruitment order is similar for the two tasks.

Reprinted from Thomas, Ross, and Calancie 1987.

recruited in all three tasks, and the order of recruitment was always the same.

Thus, motor units that take part in several different types of movements around a joint seem to be recruited for each movement according to the size principle. This seems to hold true whether or not the same units within the muscle are used for the various movements.

Isometric Contractions vs. Movements

There is reason to believe that motor units might be recruited differently during contractions that involve movements but that generate the same relative force as a corresponding static contraction. Thickbroom and colleagues (1999) showed that the functional magnetic resonance imaging (fMRI) signal from the sensorimotor cortex was considerably greater during a rhythmic, dynamic finger flexion task than when the task involved the same level of static force generation. The technique of fMRI, which yields a measurement of changes in cerebral metabolism relative to oxygen supply, is unfortunately not suited for localization of the increased activity or identification of its form (increased number of cells involved vs. increased frequency of those activated). This result does suggest, nonetheless, that signals

arriving at motoneurons are different for static and dynamic contractions.

Tax and colleagues (1989) examined the recruitment of biceps brachii motor units during isometric and slow isotonic (change in elbow joint angle of 1.5 to 3 degrees per second) contractions. Their results showed a substantially lower force threshold for motor units during slow isotonic contractions than during isometric contractions (figure 2.41). In addition, minimal firing frequency of the motor units was lowest during isotonic extensions and highest during isotonic flexions. Thus, motor units appear to be activated differently during these three different types of task. Tax and his colleagues (1990) continued their studies of this phenomenon to show that the different recruitment was due not to differing peripheral signals in the isotonic and isometric conditions, but to differences in central activation. They added an experimental condition whereby subjects experienced the same change in elbow angle as during an active isotonic contraction, but via an immovable torque motor against which they exerted a constant or slowly increasing torque. These researchers' finding that the recruitment thresholds and minimal firing rates were identical in this condition to the isometric contraction condition led them to the conclusion that

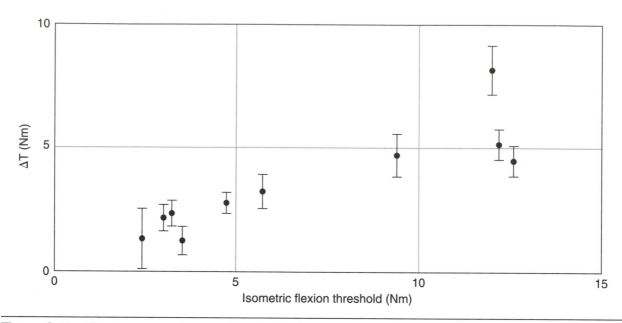

Figure 2.41   The force recruitment threshold of motor units in biceps brachii is higher during slow isotonic contractions (2 degrees of flexion per second) than during isometric contractions of the same torque. This difference (ΔT) gets larger with stronger contractions.

Reprinted from Tax et al. 1989.

central, and not peripheral, components are involved in determining the specificity of these motor unit responses.

The finding of Tax and coworkers (1989) that motor units' recruitment thresholds are reduced when contractions involve joint movements has been substantiated by Kossev and Christova (1998). These investigators recently showed that recruitment thresholds of motor units in biceps brachii were smaller during isovelocity concentric contractions than during isometric contractions. Their data are particularly interesting in that they were able to record motor units recruited at up to 54% of MVC because of the stability of their recording electrodes. Their results showed a reduction in motor units' torque threshold during concentric movements, which became more significant as the speed of the concentric contraction increased and which was more pronounced for the lower-threshold units.

Sogaard and his colleagues (1998) showed that firing rates are higher during dynamic (30 degrees per second) than during static wrist flexions involving the same torque generation. In addition, although mean motor unit firing rates are higher during an isometric contraction corresponding to 60% of MVC than one at 30% of MVC, firing rates are not different when

generating these two torque levels during a dynamic contraction. Thus, not only is the recruitment pattern different for dynamic and static contractions at the same force level, but the recruitment strategy used to increase force is also different under these two conditions.

Van Bolhuis, Medendorp, and Gielen (1997) addressed this research using a rather novel approach. They had subjects perform horizontal elbow flexions against small loads, during which they generated forces in a sinusoidal manner at various frequencies and under isometric and isotonic conditions. The researchers found that the behavior of the same motor units in brachialis and biceps muscles varied, depending on the type of contraction (isometric or isotonic) and the speed of the sinusoidal movements. They found that the phase lead of motor units (peak in burst rate as a function of peak in torque) was very reproducible among motor units during isometric contractions but varied much more among units during flexion–extension sinusoidal movements of about the same force. In addition, they found that some units showed higher firing rates when the movement changed from isometric to isotonic, whereas other units decreased in firing rate (figure 2.42). These two findings indicate that various motor units are activated differ-

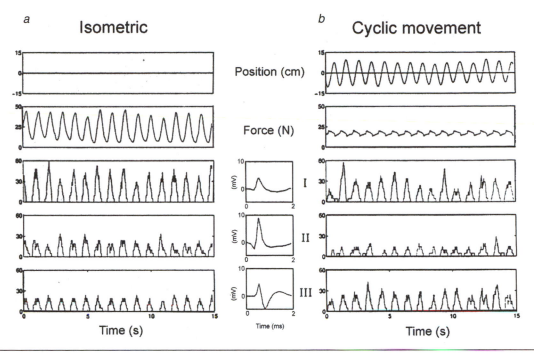

Figure 2.42 Firing rate behavior of three motor units (I, II, III) in brachialis during (*a*) sinusoidally modulated isometric contractions and (*b*) sinusoidal movements. Note the lack of consistent change in firing rates (shown in the bottom three panels) between isometric and movement conditions (firing rates decrease for units I and II and increase for unit III when changing from isometric to cyclic movement contractions).

Reprinted from Van Bolhuis, Medendorp, and Gielen 1997.

ently in isometric and movement tasks. Interestingly, these investigators found that some units had a phase lead of 180% or more during flexion–extension activity, which suggests that some flexor units are active during the extension phase. This supports the results of others that suggest that motor units are activated differently during lengthening and shortening types of contractions. These motor unit studies support whole-muscle studies that demonstrated that EMGs for concentric and eccentric movement tasks were higher than for isometric tasks (Van Bolhuis and Gielen 1997).

But do these observations indicate that the order of recruitment of motor units can be changed, depending on the isotonic or isometric nature of the task? In this case, not necessarily. While motor units' lower force threshold might indicate that some now have a higher threshold (since the total force at which these thresholds is compared is the same), which would indicate a reversal of order of recruitment for at least some units, there is some evidence that a decrease in the minimal firing frequency also occurs. Thus, in the study of Tax and colleagues

(1989), more motor units were activated during a slow isotonic contraction, but each generated slightly less force because of lower firing frequencies.

Thomas, Ross, and Calancie (1987) addressed this question of recruitment order by investigating the recruitment of first dorsal interosseus and abductor pollicis brevis motor units during the repetitive opening and closing of scissors and compared the order of recruitment to that seen in these same muscles during ramp isometric contractions. In their limited data set, they found that the recruitment according to increasing motor unit size that was seen with isometric contractions was largely preserved during the repetitive opening and closing of scissors.

Thus, there is strong evidence that motor units are recruited differently during shortening contractions as opposed to isometric contractions, even when the movement is of slow speed. The force at which recruitment occurs is lower, and the frequency of firing is higher for motor units during concentric contractions. In addition, increasing torque during an isometric contraction appears to rely more on increasing firing frequency of the motor units than is the case when torque is increased during a concentric contraction. It also appears that each motor unit may have a specific contributory role during isometric and concentric contractions, with a significant degree of variability among units as to their function in isometric versus anisometric contractions. Many of these differences are due to variations in central programming for each of these tasks.

During ballistic isometric contractions, the order of recruitment of motor units seen during slow-ramp contractions is preserved. However, as in slower anisometric contractions mentioned earlier, force thresholds are lower, and initial firing frequencies much higher during ballistic contractions (Desmedt and Godaux 1977, 1978).

Contractions involving concentric movements often include an initial high-frequency double discharge (doublet) of the motor unit, followed sometimes by a longer interpulse interval, which is evident during the initial force increase preceding the movement (Kossev and Christova 1998). The obvious advantage of this

phenomenon would be to increase force rapidly, in order to assist in overcoming the inertia of the load. The source of this phenomenon is, however, not known. As previously pointed out, it may be more pronounced in motoneurons displaying a pronounced delayed depolarization (Kudina and Alexeeva 1992; Kudina and Churikova 1990). Interestingly, this phenomenon also appears to be subject to adaptive changes with ballistic-type training (see chapter 5).

Lengthening Contractions

Nardone, Romano, and Schieppati (1989; figure 2.43) reasoned that when performing lengthening contractions in which muscle was yielding to the load, there would be an advantage to preferentially recruiting fast-twitch units, since these have faster relaxation times and therefore one would have better control over the movement trajectory. Furthermore, they hypothesized that, given the relatively large size of fast units, it would be advantageous to recruit them at low frequency and also to decrease the background force level by derecruitment of slow units. Of course, this would imply a different

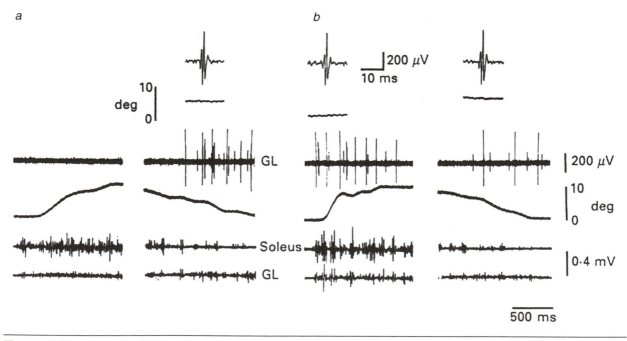

Figure 2.43   Pattern of discharge of a lateral gastrocnemius motor unit during shortening and lengthening contractions performed in (a) ramp (8 degrees per second) and (b) ballistic modes. The large-amplitude unit was not active during ramp shortening but was active during lengthening. It could be activated during ballistic shortening.

Reprinted from Nardone, Romano, and Schieppati 1984.

order of recruitment than that seen during isometric contractions of slowly increasing force. In their study, they examined the firing patterns of motor units in soleus and medial and lateral gastrocnemius during plantar and dorsiflexion (the latter achieved by gradually lowering a load on the plantar flexors, thus a lengthening, or eccentric, contraction). They found that units could be classified as S (active during shortening), L (active during lengthening), or S + L. L units (15% of all units recorded in soleus, and 50% in gastrocnemius) were often not recruited at all during S movements, but when they were, it was only at relatively high forces or during very rapid contractions. L units also had relatively large spike amplitudes and were difficult to keep activated, and their axons had relatively high conduction velocities and relatively low voltage thresholds to electrical stimulation. The authors contended that these were therefore high-threshold units. Their conclusion was that, since force requirements are certainly not greater to lower a load than to maintain it and because some units are replaced by others when changing from a shortening to a lengthening contraction, motor unit recruitment is different for shortening and lengthening contractions. The mechanism may be an increase in descending enhancement of recurrent inhibition from large to small motoneurons and presynaptic inhibition to all motoneurons, which would affect small more than large motoneurons. However, the activation of specific task groups of motor units may also be involved.

During the same submaximal load, muscle activation, as estimated using EMG, is less for the eccentric phase than it is for the concentric phase of the contraction (Moritani, Muramatsu, and Muro 1988; Nakazawa et al. 1993).

The increase in inhibitory influences on alpha-motoneurons during eccentric types of contractions is supported by the finding of Westing, Seger, and Thorstensson (1990) that torque could be increased by 21% to 24% during maximal voluntary eccentric contractions by superimposing electrical stimulation during the contraction. In 1991, Westing, Cresswell, and Thorstensson showed that EMG was lower, but torque higher, during eccentric maximum isokinetic knee extensor contractions than during concentric ones.

Similar results were later reported by J.N. Howell et al. (1995). In their study of first dorsal interosseus motor units, they found instances where motor units that normally had high force thresholds under isometric conditions were recruited during concentric–eccentric contractions of low torque and during the eccentric phase (as Nardone, Romano, and Schieppati found for plantar flexors in 1989; figure 2.44). Thus, this phenomenon is evident in small muscles of the hand as well as in larger limb muscles.

What is the mechanism for the altered recruitment of motor units during shortening and lengthening contractions? Abbruzzese and colleagues (1994) proposed that spinal cord circuits were altered in their excitability during the eccentric movement. They based their conclusion on the findings that magnetic brain stimulation (which stimulates cortical neurons transsynaptically) and electrical brain stimulation (which stimulates cortical axons) both showed a decreased motor-evoked potential in elbow flexors during eccentric contractions, and that the H-reflex was also decreased. Thus, major changes in cortical excitability are probably not as important in changing from concentric or isometric to eccentric contractions as a change at the spinal cord level, perhaps in the form of an increase in the level of presynaptic inhibition, which would have a relatively greater effect on small motoneurons.

Cocontraction, or Coactivation

Coactivation is a phenomenon worth considering when discussing recruitment, since it may under certain circumstances limit the full expression of torque around a joint. Simply put, coactivation refers to the simultaneous contraction of antagonists and agonists and is a common phenomenon that serves to stabilize the joint, aid in distributing pressure across the joint, and decrease the strain on ligaments during forceful contractions (Weir et al. 1998). Coactivation increases with increased rate of agonist shortening (Weir et al. 1998). Coactivation is under central control, via presynaptic inhibitory mechanisms (Nielsen and Kagamihara 1993), and may play a role during fatigue and in the training process. These subjects are treated in following chapters.

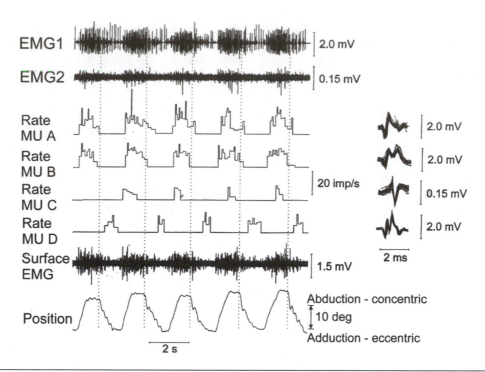

Figure 2.44   Activity of four motor units (MU A to D) during repeated abduction–adduction movements against a constant torque load. Note the differences among units in the phase of the contraction during which they are active.

Reprinted from Howell et al. 1995.

Unilateral vs. Bilateral Contractions

Reports of the bilateral deficit phenomenon, which involves a slight (about 10%) loss in maximal leg extension force when the contralateral leg is contracting at the same time, have appeared in the research literature (Vandervoort, Sale, and Moroz 1984; Schantz et al. 1989; J.D. Howard and Enoka 1991). It has been difficult to demonstrate consistently and unequivocally that this deficit, which has also been demonstrated for the upper limbs, results in a concomitant decrease in EMG during the bilateral contraction, probably because surface EMG is not sensitive enough to detect such a small difference in activation (J.D. Howard and Enoka 1991). It is probably not a limit in the amount of muscle tissue that the central nervous system can activate, since it does not occur with other than homologous muscles on opposite sides of the body (J.D. Howard and Enoka 1991) and the degree of deficit is not increased if arm exercise is added during the bilateral task (Schantz et al. 1989). Interestingly, not all subjects demonstrate a bilateral deficit; some demonstrate no difference or even a bilateral facilitation (J.D.

Howard and Enoka 1991). Since electrical stimulation of the contralateral muscle increases the performance of the ipsilateral muscle during a maximal contraction (J.D. Howard and Enoka 1991), there clearly are interlimb signals that modulate the expression of maximal force during bilateral efforts. Not all muscle groups are subject to this phenomenon; R.D. Herbert and Gandevia (1996), for example, were unable to demonstrate a bilateral difference with thumb adductor muscles. The mechanisms by which bilateral deficit occurs are not known at present.

Rhythmic Complex Contractions (Running, Cycling)

CONCEPT 6   Our primary information for recruitment during rhythmic complex contractions comes from glycogen depletion studies. Vollestad and colleagues (Vollestad, Tabata, and Medbo 1992; Vollestad and Blom 1985; Vollestad, Vaage, and Hermansen 1984) had subjects perform on a bicycle ergometer at different speeds for various periods of time and examined the disappearance of glycogen in single vastus lateralis fibers from sequential biopsies. Their results have generally confirmed a recruitment scheme consistent with the size principle. For example, during ergometer cycling corresponding to 75% of maximal oxygen consumption (figure 2.45), there was virtually no glycogen loss from IIAB and IIB fibers, corresponding to 10–30% of the muscle cross section, during the first 20 minutes of exercise. During this same period, type I and IIA fibers showed extensive glycogen loss. As exercise continued, type I and IIA fibers became depleted, and involvement of type IIAB and IIB fibers became more significant. At exhaustion, after 140 minutes of exercise, types I and IIA were severely depleted of glycogen, while a considerable proportion of type IIAB and IIB fibers were not yet depleted. This finding is consistent with the idea that the highest-threshold units require the most effort to recruit and cannot be recruited continuously, partly because of their propensity for late adaptation.

These same investigators have also shown that the number of fibers involved increases with intensity, in the order I ▹ IIA > IIAB > IIB, and that the rate of glycogen depletion also increases as a function of exercise intensity (figure 2.46).

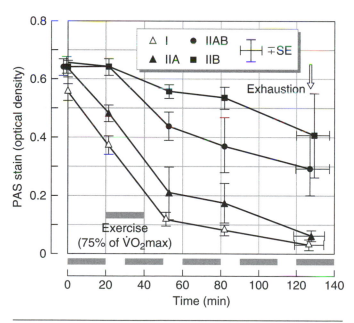

Figure 2.45 Average glycogen content, as measured by periodic acid–Schiff (PAS) staining intensity, in typed fibers of vastus lateralis during bicycle ergometer exercise at 75% of maximal oxygen consumption.

Reprinted from Vollestad, Vaage, and Hermansen 1984.

Finally, these investigators showed quite convincingly that all muscle fibers are recruited at a supramaximal intensity (up to 200% of cycling intensity corresponding to maximal oxygen consumption). Glycogen depletion rates in the different fiber types indicated that, contrary to the belief of many, type I fibers are capable, when called upon, of using significant amounts of glycogen anaerobically in order to take part in high-intensity tasks. The increase in rates of glycogen breakdown in all fiber types as intensity increases suggests that there is no significant derecruitment of type I fibers during high-intensity, primarily anaerobic tasks. Thus, even for such tasks, where the decreased anaerobic potential and lower power might render these fibers inappropriate, the size principle is, for the most part, still in force.

Of course, the limits of using the glycogen depletion to infer fiber recruitment must be understood when examining this literature. The loss of glycogen from fibers during exercise depends on, among other things, the level of oxygen delivery to the working muscle, the initial glycogen levels, and the activities of the enzymes of energy metabolism, all of which vary among individual fibers. In addition, the rate of glycogen loss tells us virtually nothing about the rate of energy turnover or of work performance of the fiber.

Do We Recruit Everything During a Maximal Contraction?

CONCEPT

7 Motor unit firing rates during maximal voluntary contractions appear to be sufficient to evoke maximal tetanic forces of their innervated muscle units in biceps brachii, adductor pollicis, and soleus muscles (Bellemare et al. 1983). This supports the view that, at least for these muscles, maximal activation is theoretically possible through voluntary effort.

Twitch interpolation is a technique whereby a muscle's nerve is stimulated while the individual is performing a maximal voluntary contraction, with the idea that if the muscle has been voluntarily recruited to maximum, one should not see additional force when the motor nerve is stimulated. This technique, first applied to the adductor pollicis, has now

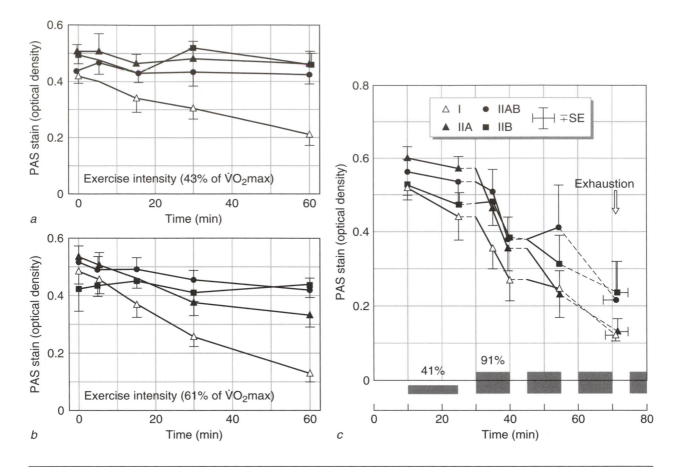

Figure 2.46   Decrease in average glycogen content of typed fibers in vastus lateralis during bicycle ergometer exercise at several intensities: (*a*) 43%, (*b*) 61%, and (*c*) 91% of maximal oxygen consumption.

Reprinted from Vollestad and Blom 1985.

been used extensively to test the "maximality" of voluntary contractions in a variety of muscles and muscle groups, with varying results.

Dowling and colleagues (1994) suggested, from their results using elbow flexors, that the relationship of interpolated twitch and percentage of voluntary force is best fitted by an exponential function. This would mean, of course, that subjects were never able to activate their elbow flexors 100%, but that the small residual twitch response becomes too small for us to measure at near-maximal effort.

It may be that many repetitions are necessary in order to permit subjects to generate a true maximal contraction. In a systematic study of the intra- and interindividual variability in degree of activation using the twitch interpolation technique, G.M. Allen, Gandevia, and McKenzie (1995) measured this in elbow flexors of five subjects, tested 10 times per day on five different days. Their results, shown in figure 2.47,

underlined the importance of several measurements in order to determine whether maximal contractions are possible in any given subject. Apparently, at least for elbow flexors, maximal activation is possible in most subjects but does not occur with all contractions.

Strojnik (1995) has shown a consistent deficit in the human quadriceps, which can be seen by superimposing submaximal electrical stimulation. This author also showed an increased rate of rise of torque during a maximal contraction when stimulation was applied.

It has been suggested that a high-frequency train of stimuli, as opposed to the single or double pulses normally used, is more sensitive in determining central failure to maximally activate muscles. For example, Kent-Braun and Le Blanc (1996) showed that a 50-hertz, 500-millisecond stimulus applied during an MVC detected 33% of control subjects who could not fully activate their dorsiflexors, whereas double

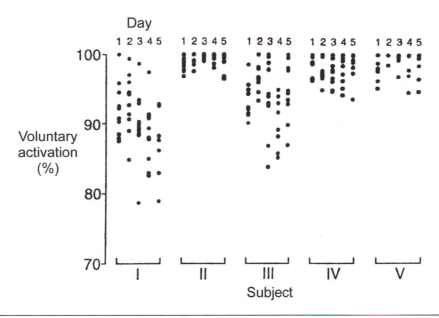

Figure 2.47   Ability to maximally activate elbow flexors during an MVC varies among trials on the same day and among subjects. Five subjects were given 10 trials on each of five days. Voluntary activation was calculated as (1 – superimposed twitch in response to electrical stimulation during MVC/twitch at rest) × 100.

Reprinted, by permission of Wiley-Liss, Inc., a subsidiary of John Wiley & Sons, Inc., from G.M. Allen et al., 1995, "Reliability of measurements of muscle strength and voluntary activation using twitch interpolation," *Muscle & Nerve* 18: 597. © 1995 Wiley-Liss, Inc.

stimuli detected only 4.8%, and single stimuli detected no cases. In spite of this, the mean ratio for control subjects was still 0.96, indicating that these subjects were capable of maximally stimulating their dorsiflexors voluntarily.

Summary

The properties of muscle units and of the motoneurons that innervate them covary in a way that appears to allow the appropriate use of motor units for specific tasks. Data on motoneuron and muscle unit properties can be applied to a simple recruitment model to demonstrate, within limits imposed by certain assumptions, how these properties interact to generate a sustained, submaximal, isometric contraction. The study of motor unit recruitment during more complex types of movements, such as those involving changing force levels, shortening and lengthening, and unilateral and bilateral efforts, shows that a certain degree of flexibility in recruitment patterns is also present. Recruitment patterns during more complex and prolonged motor tasks, such as running and cycling, are less evident, due to technical considerations. Finally, it appears that, although humans are capable of recruiting nearly all of the maximal force capability of muscles, there is a significant inter- and intra-individual variation in this capability. This might be significant in determining athletes' capacity to repeatedly perform maximal efforts in competition.

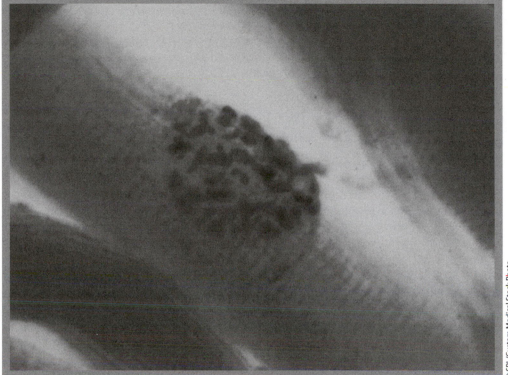

chapter 3

Neuromuscular Fatigue

"**M**uscular Fatigue affects Brain-power: severe muscular exertion may bring a disinclination and incapacity for Brain-work. Hard exercise uses up Nerve-force, and also makes Waste-products circulate in the blood; and so the section of the Brain is hindered."

F. Schmidt and E. Miles, *The Training of the Body*, 1901

Most Important Concepts From This Chapter

1 Neuromuscular fatigue can be defined as any exercise-induced reduction in maximal voluntary force or power output. Thus, fatigue is an ongoing process that occurs even under conditions where performance is maintained at a constant level.

2 Neuromuscular transmission failure, although a candidate mechanism for neuromuscular fatigue, is difficult to demonstrate unequivocally. On the other hand, decrease in motoneuron activation during exercise is a mechanism common to maintained submaximal and maximal exercise.

3 During both submaximal and maximal exercise, motoneurons experience a gradual decrease in excitatory influences and an increase in inhibitory influences. Increases in supraspinal excitation and inhibition also occur, as does an alteration in the input signal to the primary motor cortex.

4 Results from experiments with anesthetized animals show that group III and IV receptors increase their firing in the presence of several substances that change in fatiguing muscles. Increased firing of these afferents has an inhibitory effect on the firing of extensor alpha-motoneurons but a stimulatory effect on fusimotor neurons.

It is naive to think that fatigue resulting from exercise can be explained by a single mechanism. Although fatigue during certain kinds of effort involves failure of the contractile apparatus, some types of fatigue cannot be explained by a decrease in the force-generating capacity of the muscle tissue alone. Some evidence suggests that systems providing input to the muscle fiber may be involved. During maximally sustained or intermittent contractions, stimulation of the motoneuron with a supramaximal voltage stimulus may or may not evoke an increased force; the ability to increase force with an external voltage would imply that either motivation was lacking or central fatigue was occurring. Even during submaximal contractile activity, either sustained or intermittent, there are signs that changes occur in the nervous system as exercise progresses that might influence the efficacy or pattern with which the contractile machinery is activated. While contractile activity of a muscle electrically stimulated supramaximally gives us an objective measure of fatigue (i.e., the decline in maximal force response in spite of continued supramaximal

stimulus), the notion of fatigue in the exercising organism can include an increased effort necessary to maintain a submaximal contractile force even as the maintained force level is unchanging. Thus, as depicted in figure 3.1, the individual keeps exercising at the same performance level while experiencing an increase in the amount of excitation of the motor pool necessary to maintain the performance, with a simultaneous decrease in the maximal capacity of the contractile system. Whether there are limits to the length of time during which this increased effort can be sustained by central and peripheral nervous system structures is a topic of intense research.

CONCEPT

1 In keeping with these ideas, the working definition of neuromuscular fatigue that is used for this chapter is that of Vollestad (1997): "any exercise-induced reduction in the maximal voluntary force or power output."

The emphasis in this chapter is on the components, other than muscle tissue, that are involved in the fatigue process. Proposed bio-

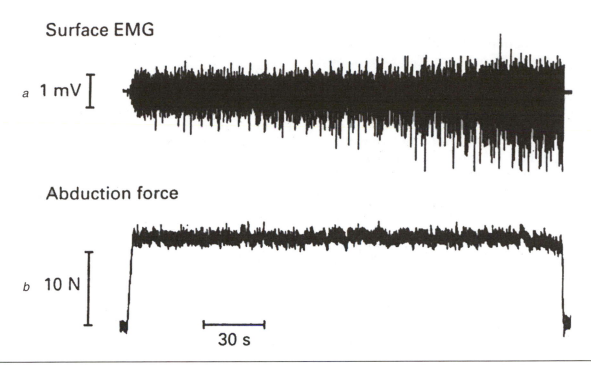

Figure 3.1 (*a*) EMG and (*b*) force of first dorsal interosseus muscle of a subject asked to maintain a target abduction force of 35% of MVC for as long as possible (210 s in this case). Note the gradual increase in EMG. MVC decreased to about 70% at fatigue.

Reprinted from Fuglevand et al. 1993.

chemical mechanisms for fatigue in muscle are the subject of several thorough reviews (Fitts 1994 and Westerblad et al. 1998, among others). Concentrating on components other than muscle provides the opportunity to focus on some less frequently considered phenomena that have been proposed as possible limiting components during exercise and that thus can be considered alongside the muscles' biochemical changes. It also allows us to consider neuromuscular fatigue as more than merely an end point beyond which continued function is not possible, but rather as a condition that develops gradually as exercise continues.

Two Basic Fatigue Mechanisms Involving the Nervous System

There are a number of physiological sites in the nervous system where neuromuscular fatigue might occur, as discussed in the sections that follow. Nonetheless, these sites can be grouped together for convenience into two major categories. The first category is the neuromuscular synapse, where fatigue occurs because the ner-

vous signals are not transmitted with fidelity to the muscle fibers. The second category involves decreased activity of the motoneurons innervating the muscle that is becoming fatigued.

Neuromuscular Transmission Fatigue

CONCEPT

2   Neuromuscular transmission failure refers to a failure of a nervous impulse to be translated into an impulse on the sarcolemma immediately beneath the motoneuron terminal. Failure at this level is notoriously difficult to determine, even in anesthetized animal preparations, but especially in the voluntarily exercising human. In many studies of fatigue, as shall be seen in later sections, neuromuscular transmission failure is eliminated as a contributing factor by the observation of an unchanged stimulation-evoked EMG response (M wave) at the point of fatigue. Caution must be exercised in trying to implicate neuromuscular transmission failure on the basis of measurements that represent propagation of the sarcolemmal action potential from the motor end plate to the recording electrode. For example, decreases in M-wave amplitude during sustained contractions can occur in the absence of decreases in evoked force (Zijdewind, Zwarts, and Kernell 1999). For that reason, I have not referenced studies in which neuromuscular propagation has been demonstrated to fail on the basis of changes in muscle EMG indices.

The research literature is replete with evidence that the neuromuscular transmission failure occurs, especially in reduced animal preparations, if the system is pushed at high enough frequencies for long enough periods of time or under conditions where factors required for optimal functioning are in limited supply. Later, I include a discussion of the most pertinent evidence for neuromuscular transmission fatigue when frequencies and durations of excitation are within a reasonably physiological range.

Probably the best evidence to date for the possibility of neuromuscular transmission involvement in fatigue emanates from studies in which muscle forces in response to direct and indirect stimulation *in situ* or *in vitro* were compared before and after fatigue induced at relatively physiological stimulation rates. Pagala, Namba, and Grob (1984) found significant differences between direct and indirect stimulation in rat extensor digitorum longus, diaphragm, and soleus after fatigue induced by stimulation at 30 hertz in trains lasting 500 milliseconds every 2.5 seconds for three to five minutes. This approach has been confirmed, especially for diaphragm, using frequencies as low as 10 hertz, which is well within the physiological range (Van Lunteren and Moyer 1996; Aldrich et al. 1986; B.D. Johnson and Sieck 1993; figure 3.2). Neuromuscular fatigue has been shown in intact rabbit (Aldrich 1987) and sheep (Bazzy and Donnelly 1993) diaphragm in response to loaded breathing by demonstrating that phrenic nerve electrical activity (ENG) continues to increase at a time when diaphragm EMG has reached a plateau. The most recent and systematic evidence of this kind has emanated from studies of diaphragm.

If we accept that this research constitutes evidence that neuromuscular transmission failure can occur under conditions that are relatively physiological, then we should consider the mechanisms. There are three main possibilities: failure of propagation of the action potential into axon branches (branch-block failure), neurotransmitter depletion, and postsynaptic membrane failure.

Axon Branch-Block Failure

Axon branch-block failure occurs when the action potential generated in the axon is not propagated into all of the branches extending to the muscle fibers. In fact, there is a significant decrease in the excitability of the smaller axon branches relative to the larger branches, so that the safety factor (the amount of excitation produced compared to that which is necessary) for action potential generation gets smaller as the caliber of the axon decreases (Krnjevic and Miledi 1959). Thus, any variable that affects this safety factor during contractile activity could result in increased blocking at these branch points. Krnjevic and Miledi (1959), in their classic studies of neuromuscular junction responses to fatigue, found evidence of branch-block failure in *in vitro* rat diaphragm preparations stimulated at 20 to 50 hertz for several minutes. They found that this blocking was not altered by substantial variations in potassium, calcium, or H^+ in the bath and concluded that the level of oxygen played a role. Evidence of branch-block failure has been shown in cat muscles subjected to maintained, rather high (80-hertz) frequencies of stimulation, which one would not expect to see *in vivo* (Sandercock et al. 1985; Clamann and Robinson 1985).

Neurotransmitter Depletion

The amplitude of end-plate potentials (EPPs) decreases with continued stimulation, due to lowered quantal content (number of vesicles), quantal size (acetylcholine [ACh] per vesicle), or both (Wu and Betz 1998; Dorlöchter et al. 1991; Reid, Slater, and Bewick 1999). As in the case of branch-block, decreases in ACh release with fatigue have generally been shown using unphysiological patterns of electrical stimulation, and therefore their implications for voluntary fatigue are unknown. The safety factor for neuromuscular transmission is probably in the neighborhood of 4 (i.e., a change in EPP of approximately 10 millivolts is required to generate an action potential, while the EPP is approximately 40 millivolts), implying that a rather drastic reduction in the EPP to 25% of its original value would be required for failure to occur due to lack of transmitter.

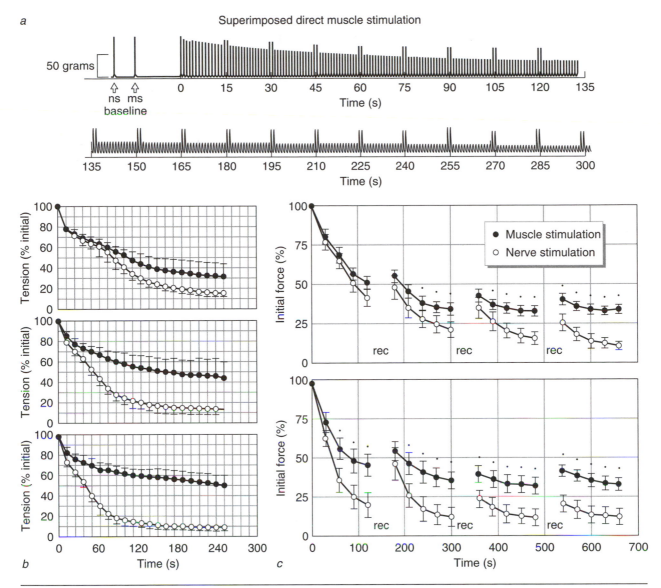

Figure 3.2   Neuromuscular transmission failure in rat diaphragm *in vitro*. (*a*) Intermittent direct muscle stimulation during phrenic nerve stimulation reveals an increasing contribution of neuromuscular transmission failure as stimulation continues. (*b*) Difference between direct (dark circles) and indirect (open circles) stimulation at 20, 40, and 70 impulses per second (*top to bottom*). (*c*) Direct and indirect stimulation at 10 Hz, intermittent protocol.

a-b Reprinted from Kuei, Shadmehr, and Sieck 1990. *c* Reprinted from Johnson and Sieck 1993.

Van Lunteren and Moyer (1996) recently demonstrated that 3,4-diaminopyridine decreases force difference between direct and indirect stimulation in fatiguing rat diaphragm *in vitro*. This substance has been shown to increase neurotransmitter release at the neuromuscular junction, most likely via an increase in the voltage-dependent calcium conductance in the nerve terminals (Van Lunteren and Moyer 1996). These researchers also found that the tetanic contractions maintained their force plateau better in the presence of 3,4-diaminopyridine. This report lends support to the hypothesis that neurotransmission failure plays a role in fatigue under reasonably realistic stimulation conditions (20-hertz intermittent stimulation for 1.5 minutes). It also suggests that calcium dynamics at the nerve terminal may limit neuromuscular transmission.

Postsynaptic Membrane Failure

Part of the problem in determining the role of presynaptic mechanisms (discussed in the preceding sections) in neuromuscular junction fatigue is that measurements taken under the postsynaptic membrane (such as EPP amplitude, used to estimate quantal content) are influenced also by changes in the sensitivity of the postsynaptic membrane. Thesleff (1959) demonstrated this quite effectively by adding pulses of ACh onto end plates at various times after stimulation of their motoneurons. He showed, for example, that the voltage response of the end plate to a given pulse of ACh was reduced immediately following a tetanic stimulation of the nerve 20 times per second for 150 milliseconds, and this reduced sensitivity lasted several 10ths of a second. This reduction was higher with higher frequencies and was similar to the reduction in the amplitude of the EPP during the train of electrical impulses, leading the author to conclude that a major proportion of the EPP decrease during trains of stimuli was due to a decrease in end-plate sensitivity and not to neurotransmitter depletion. It is now recognized that desensitization of ACh receptors is at least in part to blame for this end-plate desensitization (Magleby and Pallotta 1981; Giniatullin et al. 1986). Prolonged exposure of the receptor to ACh appears to convert it first into a desensitizable state then into a desensitized state, characterized by a smaller and slower time course associated with ACh binding (Magleby and Pallotta 1981; Giniatullin et al. 1986). Prolonged here is relative; at the frog neuromuscular junction, desensitization was demonstrated with two pulses of ACh separated by 10 milliseconds, which was insufficient time for the ACh from the first pulse to be completely removed from the synaptic cleft before the arrival of the second pulse (Magleby and Pallotta 1981). Desensitization is very easily demonstrated in neuromuscular junctions that have been treated with anticholinesterases, which do not allow the normally rapid removal of ACh from the cleft. The importance of receptor desensitization in the fatigue process, as of other processes at the neuromuscular junction, has not been established unequivocally.

In general, the role of neuromuscular transmission failure in fatigue during voluntary contractions has not been clearly demonstrated. On the other hand, adaptations that occur at the neuromuscular junction with chronically increased activity (discussed in the next chapter) allow some speculation as to the components that might limit performance in the untrained state.

Reduced Motoneuron Activity During Fatigue: "Muscle Wisdom"

During both submaximal and maximal efforts, there is abundant evidence that the way in which muscles are activated begins changing within a few seconds following the commencement of activity. This change in activation expresses itself most vividly as a decline in EMG activity, due to a reduction in the frequency of firing of the recruited motor units (Woods, Furbush, and Bigland-Ritchie 1987; Bigland-Ritchie et al. 1983; Marsden, Meadows, and Merton 1983; figure 3.3). Interestingly enough, this decreased firing frequency also occurs during submaximal contractions sustained to fatigue, even at a time when additional motor units are being recruited to maintain the force (Garland, Griffin, and Ivanova 1997). There is considerable speculation as to whether this decrease in firing frequency is a sign of failure somewhere in one or more of the mechanisms involved in increasing and maintaining the excitation of motoneurons or whether it is in fact a sensible response to changes in the contractile properties of the fatiguing muscle. A simple test of the "sensibility" of this mechanism would be to determine whether the decreasing activation of muscle resulted in a failure to express the full muscle contractile force that was available at the time. This has been tested during fatigue using supramaximal electrical stimulation, with mixed results (Bigland-Ritchie et al. 1983; Woods, Furbush, and Bigland-Ritchie 1987; Thomas, Woods, and Bigland-Ritchie 1989).

This gradual decline in muscle activation during fatigue, exemplified by the decrease in the firing frequency of motor units during rhythmic or sustained types of effort, has been given the moniker "muscle wisdom" (Marsden, Meadows, and Merton 1983). It seems "wise" for the neuromuscular system to decrease motor unit firing rates over time during continued excitation, since muscle fibers are slowing in contractile speed and thus require lower frequencies to

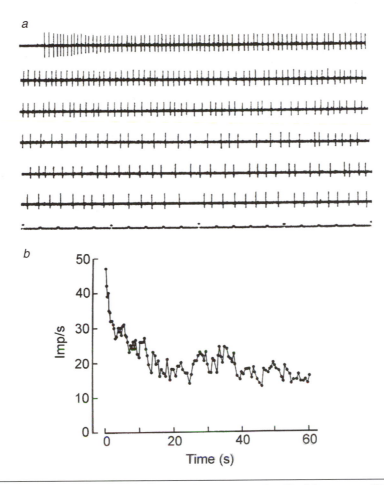

Figure 3.3 Decrease in firing frequency of a single motor unit of adductor pollicis during a maximal voluntary effort lasting one minute. (*a*) Two-second recordings; from top to bottom, at 0–2 s, 2–4 s, 8–10 s, 16–18 s, 38–40 s, and 58–60 s. (*b*) Frequency of motor unit firing in *a* during MVC.

Reprinted from Desmedt 1983.

maintain maximal force (Bigland-Ritchie and Woods 1984). Lower frequencies for the motoneuron mean less action potential generation per unit time, thus less stimulus for late adaptation, which is more pronounced at higher frequencies (Granit, Kernell, and Shortess 1963b), and less stress on the neuromuscular junction and the sarcolemma propagation mechanisms. In fact, several investigators have demonstrated that the optimal pattern for sustained force generation using electrical stimulation is via a constantly decreasing frequency of stimulation (Bigland-Ritchie, Jones, and Woods 1979; figure 3.4), thus lending some support to the concept of muscle wisdom.

What are the mechanisms underlying this muscle wisdom? How can we determine from where muscle wisdom is dispensed? There are several possible candidate mechanisms, working either individually or in combination, that could explain this decreased muscle activation with fatigue. They include (1) decreased excitability of the motoneurons, (2) decreased excitatory influence from peripheral sources, (3) increased inhibitory influences on motoneurons, and (4) decreased excitation of motoneurons from supraspinal sources.

Reduced Motoneuron Activity During Various Types of Contractions

CONCEPT

3 In the following sections, I examine each of the candidate mechanisms for muscle wisdom during maximal sustained contractions and submaximal contractions in which

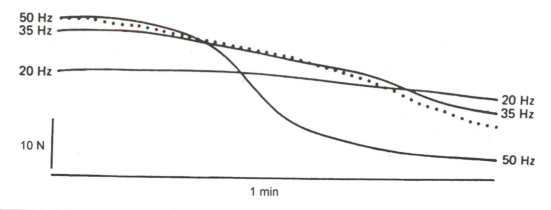

Figure 3.4 Dorsiflexion tension of the big toe evoked by electrical stimulation of the peroneal nerve at 50, 35, and 20 Hz during one minute. The dotted line shows the optimal force–time envelope that results from a continuous decrease in the stimulation rate from 50 to 20 Hz. Calibration bar = 10 N.

Reprinted from Grimby, Hannerz, and Hedman 1981.

subjects maintain a target performance as long as possible. The evolution of fatigue, in time and in intensity, is clearly different in these two types of activity. Since neuromuscular transmission failure is so difficult to establish with certainty, as discussed previously, reference to its role during maximal and submaximal contractions is confined to those circumstances where it appears not to be involved (i.e., when M-wave amplitude is unchanged).

Maximal Sustained Contractions

In experiments using maximal sustained contractions, contractile periods are relatively short (usually less than two minutes, no longer than five minutes), since maximal effort is required from the beginning of the contraction. Whole-muscle EMG begins falling immediately (Bigland-Ritchie, Dawson, et al. 1986; Bigland-Ritchie et al. 1983; Bongiovanni and Hagbarth 1990; Bigland-Ritchie, Jones, and Woods 1979; Rothmuller and Cafarelli 1995; Stephens and Taylor 1972), as do the firing frequencies of the motor units (Woods, Furbush, and Bigland-Ritchie 1987; Bigland-Ritchie et al. 1983). Some motor units cease to discharge in spite of continued maximal effort (Peters and Fuglevand 1999). The M wave in response to supramaximal stimulation of the muscle's nerve is either unchanged or slightly decreased at fatigue (D.A. Jones, Rutherford, and Parker 1989; Stephens and Taylor 1972), but its area invariably increases, which is usually interpreted as a result of decreased conduction velocity of the action

potential along the fatiguing muscle fiber membranes (Bigland-Ritchie, Jones, and Woods 1979). It is understood, nonetheless, that interpretation of changes in the M wave with fatigue is complex (Bigland-Ritchie, Jones, and Woods 1979; Stephens and Taylor 1972), especially with the recent observation that motor unit action potentials increase with fatigue in a nonsystematic way across motor units (Chan et al. 1998). Stimulation of the fatigued muscle via its motoneuron with supramaximal electrical stimulation may or may not result in a slightly increased force, signifying that, in many situations, central activation of muscle units may be less than complete, in spite of high motivation levels (Gandevia et al. 1996; Thomas, Woods, and Bigland-Ritchie 1989; Bigland-Ritchie et al. 1983; Woods, Furbush, and Bigland-Ritchie 1987).

Decreased Excitability of Alpha-Motoneurons

When alpha-motoneurons are excited with sustained or intermittent current injection above their current threshold for rhythmic firing, their firing frequency decreases gradually over several seconds and even minutes, a phenomenon known as late adaptation (see chapter 2 for a discussion of this phenomenon). This intrinsic decrease in motoneuron excitability may possibly play a role in the decreased muscle activation that occurs with sustained or intermittent effort, although this is difficult to confirm. Late adaptation is not a fatigue of the motoneuron and can be overcome via increased current injection into the motoneuron.

Motoneuron excitability is frequently estimated using the H-reflex. This technique involves stimulation of the muscle's nerve with an electric shock of a duration and intensity that preferentially excites Ia afferents (which are the largest axons in the peripheral nerve and thus the most excitable). The amplitude of the muscle EMG response that follows the stimulation at monosynaptic latency is used as an estimate of the excitability of the motoneuron pool of the muscle of interest and has been used to estimate changes in motoneuron excitation with fatigue. There are caveats to its use, however (Capaday 1997). For example, H-reflex amplitude changes can occur due to alteration in the level of presynaptic inhibition on Ia terminals, with no change in motoneuron excitability (Nielsen and Kagamihara 1993; Capaday 1997).

Following a maximal sustained effort to fatigue, H-reflex amplitude decreases, suggesting a decrease in motoneuron excitability. Garland and McComas (1990) demonstrated a 50% decrease of the H-reflex amplitude after stimulating soleus muscle at 15 hertz for 10 minutes under ischemic conditions. Duchateau and Hainaut (1993) also found a decreased H-reflex amplitude after both maximal voluntary contraction (MVC) to fatigue and electrical stimulation in first dorsal interosseus and adductor pollicis muscles. Their results were comparable to those of an earlier experiment in which researchers found a similar decrease in the stretch reflex response using a similar fatigue paradigm, thus suggesting that spindle fatigue is not involved in the decreased H-reflex amplitude (Balestra, Duchateau, and Hainaut 1992). McKay and colleagues (1995) also reported a decrease in H-reflex amplitude immediately following MVC of the ankle dorsiflexors; H-reflex amplitude recovered to control values after one minute of recovery. Thus, the excitability of the motoneuron pool (which is the cumulative effect of all excitatory and inhibitory influences acting on the motoneurons, any neurohumoral influences, and motoneuron intrinsic properties) decreases with fatigue of the muscle during a maximal effort, whether activation is voluntary or artificial. This decrease is probably not due to late adaptation of the motoneurons involved, since these H-reflex changes persist after fatigue if the muscle is maintained in an ischemic state (Duchateau and Hainaut 1993;

Garland and McComas 1990). These findings support the concept that at least part of the decrease in the H-reflex in response to maximal effort is due to a decreased excitatory or increased inhibitory influence emanating from the fatigued muscle itself.

Muscle afferents include the dynamic and static spindle afferents (groups Ia and II), afferents arising from Golgi tendon organs (Ib), mechanoreceptors (nonspindle group II and group III), and nociceptor and metabotrophic receptors (groups III and the unmyelinated group IV afferents). In view of their excitatory and inhibitory effects on motoneuron excitation, much effort has been expended in attempting to determine whether they are involved in the decreased activation of the neuromuscular system during submaximal activation. This is discussed next.

Decline in Muscle Spindle Support to Motoneurons

Spindle activity increases in fresh muscles with an increase in voluntary drive, so that new spindles are recruited and those already firing increase in their firing rate as strength of the contraction increases (Bongiovanni and Hagbarth 1990). Since muscle spindle discharge provides an excitatory influence to motoneurons, one might expect a decrease in excitation of the motoneurons innervating a fatiguing muscle if spindle discharge were to decline. Several investigators have investigated the possible involvement of alterations in spindle afferent discharge in fatigue.

Fusimotor neurons (gamma-motoneurons) innervate intrafusal muscle fibers and cause contraction of their contractile components. Thus, fusimotor neuron discharge constitutes an important component in determining the response of muscle spindle afferents during movement. Hagbarth and colleagues (1986) showed that tibialis anterior motoneuron firing rates during a maximal effort decreased significantly when nerves were partially blocked with a local anesthetic (gamma fibers are more sensitive than alpha fibers to anesthetics; figure 3.5). Reduced firing rates in their blocked preparations could be increased by stretch of the muscle or by tendon vibration during contraction, suggesting that fusimotor neuron activity is important during contraction in providing excitatory support to motoneurons during sustained effort. These researchers' finding that relatively high vibration frequencies (165 hertz) were particularly effective in countering the effects of nerve block suggests that Ia endings, which can be driven up to 200 hertz by tendon vibration (Hagbarth et al. 1986), are more important than type II endings, which seldom follow vibration frequencies higher than 100 hertz (Gydikov et al. 1976). Furthermore, they found that reductions in firing rates were exacerbated by tendon vibration of the antagonist.

Subsequent studies (Bongiovanni and Hagbarth 1990) showed that all of the decrease in EMG that occurred with a maximal sustained contraction of the dorsiflexors for four minutes could be compensated by short periods of tendon vibration. Furthermore, the effect of vibration was most pronounced in the high-threshold units (figures 3.6 and 3.7). Consistent with the

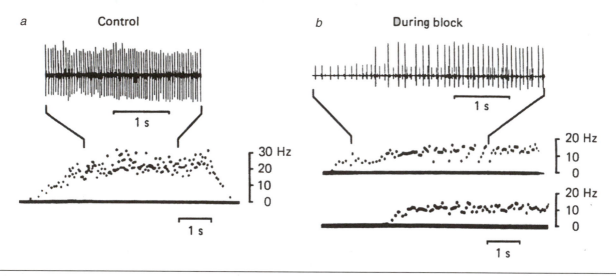

Figure 3.5   Effects of blockade of gamma-motoneurons on motor unit firing in tibialis anterior during a maximal voluntary contraction: (*a*) before blockade; (*b*) after partial block of the peroneal nerve using injection of prilocaine around the nerve. Note how firing frequencies are lower in *b* than in *a*.

Reprinted from Hagbarth et al. 1986.

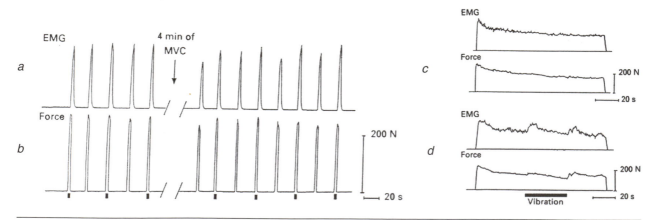

Figure 3.6   Brief (a few seconds') tendon vibration attenuates EMG and force decreases during foot dorsiflexor MVC sustained to fatigue. (*a*) EMG and (*b*) force before and after four minutes of MVC to fatigue. Vibration was applied to the tibialis anterior tendon every second contraction (indicated by dark boxes on abscissa). (*c* and *d*) Force and EMG drop-offs during a sustained MVC of foot dorsiflexors are attenuated with tendon vibration. The top two traces (*c*) are without vibration, the bottom two (*d*) with vibration of the tibialis anterior tendon where indicated.

Reprinted from Bongiovanni and Hagbarth 1990.

idea of muscle wisdom, however, the attenuation of the EMG decrease caused by vibration had no parallel effect on force loss.

More recently, Macefield and colleagues (1993) examined the recruitment of dorsiflexor motoneurons deafferentated by pharmacological blockade of the peroneal nerve. Their findings, that firing frequencies at maximal effort (18.6 hertz) when motoneurons were denied afferent excitation were lower than those of unblocked motoneurons (28.2 hertz), were consistent with the interpretations of previous investigators. However, they also noted that the firing rates of their deafferentated motoneurons did not drop during the maximal effort as much as the drop seen in experiments with the same muscle and task but intact afferents. This led them to conclude that there probably are at least two mechanisms that limit motoneuron firing frequencies during fatigue: one operating early in the task (declining spindle support) and another later (increased inhibitory influence; figure 3.8).

These results suggest that the fusimotor system provides an important source of motoneuron excitation for ongoing contractile activity during a sustained MVC and that a decrease in the functioning of this system might explain part of the decreased firing rate of motor units

during fatigue at this intensity. A reduction in the sensitivity of the spindle itself would not explain why H-reflexes are reduced even when afferents are electrically stimulated (Garland and McComas 1990; Duchateau and Hainaut 1993), thereby bypassing the spindle, or why the time course of H-reflex decrease is longer than one would expect from the rapid decrease in spindle support. A possible effect of fatiguing contractions on fusimotor activation has been proposed (discussed later); a decrease of fusimotor activation would decrease spindle activation.

Increased Inhibition of Motoneurons

Motoneurons could be increasingly inhibited during a sustained MVC, thus contributing to the decreased motoneuronal excitability and decreased firing rates. The possible sources of inhibition are shown in figure 3.9.

Undeniably, the most popular hypothesis for inhibition as a mechanism during fatigue involves the stimulation of afferents whose muscle endings are sensitive to metabolic byproducts of fatiguing contractions, which ultimately have an inhibitory influence on the motoneurons. There is a considerable amount of information of the effects of metabolic substances on muscle receptors, which are

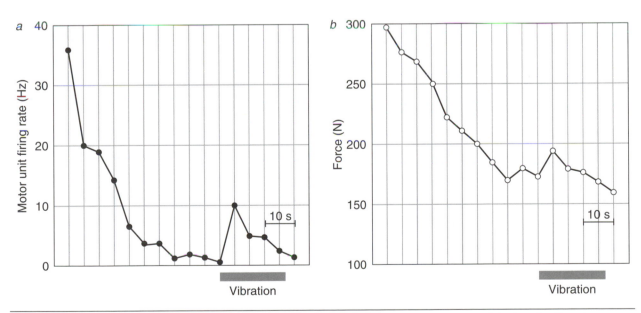

Figure 3.7 Vibration of the tibialis anterior tendon temporarily counteracts (*a*) the decline in motor unit firing rates and (*b*) muscle force that occurs during an MVC of foot dorsiflexors. Unit in upper trace is a high-threshold unit.

Reprinted from Bongiovanni and Hagbarth 1990.

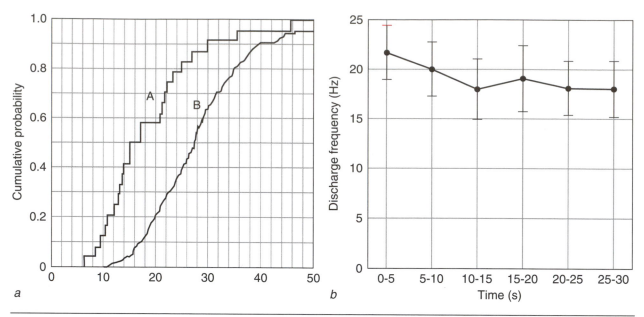

Figure 3.8   Firing behavior of single motor axons in the common peroneal nerve during an MVC in the absence of feedback from the muscle via distal pharmacological deafferentation. (*a*) Distribution of firing rates during brief MVCs of the tibialis anterior in (A) blocked and (B) unblocked preparations. Means are 18.6 and 28.2 Hz for blocked and unblocked samples, respectively. (*b*) Tibialis anterior motoneuron firing rates begin at lower rates, but fall off more slowly, during an MVC under the blockade condition.

Reprinted from Macefield et al. 1993.

discussed in a later section summarizing evidence from animal experimentation.

Bigland-Ritchie and colleagues (Bigland-Ritchie, Dawson, et al. 1986; Woods, Furbush, and Bigland-Ritchie 1987) conducted ingenious experiments that strongly pointed toward an inhibitory influence on motoneuron excitability emanating from fatiguing muscles. The design of their experiment and an example of their results are presented in figure 3.10. These investigators found that the reduced firing rates that occurred with a sustained MVC did not recover while the muscle was kept ischemic, which suggests a reflex by which information is transmitted via afferents from the fatigued muscle to the spinal cord and that has an inhibitory effect on motoneuron excitability. After all, how else would the motoneuron cell body in the ventral horn of the spinal cord, where firing rates are determined, know what was happening at the muscle metabolic level? Garland, Garner, and McComas (1988) found similar results in soleus muscle stimulated under ischemic conditions to fatigue, thus demonstrating that the decline in muscle activation does not require voluntary muscle activation and supporting the hypothesis of a reflex origin for

the decrease in firing rates. It has since been demonstrated that the decrease in EMG during MVC after fatigue still occurs to the same degree even if the large afferents are blocked using compression during the ischemic fatigue (such that the H-reflex is almost abolished, thus implicating the Ia afferents). This finding suggests that smaller afferents, not large afferents, are probably involved in this fatigue reflex.

Using this same model of stimulation of muscles under ischemic conditions, Sacco, Newberry, et al. (1997) found that EMG was depressed during an MVC in medial gastrocnemius after fatigue of the lateral gastrocnemius and thus suggested that this inhibitory reflex also affects close synergists.

The decrease in motoneuron excitability found by Duchateau and Hainaut (1993) after fatigue in intrinsic hand muscles was too slow to be attributed to motoneuron adaptation or to decreased spindle support, both of which exert their most significant effect early in the contraction; the decrease in motoneuron excitability persisted as long as ischemia was maintained. By elimination, these researchers concluded that increased inhibition involved a

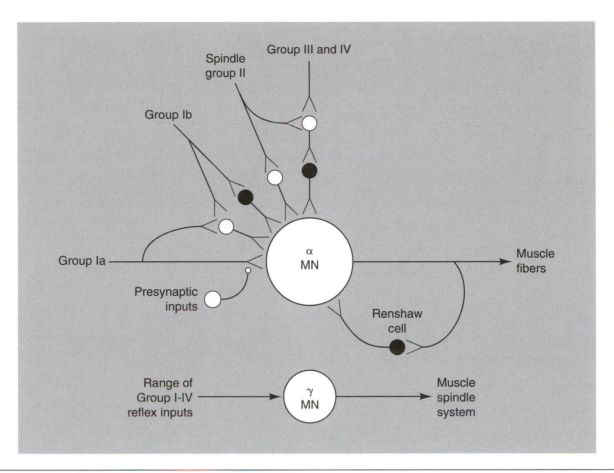

Figure 3.9 Influences on alpha- and gamma-motoneurons. Filled cells are inhibitory.

Reprinted from Gandevia 1998.

muscle-originating reflex, but like other investigators in this area, they were unable to pinpoint the exact source of the inhibition.

Ia input to motoneurons can be regulated via Ia presynaptic inhibition (Hultborn, Meunier, Pierrot-Deseilligny, et al. 1987). Presynaptic inhibition involves a depolarization of Ia afferent terminals to motoneurons, brought about by axoaxonal synapses, which reduce the size of the Ia excitatory postsynaptic potential (Hultborn, Meunier, Morin, et al. 1987; see figure 3.9). Presynaptic interneurons are subject to excitatory and inhibitory control from several brain centers (Nielsen and Kagamihara 1993; Hultborn, Meunier, Pierrot-Deseilligny, et al. 1987; Hultborn, Meunier, Morin, et al. 1987). Hultborn, Meunier, Morin, et al. (1987) found that presynaptic inhibition influencing a motoneuronal pool decreased at the beginning of a contraction of the implicated muscles and increased in muscles not involved in the contraction. They

proposed that at least a part of the increase in presynaptic inhibition of the nonrecruited muscle might be due to activation of presynaptic neurons by group I afferent information coming from the contracting muscle. However, their results indicated at least one supraspinal control mechanism of the presynaptic inhibitory system: the inhibition of motoneurons of the contracting muscle (which are disinhibited upon activation). Most probably, a tonic presynaptic inhibition is in place during the noncontracting state, and a supraspinal center or centers selectively remove this inhibition from recruited motoneurons by inhibition of these interneurons (Nielsen and Kagamihara 1993; Hultborn, Meunier, Pierrot-Deseilligny, et al. 1987). A decrease in presynaptic inhibition at the onset of recruitment ensures that sensory information from spindles is received in full fidelity by motoneurons of the activated muscle. There is some evidence, nonetheless, that

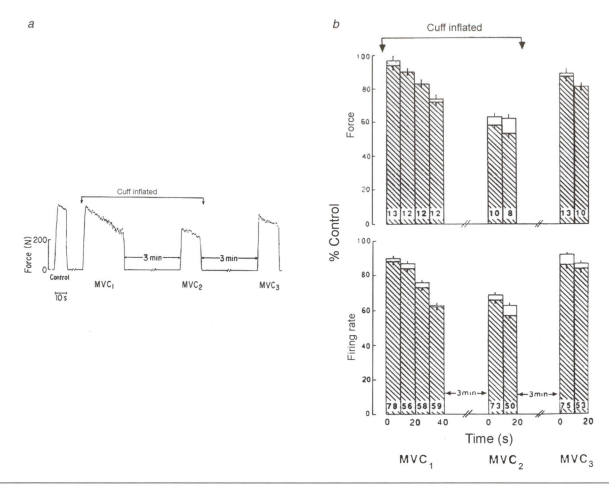

Figure 3.10   Evidence that the reduced motoneuron firing rates that occur during a sustained MVC are due to an inhibitory influence on motoneurons from muscle afferents. (*a*) The design of the experiment of Bigland-Ritchie and colleagues (Woods, Furbush, and Bigland-Ritchie 1987). Subjects performed three MVCs to fatigue (quadriceps and adductor pollicis were examined), separated by three minutes of rest. From the beginning of the first to the end of the second MVC, ischemia was induced by a tourniquet. (*b*) Example of the results obtained from these experiments (quadriceps). Motor unit firing rates did not recover until blood flow was restored (during the second rest period), implying an influence from fatigued muscles on motoneuron firing rates.

Reprinted from Woods, Furbush, and Bigland-Ritchie 1987.

presynaptic inhibition, which decreases at the beginning of contraction, may increase as the contraction continues, which would involve a gradual reduction in motoneuron excitation (Hultborn, Meunier, Morin, et al. 1987).

A possible role for the presynaptic inhibition system in maximal efforts to fatigue is supplied by the case of cocontraction, or coactivation, of agonists and antagonists. An altered activation of Ia inhibitory interneurons via supraspinal centers may explain the variable degree of agonist/antagonist coactivation that can occur with certain types of tasks. Ia inhibitory neurons appear to be controlled differently during

cocontraction than during flexion–extension movements (Nielsen and Kagamihara 1993). During repeated isometric MVCs continued to fatigue, cocontraction of agonist (quadriceps) and antagonist (biceps femoris) muscles increases, as evidenced by a decrease in EMG of the agonist with fatigue with no change in EMG of the antagonist. Increased coactivation also occurs during fatiguing isokinetic knee extensions at low and high speeds, although coactivation is higher at high speeds, even before fatigue (Weir et al. 1998). One hypothesis is that supraspinal influence reduces the Ia presynaptic inhibition of afferents that convey reciprocal inhibition to antagonists during agonist fatigue, thus increasing coactivation.

Presynaptic inhibition might also explain the phenomenon whereby prolonged tendon vibration accentuates the fall in EMG, motor unit firing rates, and force during a sustained MVC in foot dorsiflexors. This is, of course, in contrast to the effects of short periods of tendon vibration on fatiguing muscle (see figures 3.6 and 3.7). Bongiovanni, Hagbarth, and Stjernberg (1990) reported this phenomenon, noting also that this effect developed slowly and persisted for up to 20 seconds after the end of vibration, was accentuated by previous activity, and preferentially affected the ability to drive high-threshold units at high firing rates.

Kukulka, Moore, and Russel (1986) established that reciprocal inhibition, via Renshaw cell activation, might play a role in reducing motoneuron excitability during maximal activation of the soleus muscle. They showed that the inhibitory effect of a conditioning stimulus (preceding the H-reflex by 10 milliseconds) on the H-reflex of soleus was altered following fatigue, in a way that would suggest that reciprocal inhibition was increasing during the effort. Highest-threshold motoneurons contribute several times more to the excitation of Renshaw cells than lowest-threshold motoneurons (Hultborn, Katz, and Mackel 1988; Hultborn, Lipski, Mackel, et al. 1988), implying that significant reciprocal inhibition would not be unexpected when recruiting high-threshold units for these unusually long periods (up to five minutes).

Decreased Excitation of Motoneurons From Supraspinal Sources

Our knowledge concerning the possible contribution of decreased excitation of motoneurons from supraspinal sources to neuromuscular fatigue has emanated from studies using stimulation of the cortex via either transcranial electrical stimulation (TES) or transcranial magnetic stimulation (TMS) and the recording of the resulting motor evoked potential (MEP), which is a compound action potential of short latency (Taylor et al. 1996), measured at the muscle. TES excites pyramidal cortical cells at the level of their axons, while TMS stimulates these cells either directly or indirectly through intracortical connections (Taylor et al. 1996; McKay et al. 1996). The basic approach has been to deliver stimuli at submaximal intensity to the cortex in control and voluntary fatigue conditions. A change in the amplitude of the MEP yields information on the general excitability of the cortical cells and spinal cord (increased amplitude = increased excitability), whereas the silent period that immediately follows the stimulation (an interruption of the voluntary EMG lasting approximately 250 milliseconds) yields information on the degree of inhibition present in the cortical area (a longer silent period indicates more inhibitory influence).

In 1995, Mills and Thomson asked subjects to generate a sustained MVC with the first dorsal interosseus muscle and delivered stimuli using TMS during the fatigue process. The contraction lasted about 100 seconds, and isometric force decreased to about 25% of the original. They reported no change in MEP amplitude (more precisely, a slight increase that was statistically not significant) but an increase in the duration of the silent period during fatigue. They concluded that there was either no change in cortical excitability with fatigue or that the changes in cortical and spinal excitability changed in equal and opposite directions, a possibility that they were not in a position to resolve with their techniques. They attributed the increase in cortical inhibitory influence to peripheral feedback from the fatiguing muscle.

In the same year, McKay et al. (1995) found that MEP amplitude decreased immediately following a maximal sustained isometric contraction of ankle dorsiflexors and remained depressed during 20 minutes of recovery.

A similar approach was taken by McKay et al. in 1996, again using fatigue of the ankle dorsiflexors. Their results were similar to those of Mills and Thomson (1995), in that MEP amplitude did not change, but the silent period increased at fatigue. However, they also reported that the MEP of the contralateral muscle significantly increased, suggesting that cortical excitability may increase during sustained MVCs in an attempt to counter the decreased excitation from the periphery (although this was not measured). Thus, their results might suggest that the lack of change seen in MEP of the fatigued muscle during TMS is due to an equal and opposite change in cortical and spinal excitability, as was offered previously as a possible explanation for these results. They also succeeded in demonstrating that the source of the silent period change was cortical, since the increase in this period was not evident when stimulation was at the level of decussation of the pyramids (near the junction of the medulla and spinal cord).

Taylor et al. (1996) and Gandevia et al. (1996) applied similar approaches to examine fatigue following sustained MVC in the elbow flexors and combined TMS with several other techniques in an attempt to isolate the locus of fatigue. Their results (figures 3.11 and 3.12) demonstrated an increase in the amplitude of the MEP as well as an increased silent period duration at fatigue, evidence of a parallel increase in simultaneous excitation and inhibition in the cortex. When ischemia or tendon vibration were administered at the point of fatigue, the recovery of MEP amplitude and silent period was not influenced, indicating that they were not linked to peripheral events. In addition, similar to previous results, the changes in MEP amplitude were cortical and not spinal, since they were not evident with stimulation of the corticospinal tract at the cervicomedullary junction.

Perhaps more revealing, these results also demonstrated that cortical stimulation during fatigue resulted in an increased superimposed muscle twitch response, a classic demonstration of central fatigue (figure 3.13). This indicates that, in spite of increasing cortical excitation during the MVC (i.e., increased MEP amplitude), the cortex was operating at a suboptimal level for maximal force generation. This limitation was not at the level of the motoneuron, since peripheral stimulation also resulted in a larger superimposed contractile response at fatigue. The researchers also noted that when ischemia was imposed at the end of the fatigue protocol, MEP amplitude recovered rapidly, but the superimposed twitch in response to cortical stimulation remained elevated. This means that, in spite of increases during fatigue and the rapid recovery following fatigue, this cortical excitation remained suboptimal for maximal muscle force generation as long as the muscle remained in a fatigued state. The authors proposed that the suboptimal driving of motoneurons was in response to afferent information from muscles, joints, and tendons.

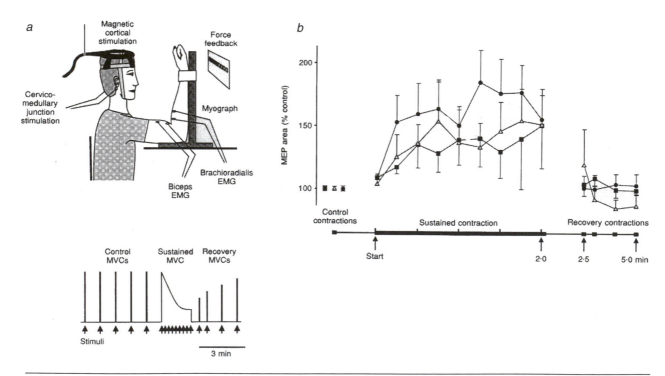

Figure 3.11 (a) The design of the Taylor et al. (1996) experiment. Stimuli were given alternately at the level of the cortex and brainstem before, during, and after an isometric MVC of the elbow flexors. (b) The area of the MEP in biceps brachii increased markedly during an MVC (filled circles), but also during contractions at 30% (filled squares) and 60% (open triangles) of MVC.

Reprinted from Taylor et al. 1996.

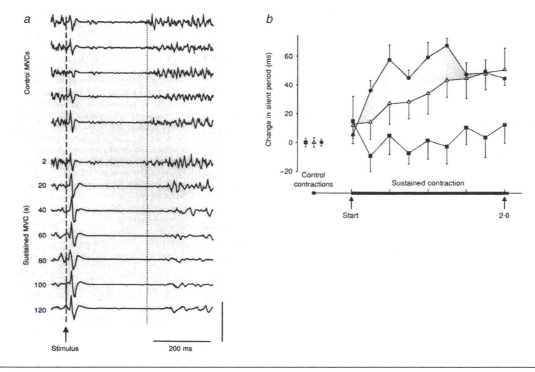

Figure 3.12 (*a*) During a two-minute MVC of elbow flexors, the silent period following the muscle potential evoked by TMS increases. (*b*) This increased silent period occurred at 60% (open triangles) and 100% (filled circles), but not 30% (filled squares) of MVC.

Reprinted from Taylor et al. 1996.

There is a gradual decrease in MEP amplitude after exercise is terminated. Bonato and colleagues (1996) asked subjects to perform rapid abduction–adduction movements of the thumb at a maximum rate for one minute and examined the MEP amplitude during the recovery period. They found that MEP amplitude decreased significantly by 15 to 20 minutes after the exercise for both ipsilateral and contralateral muscles. They also found a contraction of the motor output map of the corresponding cortex and suggested that this was due to inhibitory mechanisms.

More recently, Gandevia et al. (1999) showed a possible depression in corticospinal influence on motoneurons that occurs immediately following a 120-second MVC of the elbow flexors, using a stimulation technique that selectively activates these axons (transmastoid electrical stimulation). The time course of the effect is different from the effects of transcranial magnetic stimulation and most likely involves decreased efficacy of corticospinal–motoneuronal synapses.

These results from experiments using cortical and subcortical stimulation during a sustained MVC to fatigue can be summarized as follows: (1) The level of cortical inhibition increases; (2) the level of cortical excitation increases; (3) the cortex appears to drive motoneurons at sufficient intensity to elicit maximal contractile responses in the control condition, but perhaps not during fatigue; (4) this central fatigue is related to afferent information from the periphery; (5) decreases in corticospinal stimulation of motoneurons may also occur; and (6) since TMS results in a superimposed contractile response during fatigue, the locus of this central failure is somewhere in the sites driving the cortex. In other words, sites upstream from the motor cortex appear to limit the full expression of contractile force during a sustained MVC.

Submaximal Contractions to Fatigue

I have separated submaximal from maximal contractions to fatigue for the simple reason that they have different durations, involve different recruitment strategies, and may as a consequence involve different mechanisms. However, a comparison of responses at fatigue during

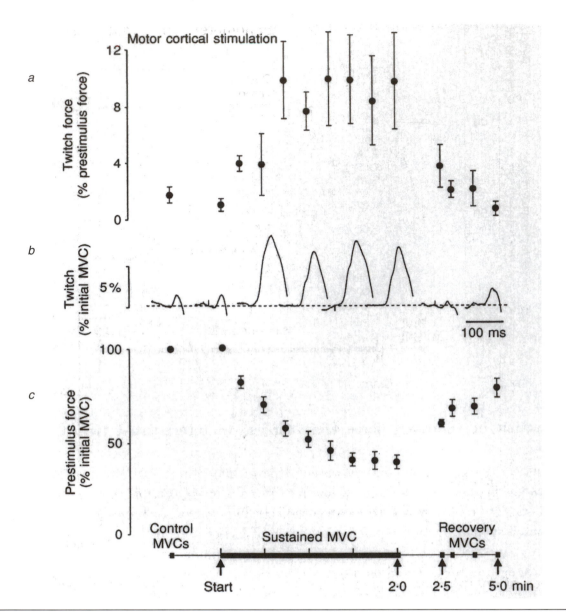

Figure 3.13   Same experimental design as in figure 3.11. (*a*) Mean values for twitches resulting from cortical stimulation. (*b*) Examples from one subject of twitches resulting from cortical stimulation every 30 s during the contraction and at periods before and after the fatiguing regimen. (*c*) Decrease in voluntary force with sustained MVC of elbow flexors.

Reprinted from Gandevia et al. 1996.

submaximal and maximal contractions reveals similarities as well. I present evidence from submaximal contractions for the same four mechanisms discussed regarding maximal contractions: decreased excitability of alpha-motoneurons, decline in muscle spindle support to motoneurons, increased inhibition of motoneurons, and decreased excitation of motoneurons from supraspinal sources.

When a subject is asked to generate a submaximal contraction until fatigue, muscle EMG increases gradually as additional motor units are recruited to maintain the force (Garland et al. 1994; Bigland-Ritchie, Furbush, and Woods 1986; Löscher, Cresswell, and Thorstensson 1996b; Macefield et al. 1991; Petrofsky and Phillips 1985; Häkkinen and Komi 1983b, 1986). This increased recruitment is substantiated by an estimated increase in the excitability of the motoneuron pool, since H-reflex amplitude increases as contraction proceeds during a maintained contraction of the triceps surae at 30%

of MVC (Löscher, Cresswell, and Thorstensson 1996b; figure 3.14). Simultaneously, and somewhat paradoxically, the majority of recruited motor units demonstrate a decline in their firing frequencies (Christova and Kossev 1998; Petrofsky and Phillips 1985; Garland, Griffin, and Ivanova 1997; Garland et al. 1994; De Luca, Foley, and Erim 1996), although some may actually demonstrate an increase (Garland et al. 1994), especially those that are newly recruited later during the task (Garland, Griffin, and Ivanova 1997; figure 3.15). Thus, as more units are recruited, at least some, and perhaps most, of those already recruited experience a decrease in firing rate.

De Luca, Foley, and Erim (1996) made some interesting observations of this phenomenon for contractions of the tibialis anterior and first dorsal interosseus maintained at various percentages of MVC. Even at relatively short periods before the onset of significant fatigue (10 to 20 seconds), motor units began to show a consistent decline in firing rates once the target force was attained. Those recruited at higher thresholds and that had lower firing rates demonstrated a greater decrease in firing rates than those with lower thresholds and higher firing rates during the contractions. This was seen even during contractions at levels at which one would find no additional recruitment, thus suggesting that force can be maintained early during constant-force contractions in spite of lack of further recruitment and declining firing rates of active motor units. The authors suggested that, at least early during these maintained contractions, lower firing rates were compensated for by contractile changes such as staircase and potentiation, which would tend to increase the force per excitation. However, this effect would last only a minute or so and does not explain the decreasing rates that occur as muscle unit forces drop with fatigue.

As motor unit firing rates decline with fatigue, electrical stimulation of the muscle may or may not result in force equal to a voluntarily generated MVC, indicating that the decline in motor unit firing frequency may limit the expression of muscle force to a greater extent than during maximal sustained contractions (Löscher, Cresswell, and Thorstensson 1996b; Sacco, Thickbroom, et al. 1997). When the target force can no longer be maintained, not only MVC but

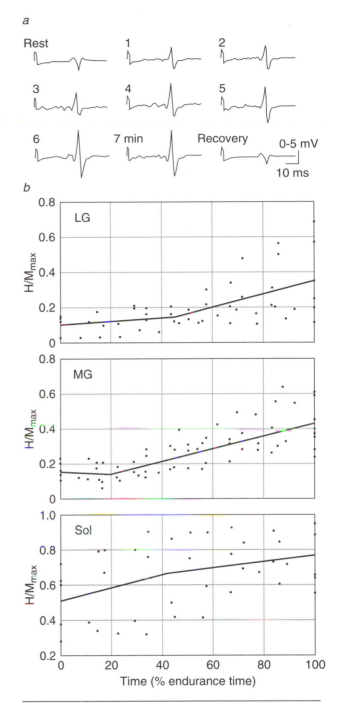

Figure 3.14   The H-reflex amplitude increases during a sustained submaximal contraction. (*a*) Time-related increase in soleus H-reflex during a contraction at 30% MVC sustained until fatigue (seven minutes). (*b*) Time-dependent increase in the H-reflex amplitude (expressed as H/M_{max} ratio, where M_{max} is maximal M-wave amplitude) in lateral gastrocnemius (LG), medial gastrocnemius (MG), and soleus (Sol) at 30% MVC sustained until fatigue.

Reprinted from Löscher, Cresswell, and Thorstensson 1996.

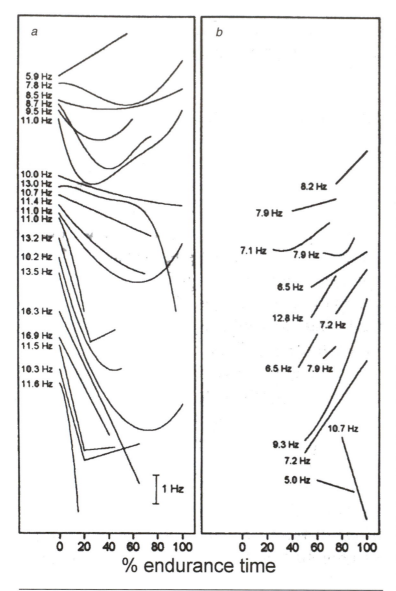

Figure 3.15 Firing rate behavior of motor units in human triceps brachii during an isometric contraction at 20% MVC sustained until fatigue (about six minutes). (*a*) Motor units that were active from the beginning of the task; (*b*) those recruited later. Beginning firing rates are shown for each unit. Units recruited at the beginning of the task tended to decrease with time, while those newly recruited during the task tended to increase with time.

Reprinted from Garland, Griffen, and Ivanova 1997.

also maximum EMG are depressed (Petrofsky and Phillips 1985; Fuglevand et al. 1993; Löscher, Cresswell, and Thorstensson 1996b). This deficit can persist for at least six hours following the exercise, and probably longer (Bentley et al. 2000). The lower the target force and thus the longer the contractile period, the greater are the depressions in maximum EMG and

motor unit firing frequencies at the point of fatigue (Bigland-Ritchie, Furbush, and Woods 1986; Dolmage and Cafarelli 1991; Fuglevand et al. 1993; Sacco, Thickbroom, et al. 1997). The depression of the voluntary EMG at fatigue, compared with a "fresh" MVC, is more than that of the M wave, suggesting that the decrease in EMG is not due (though not entirely) to neuromuscular failure (Fuglevand et al. 1993; Löscher, Cresswell, and Thorstensson 1996b; Sacco, Thickbroom, et al. 1997). These low firing rates at fatigue do not appear to substantially limit the full expression of available muscle force, however, and thus this phenomenon may constitute a demonstration of muscle wisdom. But how does this phenomenon come about?

Decreased Excitability of Alpha-Motoneurons

There is little evidence that decreased excitability of alpha-motoneurons is a major player in fatigue during sustained submaximal contractions. Löscher, Cresswell, and Thorstensson (1996b) examined H-reflex amplitude at various times during a submaximal (30% MVC) isometric contraction of plantar flexors maintained to fatigue. Their findings showed a gradual increase in H-reflex amplitude over time, until the point of fatigue (see figure 3.14).

Decline in Muscle Spindle Support to Motoneurons

During submaximal sustained contractions, spindle discharge decreases, in spite of an increased EMG as more motor units are recruited to maintain the target force (Macefield et al. 1991; figure 3.16). Macefield and colleagues (1991) demonstrated this by recording from afferent fibers in the peripheral nerve during sustained submaximal contractions of the dorsiflexors. The decrease in firing frequency was quite rapid and occurred even during the initial period of contraction as force is increasing to the target level. Spindles could be stimulated to higher frequencies by vibration of the tendon and by pressure applied over the spindle

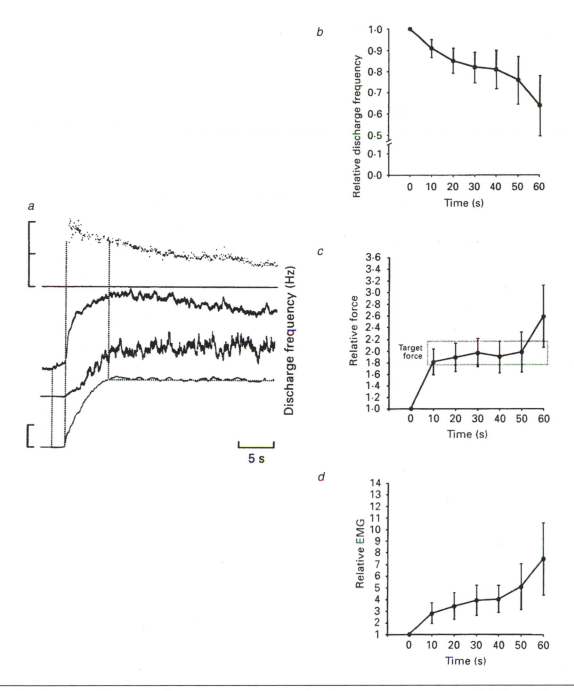

Figure 3.16 Firing of spindle afferents during constant-load isometric contractions. (*a*) Firing of a muscle spindle ending in extensor digitorum longus during the first 30 s of a contraction at 60% of MVC. Tracings are, from top to bottom, firing rate of the spindle ending, integrated neurogram (ENG) of the common peroneal nerve, muscle EMG, and force. Note how spindle firing rate declines even as force is increasing at the beginning of the contraction. (*b*) Mean spindle firing rate, (*c*) force, and (*d*) muscle EMG. During maintained submaximal contraction, mean firing of spindles decreased while muscle EMG was increasing.

Reprinted from Macefield et al. 1991.

receptive field; thus, spindles were not fatiguing. They also found decreased firing frequencies for Golgi tendon organs and cutaneous afferents during the sustained effort and decreased sensitivity to pressure applied to the receptive field (Golgi tendon organ afferents) and to maintained skin stretch (cutaneous afferents). The explanation for these decreases, according to the authors, may include a reduction in fusimotor output to spindles, a gradually reduced responsiveness of the spindles to fusimotor drive (because of axon branch-block failure or neuromuscular junction failure at the gamma-motoneuron–intrafusal fiber level), or adaptation of the fusimotor neuron that might be more pronounced than that in the alpha-motoneurons.

Avela and Komi (1998) asked their subjects to perform maximal leg extension countermovements before and after a marathon competition. They found that the short-latency stretch reflex that occurred during the eccentric phase of the contraction was virtually absent after the marathon, thus adding support to the contention that prolonged submaximal exercise may decrease spindle support.

These results are similar to those of Hagbarth, Bongiovanni, and Nordin (1995), who examined fatigue effects on finger extensor muscles. Their subjects were asked to sustain an isometric contraction at 30% of MVC for as long as possible. During the effort, the experimenters imposed torque pulses and extensor unloadings in order to study reflex responses in both extensor (unloading reflex) and flexor (stretch reflex) muscles. They found reflex responses to be less pronounced and to have a longer latency in fatigued muscles, attesting to a reduction in servocontrol with fatigue.

It does not appear that spindle sensitivity to length change decreases with submaximal fatigue, however, since the EMG response to a standardized tendon tap actually increases following a fatiguing contraction at 50% of MVC (Häkkinen and Komi 1983b; Nelson and Hutton 1985).

Increased Inhibition of Motoneurons

Reciprocal inhibition of the antagonist soleus increased after fatigue of the agonist tibialis anterior produced by a sustained contraction at 30% of MVC or by electrical stimulation (Tsuboi et al. 1995). This was expressed as a depressed H-reflex of the soleus in response to stimulation of the nerve innervating the fatigued tibialis anterior two to four seconds before.

Aymard et al. (1995) fatigued wrist extensors (extensor carpi radialis) in human subjects (sustained isometric contractions at 50% MVC, one minute on, one minute off) and examined the degree of inhibition evoked in an antagonist (flexor carpi radialis, a wrist flexor) and a transjoint synergist (biceps brachii, an elbow flexor) when group I afferents of the fatigued muscle were stimulated. As in the Tsuboi et al. (1995) study cited earlier, these researchers did this by evoking an H-reflex in the nonfatigued muscles and stimulating the afferents of the fatigued muscle at a latency that would demonstrate the inhibitory effect of the disynaptic inhibitory pathway on H-reflex amplitude. Their results, unlike those of Tsuboi et al. (1995), showed a decrease in the inhibitory response of afferent stimulation among synergists and no change between antagonists (i.e., reciprocal inhibition) with fatigue. The best explanation of their results was a selective effect of fatigue on the Ia inhibitory interneurons that govern inhibition between synergists, so that synergist inhibition was reduced.

A centrally mediated common drive between agonist–antagonist pairs results in a variable degree of agonist–antagonist coactivation during voluntary contractions (Psek and Cafarelli 1993; Rothmuller and Cafarelli 1995). The flexibility with which cocontraction can occur suggests a supraspinal mechanism for its modulation, in which reciprocal inhibitory pathways are suppressed, most likely via modulation of the Ia inhibitory system. Coactivation is particularly evident during strong and fast-displacement contractions (Psek and Cafarelli 1993; Hagbarth, Bongiovanni, and Nordin 1995; Weir et al. 1998). Consistent with this idea, gradual fatigue of an agonist during a series of submaximal, intermittent, isometric contractions results in an increased agonist EMG as more motor units are recruited to maintain the target force and a gradual and correlated increase in the recruitment of its antagonist (Psek and Cafarelli 1993). During repeated maximal contractions of knee extensors, coactivation gradually increases, even as the EMG of the fatiguing agonist decreases, since the ratio of

EMG in agonists to that in antagonists remains relatively constant, while force of agonists (but not antagonists, obviously) decreases (Weir et al. 1998). This is an example of a decrease in an inhibitory system in the spinal cord resulting from fatigue; Ia reciprocal inhibition of an antagonist decreases. The inhibition may be via a centrally controlled Ia presynaptic inhibition, which releases the inhibitory effect of agonist contraction on its antagonist. Hagbarth, Bongiovanni, and Nordin (1995) feel that at least a part of this decreasing reciprocal inhibition is due to declining Ia afferent inflow into the spinal cord.

This idea is supported by the results of Belhaj-Saïf, Fourment, and Maton (1996). They recorded from motor cortical cells in monkeys performing submaximal elbow flexion tasks to fatigue and found a significant number of cells increased in their antagonist facilitatory function.

Such coactivation during rhythmic contractions would be expected to increasingly attenuate the limb displacement against the load by the agonist as fatigue progressed. This is shown in figure 3.17: An increase in activation of biceps femoris with time would tend to oppose the mechanical consequence of activation of the vastus lateralis. This increased coactivation with fatigue could serve as a damping mechanism, however, to reduce the risk of oscillations in the agonist–antagonist pair as fatigue increases (Rothmuller and Cafarelli 1995). Finally, it might serve to limit the time to exhaustion of the agonist and thus reduce major perturbations in metabolism, morphology, or both that might result from more prolonged activation.

Decreased Excitation of Motoneurons From Supraspinal Sources

During a low-level (20% MVC), maintained, isometric elbow flexion task to fatigue, transcortical stimulation revealed a gradually increasing MEP amplitude, indicating an increasing cortex excitability (Sacco, Thickbroom, et al. 1997). Increasing duration of the silent period, indicating increased inhibition, also occurred during the second half of the contractile period, on a different time course from that of the MEP change. The MEP change occurred with TMS (stimulation of pyramidal cells directly and transsynaptically) and not TES (stimulation of

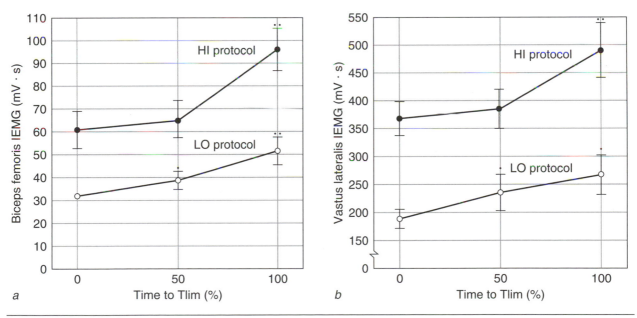

Figure 3.17   Coactivation of agonists and antagonists increases with submaximal contractions maintained to fatigue (100% Tlim). Subjects were asked to generate static, intermittent leg extensions at 30% (LO) or 70% (HI) of MVC until fatigue. (*a*) Biceps femoris integrated EMG increases with time in both protocols. (*b*) Vastus lateralis EMG. The related increases in EMG in agonists and antagonists suggests central control.

Reprinted from Psek and Cafarelli 1987.

pyramidal cell axons). This observation suggests that inhibition was in the cortex and not the spinal cord.

These results are at odds with those of a previous study of submaximal, sustained contractions of adductor pollicis (Ljubisavljevic et al. 1996). In this study, investigators found a gradual reduction in MEP amplitude and a decrease in silent period as contractions at 60% of MVC continued to the point of fatigue. Interestingly, MEP amplitude and silent period subsequently increased when contractile activity continued past the point of fatigue at a lower force. The biphasic response of MEP amplitude and silent period seen in this study, unlike the study of Sacco, Thickbroom, and colleagues (1997) cited earlier, may be due to the higher intensity and thus shorter duration of the contractile period, the use of the adductor pollicis as the muscle of study (compared with the elbow flexors studied by Sacco, Thickbroom, et al. 1997), or both.

Following exercise, two phenomena are evident using the technique of TMS: postexercise facilitation (increased MEP amplitude) and depression (decreased amplitude; Brasil-Neto, Cohen, and Hallett 1994; McKay et al. 1996; Samii, Wassermann, and Hallett 1997). Samii, Wassermann, and Hallett (1997) found that contraction of extensor carpi radialis at 50% of MVC resulted in a postexercise depression in the MEP only if the exercise was continued until the point of fatigue. For shorter times, at the same intensity, MEP amplitudes were slightly facilitated. They concluded that facilitation and depression occurred simultaneously during the contraction.

The experiments of Löscher, Cresswell, and Thorstensson (1996a) were designed to test the idea that central mechanisms, and not muscle afferents, are involved in fatigue during submaximal contractions. They asked subjects to maintain a contraction of triceps surae at 30% of MVC for as long as possible (about 400 seconds). At the point of fatigue, subjects stopped contracting, while experimenters continued stimulating the muscle electrically at the 30% torque level at a frequency of 30 hertz for one minute. They then asked subjects to generate a maximal contraction. Their principal findings were that subjects were capable, after one minute of continued stimulation of the fatigued

muscle, to resume voluntary torque generation at 30% of MVC for a further 85 seconds, even though the previously fatigued muscle had continued to work via electrical stimulation. The investigators hypothesized that the period of electrical stimulation after voluntary fatigue, while continuing to fatigue the muscle further, actually allowed the fatigued central mechanisms to rest, thus allowing the resumption of effort one minute later. This evidence would seem to speak against a peripheral source (i.e., effects of muscle state on afferent activity, which would have an inhibitory influence on motoneuron recruitment) to explain decreased muscle activation with fatigue.

Isometric vs. Anisometric Tasks

The bulk of the experimental work on fatigue, especially when it necessitates muscle or motor unit recordings during the effort, has used constant-load isometric contractions, for an obvious reason: Intramuscular electrode stability during the contractions allows unequivocal identification of the same motor unit over a prolonged period of time. Miller and colleagues (1996) investigated changes in firing rates of motor units in the triceps brachii while subjects performed a slow flexion–extension task to fatigue at a load corresponding to 17% of MVC. The researchers examined the number of spikes per contraction for the identified units recruited at the beginning of the task and found no systematic decrease with fatigue; in fact, this number decreased, increased, or stayed constant for different motor units as fatigue progressed (implying decreased, increased, or constant firing rates as fatigue progressed).

In a more recent study from this same laboratory, Griffin, Ivanova, and Garland (2000) found that motor unit discharge rate decreased when the triceps brachii performed contractions that were isometric but that simulated the torque changes that occurred during an anisometric task involving elbow flexion. The implication is that, due to the dynamic nature of this fatigue task, afferent information, blood flow, or both may be different, and thus their effects on motoneuron firing rate during fatigue may also be different. We will have to see more experiments involving dynamic types of submaximal contractions to fatigue to resolve this issue.

Spatial Variation in Muscle Activation During Fatigue: Rotation of Motor Units?

Given that continued contractile activity (sustained or intermittent) can have such detrimental effects on muscle fiber function, would it not be of some physiological advantage to be able to rotate motor units (recruit additional units to replace previously active units) during prolonged, submaximal exercise? Such a strategy would offset fatigue by giving periodic, temporary rest periods to motor units and thus would improve performance.

There is some evidence that motor unit rotation may occur during very low intensity maintained contractions. Fallentin, Jorgensen, and Simonsen (1993) demonstrated several examples of motor unit rotation in the biceps brachii muscles of individuals who were asked to maintain an isometric contraction at 10% of MVC for 1.5 to 2 hours. Furthermore, these investigators found that those subjects who were capable of lasting the longest before fatigue were also those who demonstrated motor unit rotation most consistently. Similar results have been reported by Tamaki and colleagues (1998), who found, again with very low intensity (10% MVC) contractions, that rotation among synergists involved in plantar flexion (soleus, medial and lateral gastrocnemius) occurred, especially during the latter part of the task to fatigue.

Recent studies on fatigue of the first dorsal interosseus hand muscle revealed that the recruitment of motor units may be quite heterogeneous during a task that required subjects to maintain an isometric force of 50% of MVC for as long as possible (Zijdewind, Kernell, and Kukulka 1995; figure 3.18). More recently, Westgaard and De Luca (1999) demonstrated motor unit substitution in human trapezius muscle during submaximal contractions of 10 minutes' duration. In these studies, several motor units were seen to cease being activated, only to be reactivated later in the session, while the whole-muscle activity and that of other units remained activated at a constant level. Future studies such as these may uncover other neuromuscular strategies like motor unit rotation that we can use, or perhaps learn through training to use, to improve performance.

Evidence From Reduced Animal Preparations on Mechanisms of Neuromuscular Fatigue

CONCEPT

4 The evidence cited earlier regarding fatigue in humans under various conditions is for the most part phenomenological, in

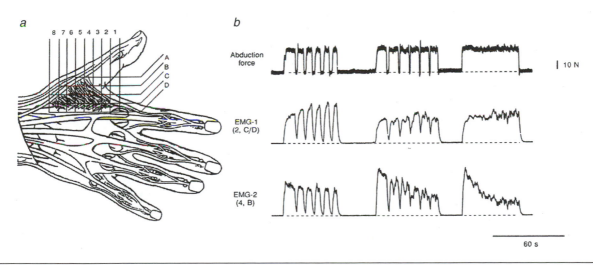

Figure 3.18 Responses of first dorsal interosseus during an isometric contraction at 50% of MVC. (*a*) Coordinate system used for various electrode positions for recording surface EMG. (*b*) Force and EMG at two electrode sites (2, C/D and 4, B) during contractions at 50% of MVC. Note the differences in repeated responses between EMG-1 and EMG-2, as well as evidence of task switching at the same electrode site.

Reprinted from Zijdewind, Kernell, and Kukulka 1995.

that specific mechanisms were not tested. Often, the mechanism was arrived at by elimination of several possibilities. This section reviews research on the previously discussed fatigue mechanisms, but which used anesthetized or decerebrate animals, in which fatigue of a muscle or muscle group was produced by electrical stimulation of the motoneuron or, in one research group, by simulation of exercise via stimulation of the mesencephalic locomotor region (MLR) in the cat. These approaches have the advantage of allowing precise measurements from such structures as alpha- and gamma-motoneurons, interneurons, and afferents. The disadvantages, on the other hand, include the fact that voluntarily generated exercise is not involved.

Behavior of Spindle and Golgi Tendon Organ Afferents

It does not appear that a major change in the sensitivity of spindle and Golgi tendon organ receptors is involved in the fatigue response. On the contrary, it appears that Ia and II spindle afferents may actually increase in dynamic sensitivity following fatigue (Nelson and Hutton 1985). Golgi tendon organs, on the other hand, may experience a diminished sensitivity with fatigue, which would probably act to increase excitation of homologous and synergistic motoneurons via a reduction in autogenic inhibition (Hutton and Nelson 1986).

When this type of experiment (one using an anesthetized animal preparation) is conducted with fusimotor neurons left intact, however, information about the sensitivity of the fusimotor–spindle system is gained. Christakos and Windhorst (1986) stimulated single cat medial gastrocnemius motor units to fatigue via single ventral root filaments and recorded the discharge from isolated muscle spindle afferents during the stimulation period. They found an increase in spindle gain as fatigue progressed; simply put, afferent firing rates did not decrease as much as force did.

The most recent work in this area (Hayward, Wesselmann, and Rymer 1991) demonstrates that spindle and Golgi tendon organ afferents show no major change in sensitivity with fatigue. In the researchers' anesthetized cats, the motoneuron was not severed, so fusimotor neurons were left intact. After fatigue of triceps surae via electrical stimulation (at a voltage below the threshold for stimulation of fusimotor neurons), spindle afferents showed no consistent change in spontaneous firing or in response to stretch. In addition, responses of Golgi tendon organs did not noticeably change with fatigue and were always related to force production.

Spinal Cord Inhibition

Hayward, Breitbach, and Rymer (1988) fatigued the medial gastrocnemius of decerebrate cats and found evidence of increased inhibition of the synergystic soleus muscle (as measured by a decrease in soleus force induced by a crossed-extensor reflex) during fatigue of gastrocnemius. These and other researchers (Lafleur et al. 1992; Hayward, Breitbach, and Rymer 1988; Zytnicki et al. 1990) found that this inhibitory influence subsided very quickly with continued stimulation during evoked contractile activity in medial gastrocnemius in decerebrate cats, however, because of the gradually reduced efficacy of transmission of inhibitory inputs to motoneurons innervating this muscle and its synergists. Declining inhibition was not due to a decrease in the firing of Golgi tendon organs. The researchers proposed that the declining inhibition was due to presynaptic Ib inhibition via IB fibers, which "filtered" information from IB fibers during the contractile period to help maintain optimal excitation of the motor pool.

The role of the inhibitory Renshaw cells in the fatigue process is uncertain. Windhorst and Kokkoroyiannis (1991) have presented evidence that Renshaw cell inhibition may be particularly strong during the first few seconds of repetitive motoneuron firing and could thus contribute to the rapid decline in firing that occurs at this time.

Excitation of Group III and IV Afferents

Group III afferents are responsive mainly to mechanical stimuli, while group IV afferents are primarily nociceptors but also include mechanoreceptors (Hayward, Wesselmann, and Rymer 1991; Paintal 1960). Group III and IV afferents exert an inhibitory influence on motoneurons innervating the muscle of origin of the afferents and their synergists (Windhorst et al. 1997; Kniffki, Schomburg, and Steffens 1981). For this reason, and because of their slow and long-lasting response (compared with that of spindles, for example), they have been targeted as the possible source of the fatigue reflex, by which muscle-related changes during fatigue act to diminish excitation of motoneurons during maximal sustained contractions (discussed earlier in this chapter).

Receptors of group IV afferents increase their firing rates in the presence of a number of substances that might increase in a fatiguing muscle, including bradykinin, potassium chloride, lactic acid, serotonin, and arachidonic acid (Djupsjöbacka, Johansson, Bergenheim, and Wenngren 1995; Djupsjöbacka, Johansson, Bergenheim, and Sjölander 1995; Sinoway et al. 1993; Windhorst et al. 1997; Djupsjöbacka, Johansson, and Bergenheim 1994; Rotto and Kaufman 1988; Kaufman et al. 1983; Kniffki, Schomburg, and Steffens 1981). The mechanically sensitive group III afferent receptors can increase their mechanical sensitivity when exposed to some of these substances, implying that they become more sensitive during fatigue (Rotto and Kaufman 1988; Kaufman et al. 1983; Hayward, Wesselmann, and Rymer 1991; Sinoway et al. 1993). Other substances released during contractile activity, which have no effect on the firing of these afferents when injected into a resting muscle, might potentiate receptor responses during contractile activity (Rotto and Kaufman 1988).

Hayward, Wesselmann, and Rymer (1991) recorded from nonspindle group II and group III afferents in triceps surae of anesthetized cats, examining their sensitivity to contraction, stretch, and surface pressure before and after fatigue produced by muscle stimulation. Their main findings were an increased sensitivity in these receptors to mechanical stimulation and an increased spontaneous discharge rate.

In 1994, Pickar, Hill, and Kaufman recorded responses from group III afferents to stimulation of the mesencephalic locomotor region (MLR) in decerebrate cats walking on a treadmill. These investigators reported an overall increase in the firing of these receptors, even at the very low forces generated in triceps surae during this locomotor task, attesting to these receptors' mechanical sensitivity. They later substantiated their findings by demonstrating that group III afferents were also excited during this rather low level of activity. These experiments were particularly important because they demonstrated, using a model that closely mimics voluntary locomotion (as opposed to electrical stimulation of the muscle), that these afferents transmit information regarding the contractile and metabolic state of the muscle not just during fatiguing activity, but also during relatively mild activity.

More recently, Hill (2000) has shown that group IV afferents from fatigued rat diaphragms show increased firing rates compared with those of fresh diaphragms.

Fusimotor Neuron Responses

Normally, flexor alpha-motoneurons are excited and extensor alpha-motoneurons inhibited by the activity of high-threshold (groups III and IV) afferents (Appelberg et al. 1983). The situation is somewhat less clear for gamma-motoneurons. Appelberg and colleagues (1983) examined the effects of stimulating group III afferents on the firing of gamma-motoneurons and found that excitatory effects outweighed inhibitory effects.

In the experiments of Djupsjöbacka, Johansson, Bergenheim, and Sjölander (1995) and Appelberg and colleagues (1983), the effects of injections into the muscle circulation of lactic acid, arachidonic acid, and potassium chloride on the firing of muscle spindle afferents were measured. Their hypothesis was that an effect of these substances on the firing level of the fusimotor neuron would be evident from the discharge of the spindle afferents when the muscle was subjected to sinusoidal stretches. They found increases in spindles' sensitivity and modulation (range in frequency in response to stretch), even when the injection was in a contralateral muscle. In addition, no such effect was evident when the muscle's nerves were cut. Thus, stimulation of chemosensitive afferents increased the excitability of fusimotor neurons.

A similar approach was undertaken by Ljubisavljevic and colleagues (Ljubisavljevic, Jovanovic, and Anastasijevic 1994; Ljubisavljevic et al. 1995), except that they examined the effects of fatiguing muscle stimulation on spontaneous firing of fusimotor neurons directly. Their various treatments (fatigue of contralateral muscles, ischemia following fatigue, denervation, etc.) demonstrated that the effects on fusimotor discharge (increased frequency) were mediated by group III and IV afferents from the fatigued muscle and were evident in nonfatigued ipsilateral and contralateral muscles. In addition, they found an early increase in firing, which they attributed to excitation by mechanoreceptors, and a late and longer-lasting increase in firing, which they attributed to the chemosensitive receptors. They concluded that this mechanism is a compensatory attempt to optimize the firing rates of the spindle afferents at a time when spindle support is declining during fatigue.

A unique approach to the role of spindle afferents in fatigue was taken in the experiments of Pedersen et al. (1998), who fatigued the medial gastrocnemius of anesthetized rabbits and examined the sensitivity of ensembles of spindle afferents (3 to 10 afferents) from the lateral gastrocnemius to stretches of different amplitudes. The analysis of groups of afferents allows researchers to derive an index of separation, or an overall idea of how well different afferents simultaneously convey distinct information about the stimulus. These researchers' statistical analysis showed that ensembles of muscle spindle afferents of lateral gastrocnemius lost the ability to discriminate the different stretch paradigms with fatigue of its synergist, the medial gastrocnemius; in a way, afferents became more similar in their responses. This effect did not occur when the medial gastrocnemius nerve was cut before stimulating it to fatigue, suggesting that the effect was transmitted to the lateral gastrocnemius via afferents from the medial gastrocnemius. They found that fusimotor drives also became more similar (less discriminatory) after fatigue by examining the effects of stretch on the modulation and frequency of firing of single afferents. Their results suggested that events in the fatigued medial gastrocnemius, probably communicated to the spinal cord via group IV afferents, influenced fusimotor neurons such that the normal variability of responses to stretch among these cells was reduced. As a result, important information regarding muscle length is lost during and immediately following fatigue.

Changes in Intrinsic Motoneuron Properties With Fatigue

Very little is known about changes in intrinsic motoneuron properties with fatigue. Stimulation of group III and IV afferents results in changes in intrinsic motoneuron membrane properties that would change their firing behavior during excitation. Specifically, in decerebrate cats, injection of substances to stimulate these afferents (bradykinin, serotonin, potassium chloride) resulted in decreases in membrane potential and cell input resistance, decreased amplitude of afterhyperpolarization, and increased synaptic noise (Windhorst et al. 1997). The authors pointed out that such changes would not necessarily contribute to adaptation during fatigue, but rather indicate a general change in parameters that are significant in cell excitability and rhythmic firing properties.

Summary

Neuromuscular fatigue includes a decrease in motoneurons' activation level, sometimes (but not always) to levels that limit the expression of muscle force. This decrease in net excitation appears to be due to a number of components, including decreased excitatory influence from spindles, increased inhibitory influence from muscle receptors that are sensitive to fatigue substances, and an altered supraspinal signal. The latter appears to involve an increased inhibitory influence, as well as an alteration in the signal coming into the primary motor cortex. It is difficult to determine the relative importance of each of these components when comparing submaximal and maximal efforts maintained to fatigue. However, we can venture a guess that muscle receptor activation by fatigue substances is probably of lesser significance and spindle support of greater significance during longer efforts to fatigue. The role of supraspinal influences during short-term versus long-term neuromuscular fatigue requires more experimentation, since most of our evidence at present is for short-term efforts.

chapter 4

Endurance Training of the Neuromuscular System

Most Important Concepts From This Chapter

1 Endurance training is very complex, due to the complexities of motor unit recruitment, and thus adaptations found using the biopsy technique in humans are descriptive. More insight into mechanisms of adaptation is afforded by examining adaptive responses and their associated mechanisms following chronic muscle electrical stimulation.

2 Chronic activity results in changes in protein synthesis at the pretranslational, translational, and posttranslational levels.

3 Several possible biochemical signals are discussed that change in concentration during acute contractile activity and that, either alone or in combination, may promote changes in protein synthesis.

4 At the level of the neuromuscular junction, morphological, biochemical, and physiological adaptations with endurance training serve to enhance endurance. These adaptations are both pre- and postsynaptic.

5 Alpha-motoneurons show biochemical evidence of adaptations to increased activity, including increased content and axonal transport of factors that seem vital for optimal adaptation. However, changes resulting from increased activity in physiological properties that would determine how they are used during exercise are unknown at present.

6 Spinal cord systems are responsive to increased usage. This is most evident in spinalized cats in which daily locomotor training with muscles distal to the lesion results in alterations in muscle usage patterns.

Most of what we know about endurance in general, and about the effects of endurance training in particular, concerns cardiovascular, respiratory, and muscular adaptations. Information regarding the latter has proliferated especially since the 1960s, due to the evolution of knowledge and techniques in histochemistry, biochemistry, and molecular biology. We are now beginning to have a clear picture of the phenotypic adaptations that characterize endurance training, although the signals that promote the adaptations and the molecular mechanisms by which these adaptations occur remain elusive.

In this chapter, my goal is to emphasize adaptations to endurance training in the muscle, at the neuromuscular junction, at the level of the motoneuron, in the spinal cord, and in the central nervous system that promote the development of neuromuscular endurance. As we move from muscle through to the central nervous system, information becomes gradually less plentiful, and examples from models other than endurance training that might involve similar mechanisms become more numerous.

Muscle Adaptations

CONCEPT 1

Exercise-training responses are complex at the level of the whole muscle. One reason for this is the complexity of the exercise being performed during the training—we obtained a glimpse of this complexity in chapter 2. Which fibers are being recruited to generate what percentage of their maximal force? Are some fibers more overloaded than others? (Our knowledge from previous chapters tells us that they are.) Do all fibers respond equally to the same degree of stress? Studies with human subjects often measure biochemical muscle responses using the muscle biopsy technique. A biopsy sample taken from vastus lateralis or gastrocnemius is in actuality a mixture of fibers of different types and, if taken from trained subjects, of different training states. We obtain an average from the fibers in the sample of the biochemical or molecular biological property being assayed, assuming that a representative sample of the muscle is possible using the biopsy technique, which is open to argument (see Lexell et al. 1983).

Chronic electrical stimulation of muscles has been used extensively during the past 30 years to demonstrate the limits of adaptability of muscles to increased activity. All indications are that the changes reported with this model may represent the ultimate endurance-trained state, which could never be achieved via voluntary activation. The results from this model have been very instructive with regard to the extent and relative time course of muscle adaptations that occur, against which we can measure endurance training–induced changes. Perhaps more important, this model has provided us with information regarding the signals that promote the adaptive changes, as well as the mechanisms of protein metabolism involved (such as gene transcription rates, translation capacity, and posttranslational modifications). This information about protein metabolism especially could not be provided using the complicated model of voluntary exercise.

Thus, despite the title of the chapter, very little discussion concerns muscle adaptations to endurance training per se. However, most of the adaptations to chronic electrical stimulation listed in table 4.1 have been reported in subjects undergoing endurance training or as differences between endurance-trained and untrained subjects, although the magnitude of the changes is less than with electrical stimulation, as one would expect.

Using Chronic Muscle Stimulation to Study Endurance Training

In this chapter, I discuss the results emanating from studies using chronic electrical stimulation conducted with animals. The changes reported with this model are in the same direction as adaptations that have been reported with endurance training, the most evident of which are changes in calcium-regulatory proteins, increased activity of mitochondrial enzymes, and changes in fiber-type proportions from type IIB toward type I. In our analysis of electrical stimulation–induced muscle changes, however, we must keep in mind the differences between electrical stimulation and endurance training:

1. Classically, chronic-stimulation experiments typically involve stimulation periods of at least 8 and often up to 24 hours per day, a period obviously longer than daily endurance-training periods, even for highly trained athletes.

2. Along the same lines, chronic-stimulation experiments often do not include rest periods. Rest periods are most likely instrumental in determining the final phenotype in response to endurance training. Rest periods allow for a higher tension/time index, whereas tension during continuous stimulation falls early in the stimulation period and may recover relatively little or not at all (Hicks et al. 1997; Green, Düsterhöft, et al. 1992). In addition, rest periods might allow changes in protein synthesis that are instrumental in determining the final phenotype.

3. Chronic stimulation does not produce the patterns of impulse activity that a muscle fiber would experience during endurance training. For example, one would expect recruitment, firing frequency, pattern of firing, and number of impulses per burst to be quite variable among motor units within a muscle, depending on the threshold and type of motor unit. This activity would also change as fatigue occurred. With chronic stimulation, the same pattern is imposed on all motor units and does not change with fatigue.

4. Finally, chronic electrical stimulation does not involve voluntary recruitment of motor units. While the significance of this is not fully known at present, it may be that voluntary recruitment involves changes in structure and function at the spinal cord and supraspinal level that are significant contributors to endurance performance. Thus, endurance training might involve muscular changes that are less marked

Table 4.1
Changes in Muscle Phenotype During Chronic, Low-Frequency Electrical Stimulation

Early period	Middle period	Late period
	INCREASED	
Activation of MAP kinases	MHC IIx/a	MHC I
Heat-shock proteins (HSP70, HSP60)	Oxidative enzymes	TnT 1s/2s
Total RNA	Na$^+$/K$^+$ ATPase	MLCs
Hexokinase II	SERCA 2a	TnI slow
Capillary density	TnT 3f/4f	TnC slow
	MLC 1f/3f	MLC 2s/2f
	Ubiquitin-proteosome system	
	Fatigue resistance	
	Twitch contraction time (CT)	
	Twitch half-relaxation time (RT $1/2$)	
	DECREASED	
T-tubular volume	MHC IIb	MLC fast
SR Ca^{2+} uptake	Parvalbumin	MHC IIx/a
Ca^{2+}-ATPase	Glycolytic enzymes	Oxidative enzymes
	SERCA 1a	TnT 1f/4f
	Calsequestrin	TnI fast
	Phospholamban	TnC fast
	DHPR content	Muscle mass
	RyR content	Muscle force
	Triadin content	
	TnT 1f/2f	

Early, middle, and late periods refer to approximately the first week, from week 1 to week 6, and after six weeks, respectively.

than, and perhaps qualitatively different from, those seen in chronic stimulation, partly because the nervous system becomes trained to alter the way muscle adaptations are translated into better performance. Examples of nervous system plasticity that might have implications for endurance training are given near the end of this chapter.

In the following section, rather than summarize the adaptations that have been reported to occur with chronic electrical stimulation, which have been the focus of several exhaustive reviews (Pette and Staron 1997; Pette 1998; Pette and Düsterhöft 1992), I have made a summary of lessons that we have learned from this technique as they most likely apply to endurance training. The phenotypic muscle changes that result from chronic electrical stimulation, and also from endurance training, are summarized in table 4.1.

Coordinated Adaptation for Several Protein Systems

CONCEPT

2 Many functional properties of the muscle fiber are determined by several proteins working in concert. When these proteins adapt to increased activity, by increasing in content or changing in isoform, the most efficient adaptive response is one in which all proteins adapt together. Teleologically, it would not be ideal for a fiber to alter its myosin heavy-chain characteristics from fast to slow contracting without also changing myosin light chains or the functional machinery involved in the regulation of calcium levels by the sarcoplasmic reticulum. This is a big order, nonetheless, since the constituent proteins of a system often vary in size, turnover rate, subunit structure, and three-dimensional complexity, which means the control of several steps in the protein synthesis and degradation machinery.

A clear example of this is the coordinated expression of type I–associated proteins, including heavy chains, light chains, and thin filament proteins, which occurs relatively late in the adaptation process (Pette and Staron 1997). Another is the coordinated changes in components of the calcium-regulatory membrane proteins of the sarcoplasmic reticulum and functionally related proteins, which include the dihydropyridine receptor (DHPR), the ryanodine release channel, triadin, sarcalumenin, and calsequestrin, early during chronic stimulation (Ohlendieck et al. 1999). Part of this coordination most likely results from common signals acting to promote the expression of the several genes that code for the proteins of a given system. In other cases, translational and post-translational modifications may operate to ensure that protein changes take place in a coordinated fashion (see examples for parvalbumin, SERCA proteins, and myosin heavy-chain IIb in later sections).

Pretranslational Control as a Major Component in Altered Phenotype

Pretranslational control refers to the control of protein synthesis by alteration of the abundance of mRNA. This increased abundance can come about by mechanisms that include increased transcription, mRNA processing, and mRNA stability (Booth et al. 1998).

During the first few days of chronic electrical stimulation, there is some evidence of an increase in mRNA stability, at least for certain proteins (Freyssenet, Connor, et al. 1999). In muscles that have been chronically stimulated for several weeks, mRNA levels generally reflect the concentration of their corresponding proteins (Pluskal and Sreter 1983; Hu et al. 1995; Brownson et al. 1988; Jaschinski et al. 1998), indicating the importance of pretranslational mechanisms. This is demonstrated in figure 4.1, from the work of Hu et al.(1995), for the proteins SERCA 1a and phospholamban and in figure 4.2, from the work of Jaschinski et al. (1998), for myosin heavy chains. In chronically stimulated rat muscle, mRNA for MHC IIb is reduced within one day of initiation of stimulation and increases rapidly after cessation of stimulation. At the same time, transcription of the IIa gene increases and decreases accordingly (Kirschbaum et al. 1990). Similarly, citrate synthase mRNA increases more than twofold after three days of stimulation of rabbit tibialis anterior (Annex et al. 1991). This shows that pretranslational processes are influenced early in adaptation. In stimulated rabbit muscles, the changes in mRNAs for myosin heavy chains reflect

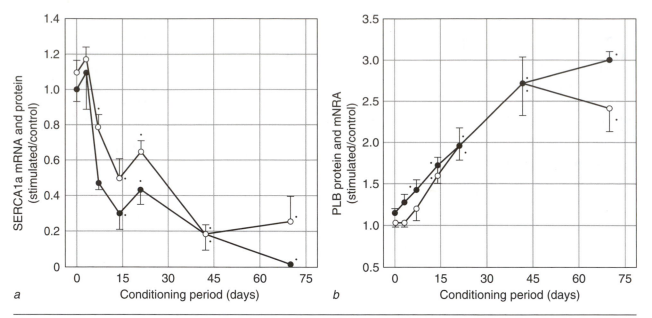

Figure 4.1   During chronic stimulation of canine latissimus dorsi at 1 Hz, the time courses for the change in protein (open circles) and for its corresponding mRNA (filled circles) are similar for (*a*) the fast-twitch calcium ATPase (SERCA 1a) and (*b*) phospholamban (PLB).

Reprinted from Hu et al. 1995.

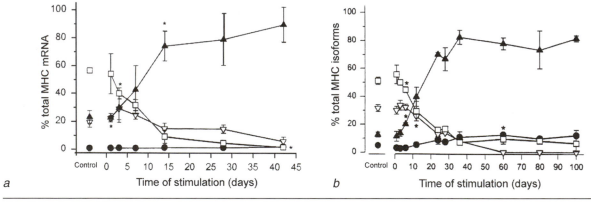

Figure 4.2   Changes in (*a*) myosin heavy-chain mRNAs and (*b*) corresponding proteins during chronic electrical stimulation (10 Hz, 10 h/day) of rat extensor digitorum longus. Symbols indicate IIb (open squares), IIx (open triangles), IIa (filled triangles), and I (filled circles) MHCs. Note the differences in timescale.

Reprinted from Jaschinski et al. 1998.

changes in the corresponding proteins after three weeks of stimulation and 12 days of recovery (Brownson et al. 1992).

This information is more readily forthcoming from stimulation studies than exercise studies because in the former the stimulus is more continuous, whereas in the latter the stimulus is short lived and intermittent. The mRNA level present in both types of study depends, among other factors, on the intensity of the stimulus, the time relative to the stimulus at which the mRNA has been measured, and the half-life of the transcript. Electrical stimulation studies, since the stimulus is more constant, allow us to make mRNA-to-protein comparisons like those depicted in figures 4.1 to 4.4. Such comparisons would be more complex with exercise training, where mRNA changes occur in pulses, with various time courses of increase and decrease (Williams and Neufer 1996).

Changes in Translational Control

Changes in translational control of protein synthesis involve changes in the amount of protein synthesized per unit of mRNA (Booth et al. 1998). This control is exerted by influences on the processes of translation: initiation, elongation of the nascent peptide, and termination. Factors that may influence these processes include adenine and guanidine nucleotide levels, NADP$^+$ and NADPH levels, or receptor-binding and mitogen-activated protein kinase (MAPK) signaling systems (Thomason 1998).

One example of changes in translational control concerns the increased translational capacity that occurs early in the adaptation period. In stimulated rabbit fast muscles, for example, total RNA, which is mostly ribosomal RNA, increases up to sevenfold during the first two weeks of stimulation (Takahashi et al. 1998; Neufer et al. 1996; Brownson et al. 1988; figure 4.3). Such a change would tend to favor increased protein production at the same mRNA level. During the initial days of electrical stimulation of rabbit tibialis anterior, citrate synthase increases faster than its mRNA, leading to the conclusion that an enhanced translation of the existing messenger was occurring (Pette and Vrbova 1992).

Hu and colleagues (1995) demonstrated another example of translational control and its consequences on muscle phenotype (figure 4.4). In chronically stimulated dog latissimus dorsi muscle, the increased concentrations of the proteins SERCA 2a and phospholamban follow a time course that is almost identical, while their mRNAs do not. Specifically, the increased SERCA 2a mRNA was almost complete by 14 days, while the protein level increased continually between 7 and 42 days. This was in contrast to all other proteins examined in this study, for which levels followed mRNA levels very closely (see figure 4.1). This suggests that, under certain circumstances, translation may be altered in order to coordinate the expression of two proteins whose functions are closely linked, as is the case with SERCA 2a and phospholamban. The au-

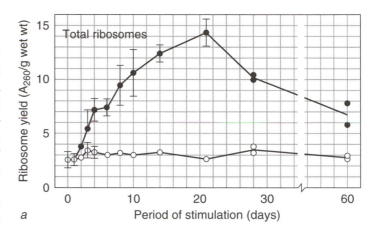

a

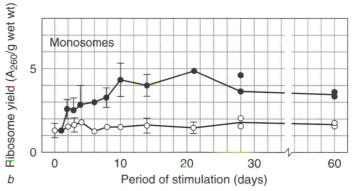

b

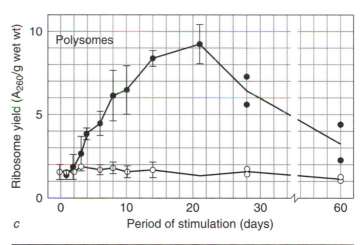

c

Figure 4.3 Increase in (*a*) total ribosomes, (*b*) monosomes, and (*c*) polysomes with chronic stimulation of rabbit tibialis anterior (10 Hz, 12 h/day).

thors presented the possibility that this control may take the form of phospholamban regulating the translation of the SERCA 2a transcript.

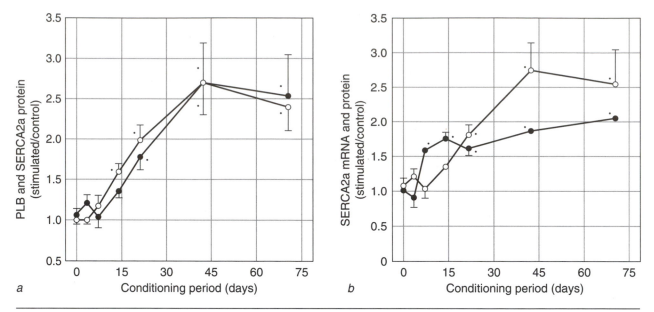

Figure 4.4 (*a*) In chronically stimulated (1 Hz) canine latissimus dorsi, the proteins SERCA 2a (slow-twitch calcium ATPase; filled circles) and phospholamban (open circles) increase with the same time course. (*b*) Although this is reflected in the increased mRNA for phospholamban (see figure 4.1), SERCA 2a mRNA (filled circles) and protein (open circles) show a different time course. Thus, SERCA 2a protein levels are also subject to posttranscriptional control. Dots indicate results significantly different from control muscles.

Reprinted from Hu et al. 1995.

This would be a means of ensuring the coordinated production of two gene products whose transcription rates differ.

Role of Posttranslational Modifications in Altering the Phenotype

Posttranslational modifications refer to any modifications of the synthesized protein, such as phosphorylation, assembly of subunits, transport, or degradation (Booth et al. 1998).

A difference between the synthesis rate of a protein (measured using radioactively labeled precursors) and its incorporation into the fiber as a functional component would indicate that posttranslational mechanisms were involved. For example, Termin and Pette (1992) noted that the rate of incorporation of labeled methionine into myosin heavy chains IIx/IIa, measured *in vitro,* increased faster than the increase in the amount of the corresponding muscle proteins during the initial period (15 days) of chronic stimulation of rat tibialis anterior (figure 4.5). Their interpretation was that synthesis exceeded the rate at which these

heavy chains replaced the IIb heavy chains that were being removed (which have a half-life of around 15 days) and thus they degraded at an increased rate (resulting in an increased turnover of the newly synthesized IIx/IIa heavy chains).

A similar situation was noted for the decrease in parvalbumin during the initial phase of stimulation-induced adaptation in rat tibialis anterior and extensor digitorum longus (Huber and Pette 1996). The rapid decline in parvalbumin mRNA (half-life of one to two days) and parvalbumin synthesis (half-life of two to three days) is not paralleled by a concomitant decrease in protein content (which does not decrease appreciably before seven days; figure 4.6). This suggests once again that protein degradation rate is altered, in this case decreased, perhaps to allow coordination of the parvalbumin decrease with the decrease in the content of myosin heavy chain IIb.

One way to influence posttranslational modifications is via the ubiquitin-proteosome system. This system is the principal protein-degradation mechanism in muscle fibers and

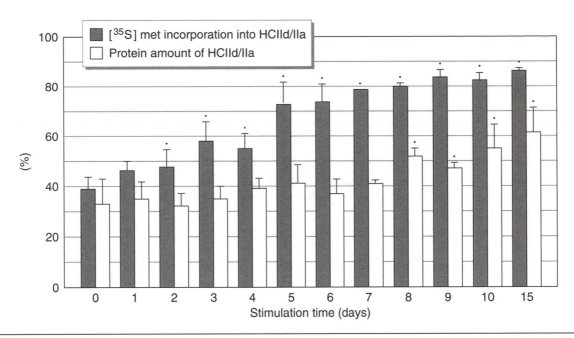

Figure 4.5 Chronic electrical stimulation (10 Hz, 10 h/day) of rat tibialis anterior. Incorporation of labeled methionine into myosin heavy-chain isoforms IIx/IIa (IId/IIa in figure; filled bars) significantly increased after two days, whereas increased protein content (open bars) is not apparent until eight days following initiation of stimulation.

Reprinted from Termin and Pette 1996.

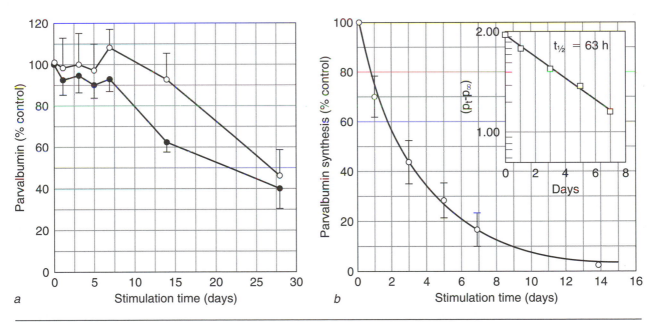

Figure 4.6 In chronically stimulated rat tibialis anterior and extensor digitorum longus, (*a*) decrease in concentration of parvalbumin is slower than (*b*) the decrease in relative synthesis rate, indicating post-transcriptional mechanisms. In *a*, tibialis anterior is shown by open circles, extensor digitorum longus by filled circles. In *b*, decline in relative synthesis rates are for tibialis anterior.

Reprinted from Termin and Pette 1992.

can be activated to provide posttranslational modification of target protein levels. Ubiquitin serves as the marker for proteins destined for degradation, while degradation is performed by the activated proteosome complex. Ordway and colleagues (2000) showed that chronic electrical stimulation of rabbit tibialis anterior results in a gradual increase in the level of proteosome protein and activity (two- to threefold) in the stimulated muscle.

Expression of the cytosolic heat-shock protein HSP70 is induced rapidly within the first day after beginning of chronic stimulation at 6 to 10 hertz and remains elevated in the muscle for up to 21 days of stimulation (Ornatsky, Connor, and Hood 1995; Neufer et al. 1996). This increased protein content can occur without a corresponding increase in its mRNA, thus implying increased protein stability as a posttranslational strategy. HSP70 is one of several molecular chaperones that are involved in the binding of polypeptides on ribosomes and the folding, stabilization, and trafficking of various proteins (Locke and Noble 1995; Ornatsky, Connor, and Hood 1995).

Along the same lines, precursor proteins are imported into the mitochondrial matrix of rat hindlimb muscles at higher rates after the first two weeks of chronic stimulation (Takahashi et al. 1998). This increase occurs as a result of modifications in the expression of components that are important parts of the import machinery.

Adapting Fibers Can Express Several Isoforms Simultaneously

The expression of several myosin heavy-chain isoforms simultaneously is unusual in control muscle fibers, but not in adapting ones. Jacobs-El, Ashley, and Russell (1993) demonstrated the presence of mRNAs for type IIx and I myosin heavy chains in 22% of rabbit tibialis anterior fibers seven days after beginning stimulation, whereas there were only 3% such fibers in controls. This has recently been confirmed by Peuker, Conjard, and Pette (1998) for stimulated rabbit tibialis anterior, in which over 50% of the fibers demonstrated mRNAs for three or four myosin heavy-chain isoforms simultaneously. This has also been demonstrated for expression isoforms of the sarcoplasmic reticulum ATPase (Leberer et al. 1989).

Evidence of the copresence of isoforms, as opposed to coexpression, is also more abundant for chronically stimulated fibers (Peuker, Conjard, and Pette 1998; Staron and Pette 1987; P. Williams et al. 1986) and fibers of humans undergoing endurance training (Klitgaard, Bergman, et al. 1990; Andersen and Schiaffino 1997). However, given the relatively long half-life of myosin heavy chains (about two weeks), one would expect a rather extended period of coexistence of several individual isoforms in the same fiber, in spite of the concurrent expression of only one.

Adaptations Do Not Require an *In Vivo* Environment

Some of the adaptations to chronic electrical stimulation seen in normally innervated muscles *in vivo* can be reproduced quite closely in denervated muscles (Windisch et al. 1998; Schiaffino and Reggiani 1996; Kilgour, Gariepy, and Rehel 1991) and in culture (Barton-Davis, LaFramboise, and Kushmerick 1996).

Adaptations Appear in a Specific Sequence

The calcium-handling proteins are among the first to be altered during chronic electrical stimulation. These include rapid decreases in the calcium ATPase of sarcoplasmic reticulum (during the first day), decreases in the proteins of the junctional triad membrane (by 10 days), and decreases in parvalbumin, calsequestrin, and SERCA 1a (by 10 to 15 days). These alterations and the corresponding effects on cellular calcium levels are accompanied by rapid changes in the muscle twitch time course, which becomes longer. The changes in cellular calcium that accompany these adaptations may provide one of the metabolic signals triggering subsequent adaptations, as discussed later (Heilmann and Pette 1979; Huber and Pette 1996; Hicks et al. 1997; Carroll, Nicotera, and Pette 1999).

The membrane enzyme Na$^+$/K$^+$ ATPase also increases quickly in chronically stimulated muscle (by five days). Green, Ball-Burnett, et al. (1992) showed that this change preceded the increased concentration of the mitochondrial enzyme citrate synthase in the same muscles (which significantly increased only after 10 days).

In stimulated rabbit tibialis anterior, the appearance of type IIx myosin heavy-chain mRNA precedes that of type I, during the first 21 days. During this time, progress from expression of type IIx to type I appears to follow increases in fiber oxidative capacity, giving the impression that a certain oxidative potential is a necessary prerequisite for fiber adaptations to continue (Jacobs-El, Ashley, and Russell 1993). The transition of myosin heavy-chain isoforms in the sequence IIb to IIx to IIa to I (Pette and Düster-höft 1992; Windisch et al. 1998; Skorjanc, Traub, and Pette 1998), based on combinations of isoforms in fibers of stimulated muscles, has been confirmed at the mRNA level (Jaschinski et al. 1998; Peuker, Conjard, and Pette 1998; see figure 4.2).

In rat, it appears more difficult to evoke the type II to I fiber transformation with chronic electrical stimulation than in the rabbit (Jaschinski et al. 1998; Peuker, Conjard, and Pette 1998). However, changes among the type II fiber types are similar in these species, once again supporting the concept of a sequential transformation of MHC from IIb to I through IIx and IIa.

In fact, changes in oxidative enzyme activities may follow the changes in myosin isoforms and may be related to the energy demands imposed on the fiber because of these isoforms.

During long-term stimulation, oxidative enzymes show an increase then a subsequent decline to levels that are still above those of control muscles (Henriksson et al. 1986; figure 4.7). In rabbit muscles stimulated for 10 months at 2.5 hertz, oxidative enzyme activities (citrate synthase, succinate dehydrogenase, hydroxyacyl CoA dehydrogenase) are higher than those of muscles stimulated at 10 hertz. In addition, muscles stimulated at 5 hertz have higher or lower activities of these enzymes depending on the degree of myosin isoform changes that have taken place (higher levels of oxidative enzymes are found in muscles with less type II–to–I transformation). These results lead us to think that ATP demands drive mitochondrial proliferation in this model and that a shift in myosin isoform to the less costly slow type results in a late decrease in oxidative potential after the initial increase (Sutherland et al. 1998; figure 4.8).

There Are Thresholds for Adaptation of Each Property or Group of Properties

As previously mentioned, the threshold for alterations in myosin heavy chains may involve an increase in oxidative capacity (Jacobs-El, Ashley, and Russell 1993).

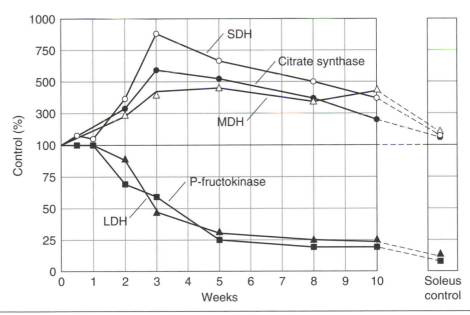

Figure 4.7 In rabbit tibialis anterior stimulated continuously at 10 Hz, oxidative enzyme levels increase, then gradually decrease to levels that are still higher than in controls. Glycolytic enzyme activities decrease to plateau levels similar to those seen in slow soleus.

Reprinted from Henriksson et al. 1986.

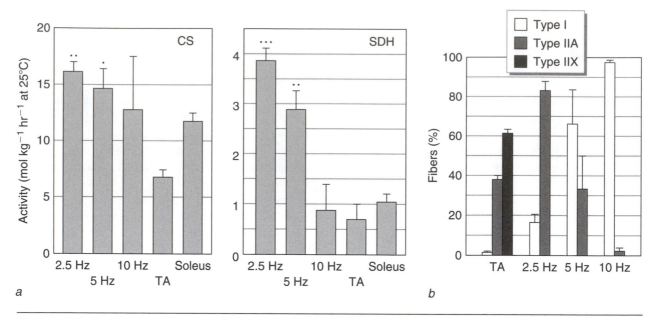

Figure 4.8   Responses of mitochondrial enzymes in rabbit tibialis anterior to 10 months of chronic stimulation depend on the frequency of stimulation. The more transformation from type II to I, the lower the oxidative enzyme level. (*a*) Citrate synthase (CS) and succinate dehydrogenase (SDH) activities. (*b*) Fiber-type proportions in the same muscles.

Reprinted, by permission of Wiley-Liss, Inc., a subsidiary of John Wiley & Sons, Inc., from H. Sutherland et al., 1998, "The dose-related response of rabbit fast muscle to long-term low-frequency stimulation." *Muscle & Nerve 21*: 1640, 1643. © 1998 Wiley-Lis, Inc.

Sutherland and colleagues (1998) stimulated rabbit tibialis anterior for 10 months at 2.5, 5, or 10 hertz in different groups of animals. Several of their findings show quite clearly that thresholds of activity exist for the adaptation of many properties. For example, while the changes in myosin isoforms and contractile speed from fast to slow were extensive in the 10-hertz group and almost absent in the 2.5-hertz group, the 5-hertz group did not show intermediate values; some resembled animals in the 2.5-hertz group, others the 10-hertz group (figure 4.9). In addition, these researchers found very few differences between the muscles stimulated at 2.5 hertz for 10 months and for 10 weeks, signifying that the muscles had stabilized at a level of adaptation corresponding to the activity level imposed.

Atrophic Effect of Chronic Stimulation on Muscle

The atrophic effect of chronic stimulation on muscle is evidenced by the often-reported finding of decreased weight, mean fiber area, and tetanic strength of the stimulated muscle (Kernell et al. 1987; Donselaar et al. 1987; Sutherland et al. 1998; figure 4.10). Although part of this decrease may be attributable to pathological mechanisms resulting in a decrease in cell number (discussed later), it appears that a major proportion may be due to missing periodic high-tension events. Even in a normally innervated muscle subjected to chronic stimulation, for example, muscles would be maintained in a chronically fatigued state, to the point that even normal activation during nonstimulation periods would be expected to produce attenuated force responses. This interpretation is supported by the finding that the atrophic effect is attenuated when low-frequency stimulation is supplemented with short, high-frequency bursts that evoke high contractile forces, even when these high-frequency episodes are of very limited total duration (0.5% of daily time; Kernell et al. 1987). Thus, the atrophic response is less when 40-hertz bursts of intermittent stimulation are used instead of 10-hertz continuous stimulation, even when the total number of impulses in the former case is four times the latter (Kernell et al. 1987).

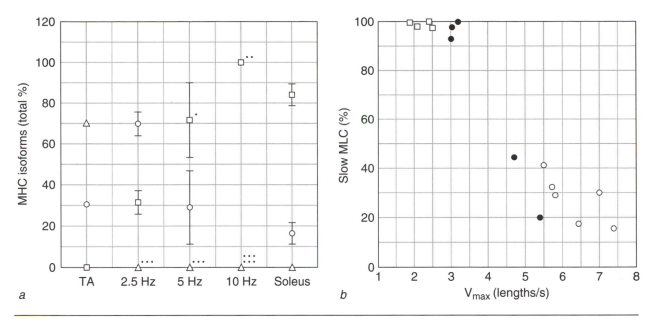

Figure 4.9 (*a*) Effect of chronic stimulation of rabbit tibialis anterior for 10 months at 2.5, 5, and 10 Hz on myosin heavy-chain isoforms. Note the large variability in the 5-Hz group, suggesting a threshold activation level for transformation to occur. (*b*) In the saae‡experiment, very few intermediate degrees of transformation were found, once again supporting the idea of a threshold for fiber transformation.

Reprinted, by permission of Wiley-Liss, Inc., a subsidiary of John Wiley & Sons, Inc., from H. Sutherland et al., 1998, "The dose-related response of rabbit fast muscle to long-term low-frequency stimulation." *Muscle & Nerve 21*: 1639. © 1998 Wiley-Lis, Inc.

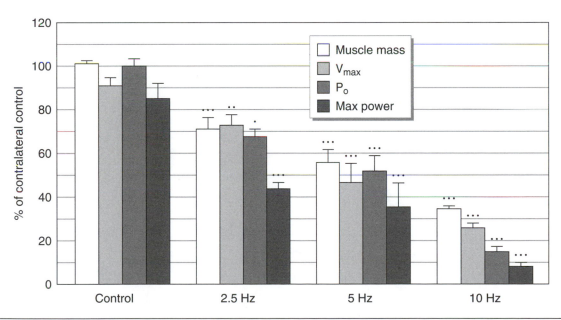

Figure 4.10 Effects of 10 months of chronic electrical stimulation of rabbit tibialis anterior at 2.5, 5, and 10 Hz on muscle mass, V_{max}, P_o, and maximum power.

Reprinted, by permission of Wiley-Liss, Inc., a subsidiary of John Wiley & Sons, Inc., from H. Sutherland et al., 1998, "The dose-related response of rabbit fast muscle to long-term low-frequency stimulation." *Muscle & Nerve 21*: 1637. © 1998 Wiley-Lis, Inc.

As we shall see in the next chapter, signals associated with stress and strain on the extracellular and cytoskeletal matrix of the muscle fiber are most likely involved in stimulating protein synthesis and thus promoting muscle hypertrophy. These signals might be virtually absent in a chronically stimulated muscle, to a level less than in even a normal, nonexercising muscle with the capacity to generate periodic high-tension contractions.

In support of these ideas, it has been demonstrated that, early during the chronic-stimulation period (up to 21 days), actin mRNA decreased in tibialis anterior and extensor digitorum longus of rabbits (Brownson et al. 1988). In humans performing aerobic exercise at 40% of maximal oxygen consumption for four hours, muscle protein degradation increases by 37% during the exercise and by 85% during a four-hour recovery period (Carraro et al. 1990).

These authors proposed that muscle mass does not gradually decrease during a training program involving this type of exercise, however, based on their observation that fractional protein synthesis rate actually increased during recovery. Muscle mass is lost in many chronic-stimulation experiments perhaps partly because of the lack of a recovery period.

Metabolic Signals That First Stimulate the Adaptive Response

CONCEPT

3 The phosphorylation potential of the adenylic acid system, $ATP/(ADP_{free} + Pi_{free})$, is persistently depressed throughout a period of stimulation extending from 15 minutes to 50 days (Green, Düsterhöft, et al. 1992; Hood and Parent 1991; figure 4.11). This contrasts with many other metabolites, which showed a recov-

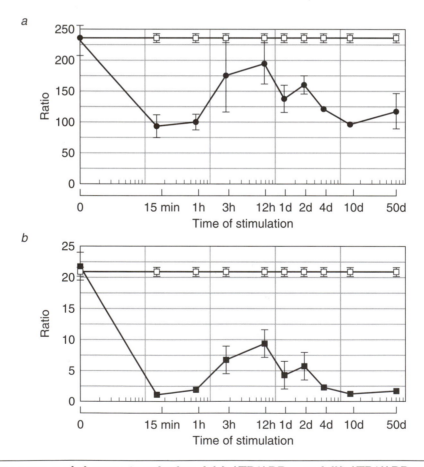

Figure 4.11   Time course of changes in calculated (*a*) ATP/ADP_{free} and (*b*) $ATP/(ADP_{free} + Pi_{free})$ in chronically stimulated (10 Hz, 24 h/day) rabbit tibialis anterior. Open squares represent unstimulated control muscles.

Reprinted from Green et al. 1992.

ery after an initial change. The phosphorylation potential of the adenylic acid system thus qualifies as a candidate for a signal that mediates phenotypic changes throughout the stimulation period. At the single-fiber level, the ATP/ADP_{free} ratio increases in the order I < IIA < IIX < IIB in normal (unstimulated) fibers, adding support to the hypothesis that this ratio may be involved in the determination of the myosin heavy-chain profile (Conjard, Peuker, and Pette 1998).

Yaspelkis and colleagues (1999) have shown the possible importance of ATP concentration in the exercise training–induced increases in the glucose transporter GLUT4 and citrate synthase in the rat. In their experiment, rats that endurance-trained while being given clenbuterol showed an attenuated increase in these two proteins, attributable to higher ATP levels at rest, and a decreased drop in ATP during exercise. When this effect was countered by giving a group of rats beta-guanidinoproprionic acid (β-GPA), ATP levels were lower at rest and after exercise, and the adaptations in GLUT4 and citrate synthase were more pronounced.

Changes in base calcium level or in calcium transients in response to stimulation have been proposed as signals that alter gene expression during increased contractile activity. This is difficult to verify, however, in an *in vivo* model. In chronically stimulated extensor digitorum longus of rabbits, intracellular calcium (measured using a calcium-sensitive electrode) increases, reaching a peak at about five times the normal level by 15 days of stimulation, after which it decreases to a base level that is slightly elevated (Sreter et al. 1987). If the stimulation is intermittent (8 hours per day instead of 24), calcium levels do not show this pattern, and conversion to slow fibers does not occur, suggesting a possible link between these two events. Interestingly, when stimulation is terminated, calcium shows a similar magnitude but shorter time course of response during the time when a reversal of histochemical profile to an increased proportion of fast fibers is occurring. Obviously, the calcium story is complicated.

Carroll, Nicotera, and Pette (1999) have recently demonstrated calcium changes in chronically stimulated rat muscles that are quite different from those previously reported in rabbit muscles. It may be that the calcium changes in rabbits are a partial reflection of the fiber in-

jury that occurs in this species during chronic electrical stimulation. In rat extensor digitorum longus, which can be chronically stimulated with very little fiber damage, intracellular calcium concentration increases during the first two hours of stimulation to a level 2.5 times that seen in control extensor digitorum longus; the concentration increase in stimulated muscle remains for at least 10 days. This increased calcium concentration is due partially to decreased calcium ATPase activity in the sarcoplasmic reticulum and decreased parvalbumin content. Interestingly enough, this increased calcium concentration is similar to that found in control slow-twitch fibers of the soleus. The constancy of calcium concentration for up to 10 days during chronic stimulation reinforces the idea that calcium may indeed be involved in altering gene expression in this model.

In myotubes in culture, increasing base calcium levels using the calcium ionophore A23187 causes a reversible increase in expression of MHC I, cytochrome c, and citrate synthase and a decrease in expression of MHC II and the glycolytic enzyme glyceraldehyde 3-phosphate dehydrogenase (Freyssenet, Di Carlo, and Hood 1999; Meissner et al. 2000). The altered gene expression of MHC associated with increased calcium occurs within 24 hours. However, the levels of calcium in these studies may be high compared with those found *in vivo*.

Abu-Shakra and colleagues (1994) demonstrated that increased intracellular calcium consequent to cholinergic stimulation induced rapid expression of an immediate early gene, *zif268,* in myotubes in culture. This effect, produced by exposing the cells to carbachol, a cholinergic agonist, was blocked by alpha-bungarotoxin, thus demonstrating that it occurred via ACh receptors. These investigators also found that the effect was modulated by ryanodine and dantrolene, which modulate calcium release from the sarcoplasmic reticulum. Thus, at least one immediate early gene is responsive to calcium released from the sarcoplasmic reticulum during at least the initial phase of chronic stimulation.

Recent evidence suggests that increased transcription of the MHC I gene can occur in rat muscles stimulated *in vitro* for six days, without changes in calcium levels (Liu and Schneider 1998). This does not mean that calcium does not play a role during more prolonged stimulation periods.

How might calcium work to alter gene expression? One possibility is via involvement of calcineurin, which is a calcium-regulated serine–threonine phosphatase. When this enzyme is activated, it activates in turn specific transcription factors that promote changes in gene expression. In transgenic mice in which expression of activated calcineurin was enhanced, an increased proportion of type I fibers was found in hindlimb muscles (Naya et al. 2000). Similarly, administration of the calcineurin antagonist cyclosporin A to rats for six weeks caused a slow-to-fast fiber transition in the soleus muscle (Chin et al. 1998). These results suggest that calcineurin is involved in the signaling required for maintenance as well as increase of MHC I gene expression. Another possible pathway for calcium involvement is via protein kinase C (PKC), which, when activated by calcium, triggers the mitogen-activated protein kinase signaling system (Freyssenet, Di Carlo, and Hood 1999; discussed next). Still another candidate is activation of calcium/calmodulin-dependent protein kinase (Baar et al. 1999).

Signals that transduce muscle activity into alterations in gene expression may involve the activation of mitogen-activated protein kinase (MAPK) signaling pathways, which are activated via a number of stresses that might include those associated with chronic muscle activity. An example of one such pathway is that of the stress-activated protein kinases, also known as the c-Jun NH_2-terminal kinases, or JNKs. These kinases, once activated by any one of a number of stressors, translocate to the nucleus, where they increase transcription of several transcription factors, among them c-Jun (Aronson, Dufresne, and Goodyear 1997). Stimulation of rat hindlimb muscle resulted in a sixfold increase in JNK activity within 15 minutes, which remained at this level for the 60-minute period that measurements were taken. Interestingly, JNK was also quickly activated and remained elevated in rats during 60 minutes of treadmill exercise (Goodyear et al. 1996). In humans, the MAPK-signaling cascade is activated in response to a 60-minute bout of ergometer exercise at 70% of maximal oxygen consumption (Aronson et al. 1998). The stress to which this system is responding is not known but could involve

autocrine mediators, metabolic changes such as the phosphorylation potential, altered cell calcium levels, or mechanical shear stresses. If the latter is involved, this signal would most likely attenuate with time, as force production decreases in the chronically stimulated muscle. Thus, it would not constitute a good candidate for a signal that would promote adaptations later in the chronic-stimulation period. It is possible that MAPK-signaling cascades (such as the extracellular signal-regulated protein kinase and p38 kinase systems) are more important in muscle adaptations to short, high-intensity activities involving higher-tension contractions. This is discussed in more detail in chapter 5.

The involvement of the activation of immediate early genes (IEGs), such as *c-Fos, c-Jun,* and *Erg-1,* in response to endurance-type training has not been established unequivocally. The IEG products c-Fos and c-Jun form dimers to bind to AP-1 sites within specific promoters, while Erg-1 is known to be activated during differentiation of bone and nerve cells (J. Michel et al. 1994). In chronically stimulated rabbit muscle, there is a transient increase in expression and concentration of these IEG products during the first 24 hours, with a secondary and sustained rise in the proteins but not the corresponding mRNAs for up to 21 days of stimulation (J. Michel et al. 1994). The relationships of these two different phases of activation of the IEGs in response to the muscle adaptations that occur during chronic stimulation have yet to be determined.

After a single 30-minute bout of endurance exercise, an increase in c-Fos and c-Jun protein and of their corresponding mRNAs are evident for several hours, with c-Fos showing a greater relative response than c-Jun (Puntschart et al. 1998).

There are a host of other intracellular changes that might constitute signals for alterations in gene expression, such as hypoxia, increased hydrogen ion concentration, and changes in concentration of a number of intermediary metabolites (R.S. Williams and Neufer 1996; Baar et al. 1999). It is probable that several signals are operant at any one time via several mechanisms and that the time course of involvement of the factors is variable, with some factors important in the early period of exercise and others in more prolonged exercise conditions (figure 4.12).

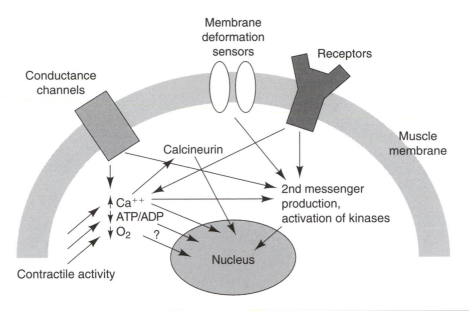

Figure 4.12 Factors that may be involved in changes in gene expression with chronic electrical stimulation or endurance training.

Degenerative and Regenerative Processes During Chronic Stimulation

Some evidence of degeneration and regeneration is present in chronically stimulated muscles, such as a transient appearance of embryonic and neonatal myosin heavy chains and the invasion of fibers by nonmuscle cells (Maier et al. 1988; Kirschbaum et al. 1990; Maier and Pette 1987). In stimulated rabbit tibialis anterior, 15% to 20% of fibers are myotubes or contain central nuclei after 30 and 60 days of stimulation, indicating an ongoing degeneration and regeneration process throughout the adaptation period (Schuler and Pette 1996). The degenerative–regenerative changes are more marked in some species than in others (Kirschbaum et al. 1990).

Sutherland and colleagues (1998) assumed that a major proportion of the decrease in specific tension of rabbit tibialis anterior after 10 months of stimulation at 10 hertz was due to degenerative loss of fibers. Their interpretation was that the loss of fibers was due to the hypoxia induced by the constant semitetanic contractions imposed by the stimulation; the loss of specific tension occurred mostly after 12 weeks of stimulation. Very little damage was seen in muscles stimulated at lower frequencies.

Lexell and colleagues (1992), based on their in-depth study of chronically stimulated rabbit tibialis anterior, concluded that the contribution of degeneration and regeneration to muscle fiber-type conversion is insignificant. More recently, however, studies of chronically stimulated rabbit extensor digitorum longus suggest that degeneration and regeneration may play a substantial role, depending on the muscle (Schuler and Pette 1996).

Summary: Muscle Adaptations to Chronic Electrical Stimulation and Endurance Training

A summary of the changes that occur with chronic electrical stimulation is presented in table 4.1. The early, middle, and late periods are relative and differ among the species that have been studied and among continuous stimulation, intermittent stimulation, and endurance training. As previously pointed out, the acute stimuli associated with the onset of chronic stimulation would not be expected to be the same as during endurance training. One might expect, for example, that some of the signals that exert their influence during the first minutes of electrical stimulation would be of special significance during endurance training, in which the repeated rest-to-exercise transitions would reactivate these pathways at each training session. In addition, although the total tension/time integral would be expected to be larger for electrical stimulation (8 to 24 hours of maximal contractile activity per day) compared with endurance training (2 hours of submaximal activation per day), a higher integral-per-unit time would be characteristic of endurance training, which would thus deliver a more intense and most likely different stimulus than chronic stimulation.

Not all of the muscle responses to endurance training are included in table 4.1. For example, endurance training appears to result in an increase in the antioxidant enzymes superoxide dismutase and glutathione peroxidase and an increase in the concentration of intracellular glutathione. The resulting increased capacity to reduce the persistence of free radicals during and following exercise is probably significant in reducing the cellular damage that these substances can produce (Powers, Ji, and Leeuwenburgh 1999). This is most likely a middle-period adaptation, although this response has not been studied in the electrical-stimulation model. It is certainly an important adaptation and one that needs to be investigated in more detail.

In spite of the many differences between electrical stimulation and endurance training, it appears that the pattern and temporal order of adaptations is very similar. Thus, the chronic-stimulation model will most likely continue to provide us with information regarding the adaptive capacity of skeletal muscles to increased activity in general, which we can apply to the infinitely more complex model of endurance training.

The Neuromuscular Junction

CONCEPT 4 Neuromuscular junctions are different in slow and fast muscles. Morphologically, junctions in the fast muscles are larger and demonstrate some systematic differences in the terminal axons (shorter, less numerous, and of larger diameter than in fast fibers) and the synaptic grooves and junctional folds (less extensive than in fast fibers; Ogata 1988). Nonetheless, there is not total agreement as to the universality of these differences across muscles,

fiber types within muscles, and species (Galvas, Neaves, and Gonyea 1982; Ogata 1988; Prakash et al. 1996; Wood and Slater 1997). Physiologically, there is no doubt as to the difference between fast and slow neuromuscular junctions; end-plate potentials (EPPs) are larger in fast fibers but run down more quickly in response to repetitive stimulation (Gertler and Robbins 1978). This difference gives the fast fiber a higher safety factor for transmission, defined as $EPP/(E_{ap} - E_m)$, where E_{ap} is the threshold potential for generating an action potential and E_m is the membrane potential (Ruff 1992). The safety factors for rat soleus and extensor digitorum longus neuromuscular junctions have been estimated as 3.5 and 5, respectively, from electrophysiological experiments (Wood and Slater 1997).

Several mechanisms can be evoked to explain this relatively higher safety factor of fast, as compared with slow, neuromuscular junctions. Quantal release is higher from fast terminals (Wood and Slater 1997). Fast postsynaptic membranes may also have higher densities of acetylcholine receptors (Sterz, Pagala, and Peper 1983). The sodium current density at fast end plates is up to four times higher than that at slow end plates, due primarily to a correspondingly higher sodium-channel density (Ruff 1992). Sodium channels at end plates appear also to be qualitatively different between fast and slow fibers. The voltage dependences of activation and of fast and slow inactivation of the sodium current are different between these two fiber types; they are at more negative potentials in fast fibers (Ruff and Whittlesey 1993). This may indicate that fast fibers are more susceptible to a use-related reduction in membrane excitability, which would tend to limit the duration that these fibers could fire at high rates (Ruff 1996).

Neuromuscular junctions in fast muscles are also characterized by larger amounts of the asymmetric form of acetylcholinesterase (AChE) found attached to the postsynaptic membrane via a collagen tail and of the globular G_4 form that is found in the perijunctional region (Gisiger, Bélisle, and Gardiner 1994; Sketelj et al. 1998). Whether fast and slow fibers within a muscle containing a mixture of fiber types exhibit these same differences is unknown.

Evidence From Chronic Stimulation

Chronic electrical stimulation has been used to determine the adaptability of the neuromuscular junction in much the same way that it has been used to demonstrate muscle adaptability. Stimulation changes neuromuscular junctions of fast muscles to those resembling, at least morphologically, junctions normally found in slow muscles. For example, Waerhaug and Lomo (1994) found that junctions of chronically stimulated extensor digitorum longus muscles were reduced in area and had decreased density of terminal varicosities. Chronic stimulation of rat extensor digitorum longus resulted in an AChE profile resembling that found in soleus (lower A_{12} and G_4 AChE molecular forms; Jasmin and Gisiger 1990). On the other hand, chronically stimulated (for 20 days) extensor digitorum longus muscles exhibited an increased quantal content while maintaining the property of rapid decrease in EPP amplitude with increasing frequency of stimulation, suggesting that a change to a slow-type junction (which would imply a decrease in EPP amplitude and an increase in fatigue resistance of transmitter release) was not complete.

The research team of Harold Atwood at the University of Toronto has contributed valuable information to this issue. During the past 20 years, they studied synaptic adaptations in the neuromuscular junctions of the crayfish claw and abdominal muscles in response to chronic stimulation. The motoneurons innervating these muscles can be distinguished as phasic or tonic based on their morphological and physiological characteristics. With continued stimulation, phasic motoneurons show a high initial neurotransmitter release and a faster drop-off in transmitter release (lower fatigue resistance) than tonic motoneurons. This phasic–tonic distinction is thus similar functionally to the fast–slow distinction seen in mammals.

By chronic stimulation of phasic motoneurons for as little as three days, for two hours per day, these researchers showed a phenomenon that they termed long-term adaptation (LTA) of the neuromuscular junction. This adaptation is characterized by an increased synaptic mitochondrial content, lower initial neurotransmitter release, and higher fatigue resistance in the phasic motoneuron, which thus becomes more like a tonic motoneuron (figure 4.13).

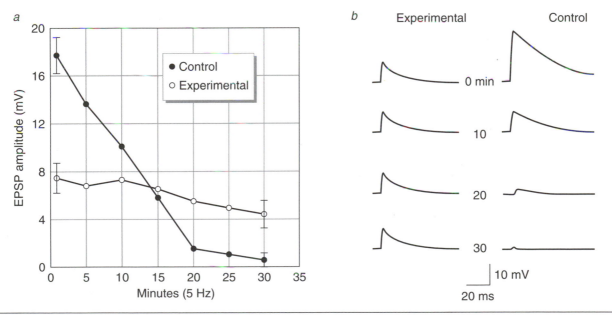

Figure 4.13   Long-term adaptation of synaptic transmission in crayfish claw closer muscle. Electrical stimulation of the innervating nerve was at 5 Hz, 2 h/day, for three consecutive days. (*a*) Amplitude changes in excitatory postsynaptic potential (EPSP) in stimulated and contralateral claw muscles during stimulation at 5 Hz, measured two days after cessation of the adaptation protocol. (*b*) Intracellular measurements of EPSP amplitudes during stimulation at 5 Hz from a typical experiment.

Reprinted from Nguyen and Atwood 1990.

These researchers' systematic search for the mechanisms involved in this adaptive response resulted in the following findings:

• Neurotransmitter release is not necessary for LTA. The adaptation was seen when axons were blocked with tetrodotoxin and stimulated central to the blockage so that the impulse did not even reach the terminal (Lnenicka and Atwood 1986).

• The production of action potentials by the stimulated motoneuron is not necessary for the expression of LTA. This conclusion was based on results of experiments in which LTA occurred after stimulation of sensory afferents that caused subthreshold depolarization in the phasic motoneuron (Lnenicka and Atwood 1988).

• LTA does not require a change in protein synthesis by the adapting motoneuron. LTA still occurred when the protein synthesis blocker cycloheximide was injected at the beginning or at the end of each stimulation session (Nguyen and Atwood 1990).

• Examining different frequencies and total amounts of stimulation showed that the increase in neurotransmitter fatigue resistance had a lower threshold than the decrease in initial neurotransmitter release. Thus, these two adaptive responses were distinct (Mercier, Bradacs, and Atwood 1992).

• The increase in synaptic fatigue resistance with LTA depends more on axonal transport than does the decline in initial transmitter release. Thus, after establishment of LTA, axotomy resulted in a decrease in fatigue resistance more rapid than change in initial transmitter release (Nguyen and Atwood 1992).

• The decrease in initial transmitter release, which occurs *in vitro* after five hours of depolarization of the cell body or axon, does not occur with simultaneous application of a calcium-channel blocker. Thus, calcium influx during motoneuron depolarization is involved in this adaptive response (Hong and Lnenicka 1993).

Endurance-Training Effects on the Neuromuscular Junction

Research reports concerning endurance training–induced morphological adaptations at the neuromuscular junction are not numerous, and their conclusions are far from consistent (Deschenes et al. 1994). This is most likely a reflection of the difficulties involved in making quantitative morphological measurements, which is most likely the reason for a lack of consensus, referred to above, regarding the morphological differences between fast and slow neuromuscular junctions. The most consistent responses to endurance training to date appear to be an increased neuromuscular junction area and an increased complexity of nerve terminal branching (Deschenes et al. 1993, 1994).

Biochemically, endurance training induces increased AChE activity in fast muscles, which is evident in the end-plate region of the muscle (Gisiger, Sherker, and Gardiner 1991; Jasmin and Gisiger 1990; Fernandez and Hodges-Savola 1996; Fernandez and Donoso 1988; Crockett et al. 1976). This response has several notable and highly intriguing characteristics. First, it affects the globular 4 (G_4) form of AChE selectively (Jasmin and Gisiger 1990; Fernandez and Donoso 1988). Correspondingly, in slow muscles such as the soleus, which contain very small amounts of G_4, this response is absent. Second, the response is very rapid and reflects the level of activity within the preceding few days (Hubatsch and Jasmin 1997; Gisiger, Bélisle, and Gardiner 1994; Fernandez and Donoso 1988; figure 4.14). Third, the patterns of adaptation in ankle flexors and extensors to treadmill running, compensatory overload, and swimming have led to the hypothesis that the adaptation may depend on the tonic (very little change in G_4 AChE) versus phasic (increased G_4) nature of the training stimulus (Gisiger, Bélisle, and Gardiner 1994; Jasmin, Gardiner, and Gisiger 1991; Gisiger, Sherker, and Gardiner 1991; Jasmin and Gisiger 1990). G_4 appears to be located in the perijunctional sarcoplasmic reticulum, where it forms a conjunctional compartment that bathes the end plates in an AChE-rich environment (Gisiger and Stephens 1988). It has been proposed that G_4 may play a significant role in removing acetylcholine (ACh) from the synaptic cleft during high-frequency, repetitive (i.e., phasic) activity and thus help avoid ACh receptor desensitization, which would continue to activate the end-plate refractory phase. It is worth noting that, as discussed in chapter 3, end-plate desensitization is considered a pos-

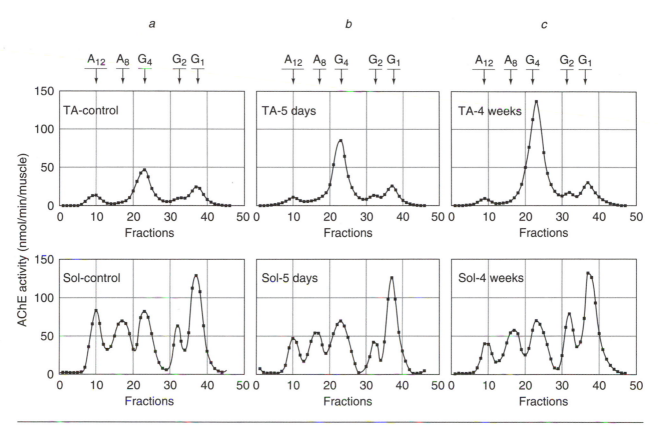

Figure 4.14   Acetylcholinesterase (AChE) molecular form content in tibialis anterior (TA) and soleus (Sol) muscles in (*a*) control rats and in (*b*) rats subjected to five days and (*c*) four weeks of exercise in voluntary wheel cages. Note the difference in the AChE profiles of soleus and tibialis anterior and the G_4 response to exercise in tibialis anterior.

Reprinted from Gisiger, Bélisle, and Gardiner 1994.

sible mechanism for fatigue of the neuromuscular system under certain conditions. The mechanism for the increased AChE in response to training is not known but may involve factors such as CGRP (calcitonin gene-related peptide) or ARIA (acetylcholine receptor–inducing activity), which are released from nerve terminals during activation and which influence the composition of the postsynaptic membrane (Fernandez and Hodges-Savola 1996; figure 4.15).

Consistent with the proposed increase in the area of the neuromuscular junction with endurance training, the number of acetylcholine receptors also increases in several muscles, including diaphragm (Desaulniers, Lavoie, and Gardiner 1998).

It appears that endurance training has consequences for the electrophysiology of the neuromuscular junction. Dorlöchter and colleagues (1991) examined single end-plate potential (EPP) amplitudes and EPP responses to repetitive stimulation in *in vitro* extensor digitorum longus muscles of mice that endurance-trained in voluntary wheels for two to eight months. Their results showed a doubling of EPP amplitude, which remained above control values during stimulation at 100 hertz for one second (figure 4.16). They also demonstrated that preparations from trained mice exhibited a higher resistance to both presynaptic (high Mg^{2+}, low Ca^{2+}) and postsynaptic (curare) blockade, suggesting that the adaptation involved primarily increased neurotransmitter release.

Thus, a picture is slowly emerging in which the neuromuscular junction appears to adapt to endurance training by increasing both its presynaptic capacity (increased and more sustained transmitter release) and its postsynaptic efficiency (higher amount of G_4 AChE and number of ACh receptors). The mechanisms for these adaptations are not yet known but may include neurotrophic and motoneuron-derived

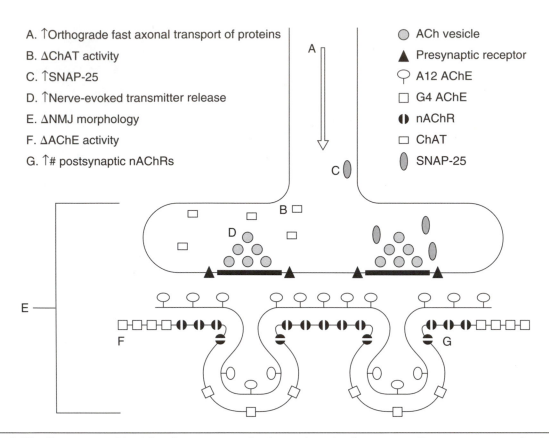

A. ↑Orthograde fast axonal transport of proteins

B. ΔChAT activity

C. ↑SNAP-25

D. ↑Nerve-evoked transmitter release

E. ΔNMJ morphology

F. ΔAChE activity

G. ↑# postsynaptic nAChRs

○ ACh vesicle

▲ Presynaptic receptor

♀ A12 AChE

□ G4 AChE

◑ nAChR

□ ChAT

⬭ SNAP-25

Figure 4.15 Summary of loci for demonstrated adaptations in the mammalian neuromuscular junction with increased chronic activity.

Reprinted from Panenic and Gardiner 1998.

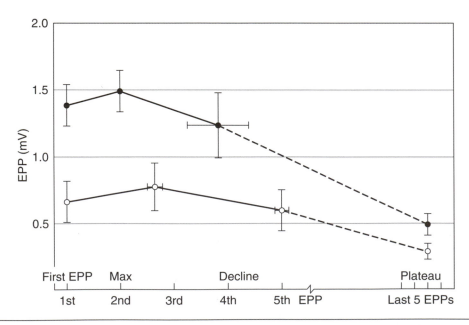

Figure 4.16 Effects of two to eight months of training in voluntary exercise wheels on end-plate potential (EPP) at the neuromuscular junction of mouse extensor digitorum longus. Trained values (filled circles) were elevated compared with nonexercised controls (open circles) throughout a 100-Hz, 1-s train of nerve stimulation.

Reprinted from Dorlöchter et al. 1991.

influences that are released at the neuromuscular junction and influence pre- and postsynaptic structures (figure 4.15). The physiological consequences of these changes are also not yet known; technical constraints to date have prevented even the ability to determine whether the neuromuscular junction limits performance. However, teleologically speaking, adaptations found at this level suggest the strengthening of a potential weak link and thus shed some light on possible limiting components in the untrained state.

Motoneuron Adaptations to Endurance Training

CONCEPT

5 Anterograde and retrograde axonal transport are functional properties of motoneurons by which materials are transported toward the nerve terminals and the soma, respectively. Anterograde axonal transport is classified as fast or slow based on the speed with which materials are transported. Fast and slow anterograde transports are also distinguishable by the materials that are moved and the energy requirements for transport. Fast transport (200 to 400 millimeters per day) involves transport of primarily membrane proteins and lipids, as well as neurotransmitters, while slow transport (0.2 to 10 millimeters per day) includes neurofilaments, microtubules, microfilaments, and metabolic enzymes (Siegel et al. 1989). We have some evidence that these fundamental nerve cell properties may be responsive to increased chronic activity. Adaptive changes have been found in fast axonal transport in motoneurons of endurance-trained rats (Kang, Lavoie, and Gardiner 1995; Jasmin, Lavoie, and Gardiner 1988). This experiment was performed by injecting radioactively labeled leucine into the ventral horn of the spinal cord of anesthetized rats and measuring the appearance of the radiolabeled proteins in the sciatic nerve four hours later. Peak and average transport velocities increased, as did the total amount of transported protein. This response may indicate elevated protein synthesis of the trained motoneuron, enhanced loading of protein onto the transport machinery, or both. We can postulate that motoneurons need more material in the nerve terminals of adapt-

ing neuromuscular junctions or that they need to enhance the delivery of trophic substances involved in pre- or postsynaptic adaptations.

It may be that certain proteins are preferentially targeted for rapid delivery to endurance-training junctions because of their special roles in the adaptive process. Kang, Lavoie, and Gardiner (1995) found that a synaptosome-associated protein, SNAP-25, was transported in higher amounts than other axonally transported proteins in endurance-trained motoneurons (figure 4.17). This protein plays an important role in the interaction of the synaptic vesicle with the presynaptic membrane of the nerve terminal and thus might be associated with the increased transmitter release found in trained nerve terminals. This protein would presumably also need to be up-regulated to accommodate any neuromuscular junction remodeling that involves increased nerve terminal size or extent of branching.

More recently, Gharakhanlou, Chadan, and Gardiner (1999) demonstrated that motoneurons from endurance-trained rats demonstrate increased content and anterograde transport of calcitonin gene-related peptide (CGRP; figure 4.18). CGRP, produced by motoneurons and released at terminals, has been shown to stimulate acetylcholine synthesis and may be involved in the control of muscle AChE.

The increase in fast axonal transport in general and in proteins such as SNAP-25 and CGRP in particular might explain why the sprouting response of motoneurons in the presence of denervated fibers is more robust in rats with increased daily activity levels (Seburn and Gardiner 1996; Gardiner, Michel, and Iadeluca 1984) and why the more-active slow motoneurons appear to regenerate and reinnervate muscles faster than fast motoneurons (Desypris and Parry 1990; Foehring, Sypert, and Munson 1986).

There is some evidence that the soma of the motoneuron shows adaptations to endurance training. Motoneurons show signs of elevated protein synthesis (Edstrom 1957) and less stress in response to acute exhaustive exercise (Jasmin, Lavoie, and Gardiner 1988; Gerchman, Edgerton, and Carrow 1975). Whether metabolic enzyme systems in motoneurons are altered by endurance training is controversial. Since mitochondrial enzyme activity is inversely related to soma size (Ishihara, Roy, and Edgerton 1995;

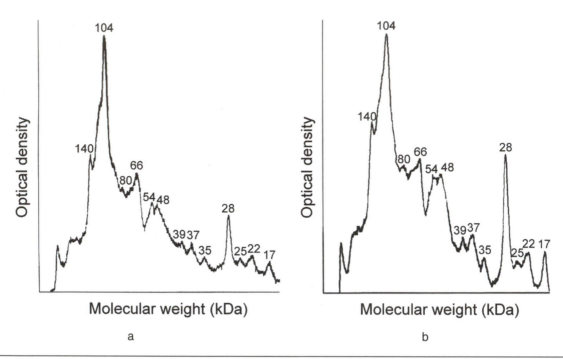

Figure 4.17   Endurance training increases the abundance of SNAP-25, a synaptic protein, among proteins transported by fast axonal transport in rat sciatic nerve. Densitometric scan of fluorograms of nerve axon extracts on 12.5% gels from (*a*) control and (*b*) endurance-trained rats. Note that the concentrations of proteins of all molecular weights (height of scan tracing) are slightly (30%) increased, while that of the 28-kDa protein (SNAP-25) is disproportionately increased.

Reprinted from Kang et al. 1995.

Suzuki et al. 1991) and slow fibers are innervated by smaller motoneurons, it seems reasonable to assume that endurance training might increase mitochondrial enzyme activity and perhaps even reduce motoneuron size. An early study (Gerchman, Edgerton, and Carrow 1975) showed that the activity of the enzyme malate dehydrogenase (MDH) was modestly elevated in the somata of hindlimb motoneurons of endurance-trained rats. More recent evidence, using more precise quantitative techniques and a different enzyme (succinate dehydrogenase, SDH), seems to suggest that if motoneurons do become more oxidative in enzyme profile with endurance training, the effect is slight (Nakano et al. 1997; Seburn,

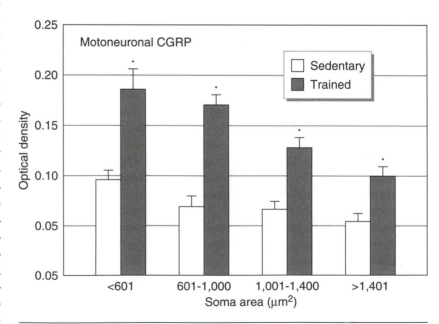

Figure 4.18   Endurance training increases the concentration of the neuropeptide calcitonin gene-related peptide (CGRP) in rat lumbar motoneurons.

Reprinted from Gharakhanlou et al. 1999.

Coicou, and Gardiner 1994; Suzuki et al. 1991). Motoneurons innervating muscles that are chronically stimulated to evoke classic muscle adaptations (such as slowing of contractile speed and increased oxidative enzyme activities) are not changed in size or soma SDH activity (Donselaar, Kernell, and Eerbeek 1986). Since stimulation of the muscle via the motor nerve evokes action potentials in the cell body via antidromic excitation, this result demonstrates clearly that the increase in the number of action potentials generated at the level of the soma does not constitute a stimulus for increased mitochondrial enzyme activity.

Does endurance training change the electrophysiology of motoneurons? Such changes might prove to be quite meaningful if they did occur. As we saw in chapter 2, fundamental properties of motoneurons such as rheobase, input resistance, duration of afterpotentials, and susceptibility to such phenomena as late adaptation and bistability significantly influence how motor units are used during movement. Since these properties also influence the conscious effort necessary to drive motoneurons efficiently during effort and might affect fatigue, training-induced adaptations would prove meaningful. Unfortunately, there is currently no information on this particular topic. There is some evidence, however, that use of motoneurons differs between muscles in dominant and nondominant hands, which might reflect such motoneuronal changes. Adam, De Luca, and Erim (1998) examined motor unit recruitment and firing behavior in first dorsal interosseus muscles of dominant and nondominant hands of human subjects (table 4.2). Their finding that motor unit firing rates at a submaximal target force were lower in dominant hands was attributed to the effect of prolonged use on the contractile properties of the muscle fibers. Thus, slower fibers would require a lower frequency of firing to obtain a given absolute force. They also found a lower mean force threshold, lower initial firing rates, and lower discharge variability in motor units of dominant hands. These findings suggest a change in fundamental motoneuronal properties that would tend to optimize performance. For example, lower average force threshold suggests more motor units recruited to satisfy the same force, which would indicate a lower firing frequency on the part of each unit, thus avoiding the problems associated with increasing firing rates (e.g., adaptation, neuromuscular junction and branch-block failure, muscle fatigue). As seen previously, more closely spaced recruitment thresholds are characteristic of motoneurons that innervate fatigue-resistant, as opposed to fatigue-sensitive, muscle units (Bakels and Kernell 1994). Lower initial firing rates and reduced variability in firing may indicate fundamental changes in the components that control the afterhyperpolarization (see chapter 2). These interpretations from the work of Adam, De Luca, and Erim (1998) are based on the premise that handedness is a model of chronically increased neuromuscular usage.

One might think that, like muscle fibers, motoneurons might become more like slow types in their electrophysiological properties in response to chronic stimulation and endurance training. The changes in motoneurons' excitability and their firing frequencies during activation during such conditions is clear (see chapter 2). The closest that we can come to resolving this issue is to examine the work of Munson and colleagues (1997), who investigated the electrophysiological properties of cat hindlimb motoneurons following chronic electrical stimulation to promote the classic adaptations in their innervated muscles. In that study, after two to three months of chronic electrical stimulation of the medial gastrocnemius nerve, virtually all of the fibers of the medial gastrocnemius were converted to type I fibers. Motoneurons innervating this muscle also showed tendencies to become more like slow motoneurons: Afterhyperpolarization duration and input resistance increased, and rheobase decreased (figure 4.19). In addition, excitatory postsynaptic potentials (EPSPs) in response to high-frequency stimulation changed from positive modulation (facilitation, characteristic of fast motoneurons) to negative modulation (depression, characteristic of slow motoneurons). The mechanisms involved in this response have not been determined but involve pre- and postsynaptic mechanisms.

Finally, the authors of this study found that these changes also occurred when the nerves that were stimulated innervated skin, as opposed to muscle. Thus, it may be that stimulation of the peripheral nerve per se, involving

Table 4.2
Dominant (D) vs. Nondominant (ND) Hand Differences in First Dorsal Interosseus Muscle and Motor Unit Properties

WHOLE-MUSCLE PERFORMANCE	
MVC	D = ND
Force variability during sustained contraction at 30% MVC	D < ND**
MOTOR UNIT PROPERTIES	
Recruitment thresholds	D < ND*
Average firing rates	D < ND**
Initial firing rates	D < ND**
Discharge variability	D < ND**

* Significant differences at $p < 0.05$

** Significant differences at $p < 0.01$

Reprinted from Adam, De Luca, and Erim 1998.

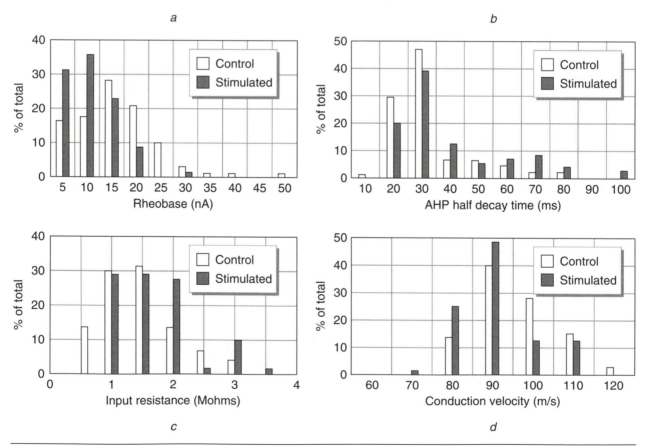

Figure 4.19 Electrophysiological properties of motoneurons of cat medial gastrocnemius tend to become "slower" after slowing of the innervated muscle fibers resulting from several months of chronic low-frequency stimulation. This is shown by a shift in the distribution of chronically stimulated motoneurons (solid bars) toward (*a*) lower rheobase, (*b*) longer AHP, and (*c*) higher input resistance.

Reprinted from Munson et al. 1997.

afferent as well as efferent stimulation, evoked the recorded responses. However, it is exciting to think that the changes in muscle exerted a retrograde effect on the properties of the motoneurons, an interpretation which is supported in these results by the motoneuron adaptations' apparent lagging behind muscle adaptations. Such a retrograde effect could be exerted by neurotrophins produced by chronically active muscle fibers, such as neurotrophin-4 (NT4; Funakoshi et al. 1995). The results of experiments in which motoneurons innervating skin also demonstrated adaptations might be explained by a common or similar neurotrophic mediator that is produced by both slow or active muscle and skin and that has slowing effects on motoneuron properties.

Spinal Cord Adaptations to Endurance Training

CONCEPT

6 Although endurance-type activities cannot generally be considered as highly skilled as, for example, activities demanding power (power lifting) and perceptuomotor coordination (golf), there is a case for possible beneficial spinal cord adaptations to endurance training. For example, the highly developed spinal circuitry governing the sequential activation of flexors and extensors must also be flexible enough to be capable of responding successfully to changes in afferent feedback, which are destined to occur during long-distance endurance-type activities due to changes in the muscle properties and in the functioning of the receptors themselves (see chapter 3). In addition, supraspinal inputs to the locomotion-generating spinal circuits change as overall effort increases during dropout of muscle fibers during fatigue. The outcome measure for successful spinal cord response to these changes during an endurance event is the capacity to maintain successful muscle activation patterns throughout the activity and to respond to unexpected perturbations in the locomotor system (such as tripping, stepping over a barrier, adjusting stride length, and the like). Chronic spinal cord adaptations (besides changes at the motoneuron level mentioned in the previous section) might include changes in the strength and proliferation of existing synaptic contacts with motoneuron pools or the uncovering of previously unused pathways.

In the following sections, I summarize several sources of information that give evidence of the plasticity of the spinal cord in response to repetitive activation (training).

Training of the Monosynaptic Reflex

Monkeys can be trained to either increase or decrease the amplitude of the monosynaptic stretch reflex response. This finding is the result of experimentation performed by J.R. Wolpaw and colleagues since the early 1980s (figure 4.20). This series of experiments demonstrated hard-wired changes in spinal cord circuitry that have resulted from behavioral conditioning. For example, these investigators found that changes in the amplitude of H-reflexes were apparent when animals were anesthetized and for up to three days following removal of supraspinal influences by spinalization (Wolpaw and Lee 1989). They have also ascertained that at least a portion of the adaptation may involve changes in intrinsic motoneuron properties. For example, in monkeys trained to decrease the amplitude of their triceps surae H-reflex, motoneurons had a more positive firing threshold and thus needed more depolarization to reach that threshold (Carp and Wolpaw 1994). The authors proposed that this change may be caused by a positive shift in the voltage for sodium-channel activation in the involved motoneurons, due in turn to a change in activation of intraneuronal protein kinase C (Halter, Carp, and Wolpaw 1995). The spinal changes produced by conditioning to up-regulate and down-regulate the amplitude of the H-reflex are not mirror images, since the latter but not the former are accompanied by changes detectable at the single-motoneuron level (Carp and Wolpaw 1995). Thus, these adaptive responses most likely include adaptations at several levels: motoneuron, afferent–motoneuron synapse, and interneurons.

Locomotor Training Following Spinalization

During the past two decades, a general picture has been emerging that suggests that spinal

a

b

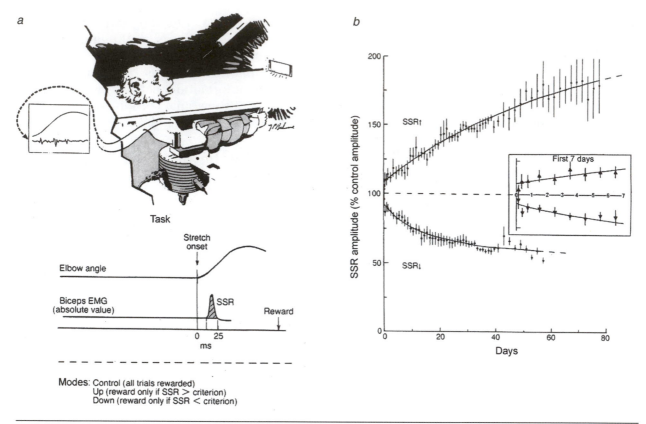

Figure 4.20   Experiments of J.R. Wolpaw and colleagues demonstrated that monkeys can be trained to increase or decrease the amplitude of the monosynaptic reflex. (*a*) An extension torque is superimposed on a low-level tonic elbow flexion, evoking a spinal stretch reflex (SSR) at monosynaptic latency. (*b*) Time course of changes in SSR amplitude in animals trained to increase (upper line) or decrease (lower line) the SSR amplitude.

Reprinted from Wolpaw and Carp 1990.

cord plasticity is possible in the presence of repeated activation in the form of training. The preferred subject in this research has been the cat, which is capable of demonstrating coordinated activation of hindlimb locomotor muscles that is surprisingly similar to a normal pattern after spinal cord transection (Lovely et al. 1990; Barbeau and Rossignol 1987; J.L. Smith et al. 1982; Bélanger et al. 1996). The quality of the locomotion evoked in the weeks and months following spinal cord transection (as measured by treadmill speed, stride length, amplitudes and temporal coordination of extensor and flexor bursts, incidence of failed steps) is greatly improved if daily training is imposed, that is, if the cat is supported on the treadmill and the hindlimb locomotor pattern is evoked by passive movements of the limbs or various means of sensory stimulation (Lovely et al. 1986; Chau, Barbeau, and Rossignol 1998; De Leon et al.

1998; figure 4.21). Furthermore, the beneficial response to training is somewhat specific to the training modality; cats trained for weight support stand but do not walk better, while those trained for walking improve in walking more than in weight support (Hodgson et al. 1994; figure 4.22).

Similar findings, again in cats, describe the effect of axotomy of the nerves innervating ankle flexors on treadmill locomotor gait before and after spinalization (Carrier, Brustein, and Rossignol 1997). When these nerves, which are important in lifting the foot during the swing phase of locomotion, were cut in the otherwise intact normal cat, compensation for the deficit took the form of increased activation of knee or hip flexors, or both. When these cats were spinalized one month later and then made to demonstrate hindlimb locomotor patterns on a treadmill, their locomotor patterns were

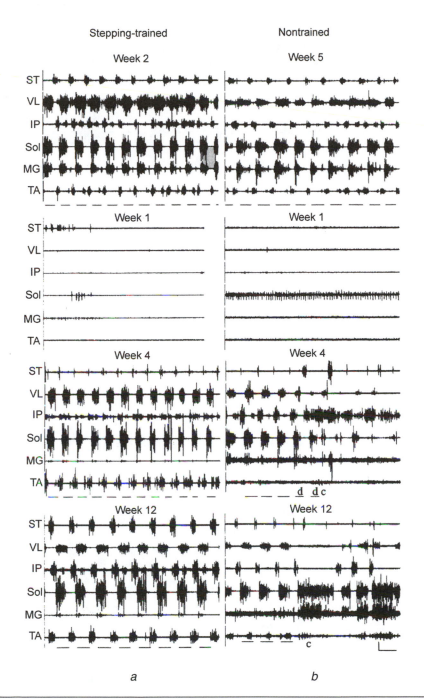

Stepping-trained · Nontrained

Figure 4.21 EMG activity of hindlimb muscles during bipedal stepping at 0.4 m/s in (*a*) a stepping-trained cat and (*b*) a nontrained cat. From top to bottom, records are from before, 1 week after, 4 weeks after, and 12 weeks after spinal cord transection at T12-T13. Training was 30 min per day, five days per week, after spinal transection in the trained cat. Muscles are semitendinosus (ST), vastus lateralis (VL), iliopsoas (IP), soleus (Sol), medial gastrocnemius (MG), and tibialis anterior (TA). Notice how the EMG pattern of the trained cat resembles the pretransection condition more closely than that of the nontrained cat.

Reprinted from De Leon et al. 1998.

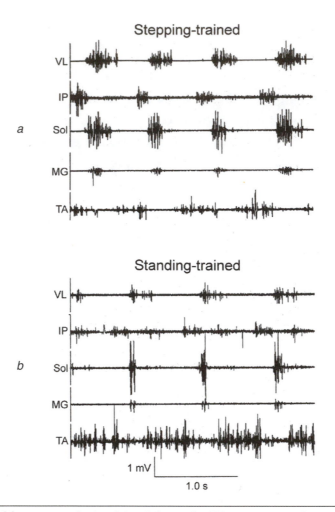

Figure 4.22 EMG record from a cat during bipedal treadmill locomotion after spinal cord transection at T12-T13. (*a*) Records after three months of daily stepping training, and (*b*) records after stepping training was replaced by weight-support standing training for three months.

Reprinted from Hodgson et al. 1994.

disorganized and asymmetrical, with exaggerated knee flexion on the side of the axotomy. This locomotor pattern contrasts with the symmetrical locomotor pattern seen in spinalized cats that were not axotomized and in spinalized cats in which the axotomy procedure was performed following the spinal cord transection. The best interpretation of these results is that the cat made supraspinal adjustments (higher recruitment of knee or hip flexors) during locomotion to compensate for the deficit caused by axotomy, which gradually resulted in altered pathways in the spinal cord. The altered pathways were revealed when the supraspinal influences were removed via spinalization.

These findings may have more significance for rehabilitative treatment than for endurance training; time will tell. They do, however, tell us that adaptations at the spinal cord level can result in alterations in the way that locomotor muscles are used. These demonstrated adaptations at the spinal cord level also suggest that more central adaptations may occur in response to increased activity. Central nervous system plasticity has been demonstrated following skill learning (Kleim, Barbay, and Nudo 1998; Bernardi et al. 1996; Pearce et al. 2000); voluntary endurance running (Van Praag et al. 1999); and peripheral nerve injury, spinal cord injury, and stroke (Liepert et al. 1998; Barbeau et al. 1998; T.A. Jones, Kleim, and Greenough 1996; Nudo et al. 1996; Jain, Florence, and Kaas 1998; T. Jones et al. 1999). These types of plasticity may eventually prove to be similar to adaptive responses to endurance training by which perception of fatigue is offset and rhythmic locomotor patterns kept more efficient during long-duration exercise.

Summary

For many years, it has been known that endurance training does a number of things to skeletal muscle: Mitochondrial enzyme content increases, the proteins that control cellular calcium levels are altered, and type IIB fibers tend to decrease while types IIA and I (and IIC) correspondingly increase. Investigations of chronically stimulated muscles have provided us with the opportunity to consider mechanisms as opposed to mere descriptions. Using the combination of electrical stimulation of animal muscles and molecular biological techniques, investigators have come a long way in the past 15 years in determining signaling pathways and the initial signals that promote these adaptations to increased activity. The task is now to attempt to put this information into a framework that takes into consideration the complexities of human motor unit recruitment during endurance-type exercise and the differences in adaptive responses of the different fiber types. The adaptations within the nervous system (neuromuscular junction, alpha-motoneuron, spinal cord systems, and supraspinal mechanisms) are less well known but may prove important in explaining at least a portion of the increased performance that occurs as a result of training. Perhaps more important, these nervous system adaptations, once fully described, might be relevant for the use of endurance-type training in certain pathological states with nervous system involvement.

chapter 5

Strength Training

"**T**he first requisite for healthy and vigorous muscular action is the possession of strong and healthy muscle fibres. In every part of the animal economy, the muscles are proportionate in size and structure to the efforts required from them; and it is a law of nature, that whenever a muscle is called into frequent use, its fibres increase in thickness within certain limits, and become capable of acting with greater force and readiness."

A. Combe, *The Principles of Physiology Applied to the Preservation of Health and to the Improvement of Physical and Mental Education,* 1843

Most Important Concepts From This Chapter

1 Protein synthesis and degradation both increase within hours of an acute strength-training bout. These changes involve primarily posttranscriptional mechanisms, including changes in initiation, elongation, or termination of protein synthesis. Protein degradation pathways that are activated include calcium-activated neutral proteases (calpains), the ubiquitin–protease system, and several lysosomal enzymes.

2 While signals that promote the adaptation of muscles to endurance training appear to be related to the altered intracellular environment caused by prolonged contraction, the signals associated with muscle stretch and strain seem fundamental in the responses to strength training.

3 Passive stretch or strength-type overload results in activation of several signals, including the production of insulinlike growth factor-1 (IGF-1), activation of phospholipases A and C, and activation of the kinase systems protein kinase C (PKC), focal adhesion kinase (FAK), and mitogen-activated protein kinase (MAPK). These signals promote changes in protein synthesis via activation of nuclear transcription factors, altered translation mechanisms, and activation of satellite cells.

4 The production of chemical substances by invading cells consequent to the muscle damage produced during the exercise may also provide signals that promote an adaptive response to strength training.

5 Phenotypic responses to strength training include the effects on fiber size, fiber ultrastructure, fiber number, fiber-type proportions, and muscle functional properties.

6 There is evidence of a neural effect of strength training, which can influence performance in the absence of any measurable morphological response. While some of this evidence borders on the anecdotal, other evidence consists of unequivocal neurophysiological results.

The development of increased muscle mass and maximal voluntary muscle strength involves different primary signals at the muscle level than are involved during the development of neuromuscular endurance. The emphasis here is on primary signals, since, qualitatively speaking, many of the signals are probably similar. There are several factors that distinguish activities that promote increased neuromuscular strength and mass from those that promote neuromuscular endurance. The main factor, of course, involves the forces to which structures are subjected during neuromuscular activation. Stress and strain on muscular structures can constitute a factor that, through mechanotransduction via several pathways, provides biochemical signals that can alter gene expression and ultimately phenotype, examples of which are included in this chapter. A second factor that distinguishes strength from endurance training is the degree and pattern of recruitment of motor units during the training to induce hypertrophy or increased strength. This factor influences the outcome of the phenotype, not merely by virtue of the high forces and firing frequencies to which the individual muscle fibers are subjected, but also by the apparent influence that recruitment patterns during this high degree of voluntary effort have on the associated nervous pathways.

In this chapter, I summarize what I believe are the signals that are important in modulating protein synthesis to produce the phenotypic response to resistive overload. In doing so, I call upon several models of hypertrophy, including chronic stretch of animal muscles *in vivo,* chronic and cyclic stretch of isolated cells (sometimes other than muscle cells, such as myocytes and epithelial cells) *in vitro,* chronic compensatory overload of muscles induced by ablation of synergistic muscles, and actual strength training in both animals and humans. Clearly, some aspects of the models other than strength training per se are of limited value in furthering our knowledge of the latter. Consequently, I decided to leave out some information that others might think should be included in this chapter, while including other information for which the application to strength training in humans might be considered tenuous. As in the case of endurance training, a great deal of information can be gleaned concerning mechanisms by examining what happens in the minutes, hours, and days following the delivery of the first training stimulus.

Next, I summarize the phenotypic changes with strength training. I put some emphasis on issues such as hyperplasia as a mechanism and the specific effects of eccentric training, since these seem to be somewhat controversial subjects. Finally, I discuss the neural influences that seem to be so important in the improved performance that occurs with resistance training.

Acute Effects of Strength Training on Protein Synthesis and Degradation

CONCEPT **1** Compensatory overload of rat hindlimb muscles induced by tenotomy of synergists results in increases in both protein synthesis and degradation (Goldspink, Garlick, and McNurlan 1983). This response is also evident in resistance training in humans. Within three hours of a bout of resistance exercise involving

concentric and eccentric contractions, fractional synthesis of mixed muscle proteins more than doubles and remains elevated, but at a lower level, for at least 48 hours (MacDougall et al. 1992; Chesley et al. 1992; Biolo et al. 1995; figure 5.1). The rapid increase is accompanied by an unchanged RNA content (thus increased RNA activity), signifying that alterations in post-transcriptional mechanisms are important contributors to the increased protein synthesis. Fractional degradation of proteins is also elevated in the hours following a bout of resistance exercise (Phillips et al. 1997, 1999; Biolo et al. 1995). The increased muscle protein synthesis is accompanied by increased passage of the amino acids leucine, lysine, and alanine from plasma into muscle, indicating that increased amino acid transport may constitute a significant promoter of this phenomenon.

Wong and Booth (1990a, 1990b) approached this problem of acute changes accompanying resistance training by electrically stimulating hindlimbs of anesthetized rats so that muscles worked against a specified load. They then measured various components of protein synthesis at periods following the "exercise." They reported several results: (1) Protein synthesis rates were higher than in control muscles 12 to 17 hours following activity involving 192 unloaded or loaded (concentric) contractions; (2) this elevated protein synthesis was still present 36 to 41 hours following the exercise; (3) protein synthesis rate increases were greater than increases in RNA; and (4) there was no relationship between the ability of the stimulation protocol to stimulate protein synthesis acutely and to increase muscle mass chronically.

The findings in general show that protein synthesis and degradation are both elevated in muscles as a result of resistance overload and that these changes, induced by a relatively short period of exposure to the contractile stimulus, last for several days. Several mechanisms are involved in these changes; the major ones are discussed in the following sections.

Stretch as an Important Signal

CONCEPT **2** One way that resistance training differs from endurance training is the stress and strain to which contractile and noncontractile tissues are exposed during an acute training

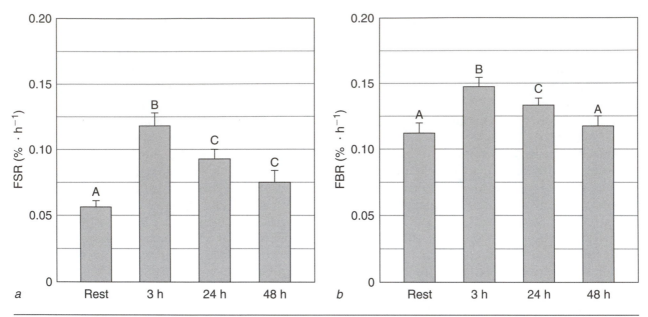

Figure 5.1   Effect of a single bout of resistance exercise (eight sets of eight concentric or eccentric contractions at 80% of concentric 1-RM) on (*a*) mixed muscle protein fractional synthesis (FSR) and (*b*) fractional breakdown (FBR) rates. Means with different letters are statistically different.

Reprinted from Phillips et al. 1997.

bout as a consequence of the relatively high forces involved. Passive stretch constitutes a strong signal for changes in protein synthesis. When rabbit muscles are immobilized in a lengthened or stretched position, for example, muscle weight increases, but primarily as a result of increased muscle fiber length (Dix and Eisenberg 1990). Under these conditions, much of the increased protein synthetic activity occurs in the distal end of the muscle, where sarcomeres are being added in series. During the first six days, increases in the concentration of polysomes are seen in the distal muscle region, attesting to increased protein synthesis. Molecules that are involved in anchoring the muscle fibers to the myotendinous junction, such as vinculin and talin, increase in expression. When electrical stimulation is added to the stretch, changes are more profound and are usually greater than the sum of electrical stimulation and stretch individually (Osbaldeston et al. 1995). This observation is indirect proof of the importance of stretch of noncontractile tissues on the adaptive response to resistance training.

The *in vivo* passive-stretch model just described is somewhat limited in its value as a model for resistance training–induced adaptations, since it usually involves immobilization of the muscle to fix its length. In short-term experiments, however, before the atrophy of inactivity begins to express itself, it does reveal that activation of components other than the contractile apparatus can have a mitogenic (i.e., growth-promoting) effect on muscle fiber protein synthesis. In this effect is found the fundamental difference between the signals involved in the adaptations to resistance and endurance training. While the signals for endurance-training adaptations are probably linked to metabolic demand and supply, the signals for resistance-training adaptations are linked to forces exerted on structures that support the contractile machinery and perhaps on the contractile machinery itself.

CONCEPT

3   There is converging evidence that stretch per se, independent of increased contractile activity, can stimulate protein synthesis via several pathways. The stretch of extracellular matrix generates intracellular signals that eventually lead to altered gene expression. Much of this evidence comes from experiments in which cell types other than skeletal muscle—such as cardiac myocytes, epithelial cells, and vascular smooth muscle cells—

are exposed *in vitro* to controlled sustained or cyclic stretches. A general scheme of the intracellular signals is shown in figure 5.2.

Activation of Mitogen-Activated Protein Kinases

Mitogen-activated protein kinases (MAPK; Alberts et al. 1994) constitute important communication links between the events occurring at the cell surface membrane and transcriptional events in the nucleus. These kinases are activated by a variety of signals resulting from activation of G protein kinase–based or tyrosine kinase–based transmembrane receptors. These receptors in turn phosphorylate transcription factors and are thought to be involved in the induction of some of the immediate early genes, such as *c-Fos* and *c-Jun* mentioned further on (Seedorf 1995). An increasing number of MAPK-regulated transcription factors are being discovered (Sugden and Clerk 1998). MAPK may also play an important role in the posttranscriptional control of protein synthesis, allowing a more rapid response in protein synthesis than would be possible by relying on transcriptional mechanisms alone.

One hypothesis (Watson 1991) that might pertain to resistance training maintains that passive muscle stretch, which would cause a physical deformation of the three-dimensional configuration of the transmembrane receptors, results in an activation of that receptor, much the same as if by attachment of its ligand. In this way, passive muscle stretch would supply the intracellular compartment with mitogenic signals. This mechanism could help to explain how passive muscle stretch increases protein synthesis but has yet to be substantiated.

It has been suggested that one site where stretch is transduced into changes in protein synthesis is the integrin molecule (Carson and Wei 2000). Integrin is a transmembrane protein that is attached on the extracellular surface to components of the extracellular matrix and on the intracellular side to components of the cytoskeleton, including talin, vinculin, and alpha-actinin (Longhurst and Jennings 1998; Alberts et al. 1994). Integrins are clustered at sites termed focal adhesions (Burridge and Chrzanowska-Wodnicka 1996). Intracellular signaling via integrins is accomplished by protein phosphorylation via a focal adhesion kinase

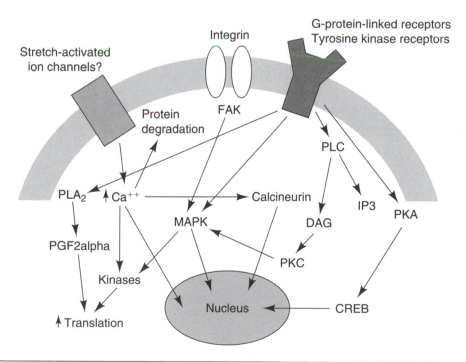

Figure 5.2   Intracellular signals that may be involved in overload-induced muscle hypertrophy. FAK = focal adhesion kinase; PLC = phospholipase C; DAG = diacylglycerol; IP3 = inositol triphosphate; PKC = protein kinase C; PKA = protein kinase A; CREB = cyclic AMP–responsive element binding protein; PLA_2 = phospholipase A2; PGF2alpha = prostaglandin F_{2alpha}.

(FAK), which can lead to MAPK activation, among other downstream events. Thus, the stretch signal that activates protein synthesis may be mediated via deformation of the integrin molecule during stretch of the membrane, which activates FAK and, as a result, MAPK and gene expression (E. Clark and Brugge 1995).

Very little information is available at present regarding activation of MAPK in muscle; available information comes from *in vitro* studies with stretched myocytes (Sadoshima and Izumo 1993), epithelial cells (Papadaki and Eskin 1997), and osteoblasts (Schmidt et al. 1998). However, one recent report shows that MAPKs are activated robustly in intact muscles with passive stretch (Martineau and Gardiner 1999). One of the MAPKs, c-Jun NH_2-terminal kinase (JNK), is activated by 60 minutes of bicycle ergometer exercise, with a corresponding increase in the expression of its downstream nuclear target, c-Jun (Aronson et al. 1998). More recently, it has been shown that JNK is particularly activated by eccentric contractions with knee extensors in humans (Boppart et al. 1999). In addition, electrical stimulation of rat muscles to perform strong, intermittent, isometric contractions (500-millisecond-duration contractions at 100 hertz, once per second) results in activation

of both JNK and extracellular signal-regulated kinase (ERK), to different magnitudes and with different time courses (figure 5.3). Thus, the signals resulting in activation of these MAPKs and their targets are distinct to some degree.

Although the idea of a signal or group of signals that activate the kinase system with stretch is appealing, we must consider the possibility that deactivation via a phosphatase system may also constitute a mechanism involved in increased protein synthesis and subsequent hypertrophy. For example, calcineurin, a calcium-regulated phosphatase, is necessary for the rapid gain in muscle mass and the severalfold increase in type I fibers that occur during four weeks of compensatory overload in rat plantaris (Dunn, Burns, and Michel 1999).

Insulinlike Growth Factor-1

Insulinlike growth factor-1 (IGF-1), which is produced primarily by the liver, is also produced by muscle and has stimulatory effects on protein synthesis, hyperplasia, and the proliferation of myonuclei *in vitro* (DeVol et al. 1990). Most effects are mediated via the IGF-1 receptor, which, like the insulin receptor, is a tyrosine kinase. In cultures of avian myotubes, IGF-1

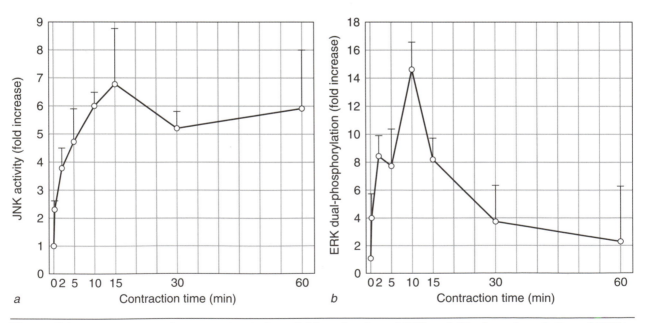

Figure 5.3   Time course of contraction-induced (*a*) activation of JNK and (*b*) phosphorylation of ERK in rat gastrocnemius. Muscles in the anesthetized animal were stimulated electrically via the sciatic nerve with one 500-ms train of impulses at 100 Hz, applied once per second.

Reprinted from Aronson, Dufresne, and Goodyear 1997.

stimulates cell hyperplasia and cell hypertrophy (insulin also does, but at a much higher, nonphysiological concentration, and IGF-2 stimulates hyperplasia but not hypertrophy under these conditions). One way in which IGF-1 exerts its positive effect on protein synthesis is via enhancement of the formation of a complex of initiating factors (eIF4E–eIF4G) (Vary, Jefferson, and Kimball 2000).

Many sources of evidence converge to suggest that the increased expression of IGF-1 by stretched or overloaded muscle fibers has significance for the proliferation of myonuclei. Yang and colleagues (1996) found that IGF-1 was upregulated within two hours after initiation of stretch in rabbit tibialis anterior and was found primarily in small fibers containing neonatal MHC. Thus, the ends of normal fibers are the regions where new longitudinal growth takes place, and IGF-1 takes part in this process.

However, the effect of IGF-1 is not isolated to myonucleus proliferation. Barton-Davis, Shoturma, and Sweeney (1999) showed that IGF-1 still exerts a significant hypertrophic effect (50% of the total effect) in mouse muscles in which increased myonucleus number is prevented by previous irradiation.

The stretch component during strength training, and not merely the increased metabolic rate during exercise, is clearly the signal promoting IGF-1 production in muscle. Goldspink and colleagues (1995) found that the response of IGF-1 to stretch plus stimulation was different from the response to stimulation alone. Static stretch alone increased IGF-1 mRNA and stimulated protein synthesis. Low-frequency stimulation amplified this response considerably, whereas stimulation alone had no effect.

Adams and Haddad (1996) showed that compensatory hypertrophy of rat plantaris resulted in increased IGF-1 mRNA and protein that peaked at three days at levels that were sixfold and fourfold, respectively, that of controls. Increases in IGF-1 preceded the hypertrophic response. They also found an increase in DNA content that was proportional to the increased muscle mass at 3, 7, 14, and 28 days of overload and an increase in mitotically active nuclei, indicating activation of satellite cells. They concluded that IGF-1 upregulation by muscle fibers appears to be involved in the increase in DNA content, perhaps by activation and incorporation of satellite cells.

Their more recent work (Adams, Haddad, and Baldwin 1999) supports this by showing a close temporal relationship between increased IGF-1 expression and markers of satellite cell activation during the early period of overload.

In related work, Adams and McCue (1998) found that localized infusion of IGF-1 directly onto a rat tibialis anterior muscle resulted in increased total muscle protein and DNA content.

Nonetheless, IGF-1 production by overloaded muscle is only part of the puzzle. Yan, Biggs, and Booth (1993) examined the effect on IGF-1 expression of one acute bout of 192 electrical stimulation–evoked eccentric contractions of the rat tibialis anterior. IGF-I protein became significant at levels three times higher than the control value by the fourth day after eccentric contractions. IGF-1 thus peaks *after* the increase in protein synthesis and therefore is not responsible for it.

Activation of Protein Kinase C

Protein kinase C (PKC) is a widely distributed kinase with broad substrate specificity that is activated by calcium, phospholipids, and diacylglycerol and is therefore activated upon stimulation of phospholipase C (Haller, Lindschau, and Luft 1994). Increased calcium levels result in increased binding of PKC to membrane lipids. Among its functions is negative feedback, via phosphorylation of tyrosine receptors, that decreases these receptors' signaling activity (Haller, Lindschau, and Luft 1994). PKC also increases calcium flux from the extracellular space and from intracellular stores and regulates several calcium-regulating enzymes, including calcium ATPase. PKC is also involved in mitogenic signaling during cell growth. Certain isoforms of PKC are translocated to the nucleus, where they may be involved in gene transcription (Haller, Lindschau, and Luft 1994).

Compensatory hypertrophy in the rat results in an increase in PKC that peaks at double the control value four days after the initiation of overload. PKC activity subsequently declines to control values by about nine days (Richter and Nielsen 1991).

Production of Prostaglandins

Vandenburgh and colleagues (1995) contributed significantly to our knowledge regarding prostaglandin synthesis during stretching of myotubes. The techniques used in their series of studies include cyclical stretching of avian myofiber cultures embedded in an extracellular matrix and maintained *in vitro* for several days (Vandenburgh, Swasdison, and Karlisch 1991). Stretched cultures respond by producing two forms of prostaglandins with different time courses. Prostaglandin PDF_2 is the form that stimulates protein synthesis, while prostaglandin PGE_2 stimulates protein degradation. PDF_2 production increases gradually from the initiation of cyclic stretching for at least four days, while PGE_2 increases rapidly, then declines to baseline within 24 hours (Vandenburgh et al. 1990, 1995). These events are blocked by indomethacin, a blocker of prostaglandins synthesis. Thus, the increased production of these prostaglandins by muscle is not due to mobilization of presynthesized intramuscular stores. Prostaglandin production is also blocked by inhibiting phospholipases (Vandenburgh et al. 1993).

These changes are related to G proteins, since treatment of cultures with pertussis toxin inhibited PDF_2 production, the activation of cyclooxygenase (the enzyme that converts arachidonic acid to prostaglandins), and phospholipase activation (Vandenburgh et al. 1995). In addition, activation of these pathways does not require electrical activation of membranes, since it occurred in the presence of tetrodotoxin, a sodium conductance channel blocker (Vandenburgh et al. 1993, 1995). Thus, it appears that mechanical stretch stimulates a G protein–linked cascade that results in prostaglandin synthesis, although how this transduction takes place and the species of G proteins involved are not known. PDF_2 has been shown to stimulate the phosphorylation of the 40S ribosomal protein S6 via stimulation of the implicated kinases (p70^{S6K} and p90^{RSK}; M.G. Thompson and Palmer 1998). It may also stimulate phosphorylation of PHAS-1, a protein that, when not phosphorylated, is bound to the mRNA cap-binding protein eIF4E and prevents it from initiating translation (M.G. Thompson and Palmer 1998).

Vandenburgh and colleagues (1993) presented several possibilities as to how stretch might activate phospholipases. Stretch may increase the sensitivity of the muscle cells to growth factors that either activate phospho-lipases, transducers such as G proteins that activate phospholipases, and calcium-sensitive lipases via alterations in intracellular calcium levels, or increase the accessibility of the phospholipase to its membrane substrate.

Stretch activates the phospholipases PLA2, PLD, and the phosphatidylinositol-specific phospholipase C, which is activated by electrical rather than mechanical events (Vandenburgh et al. 1993). PLA2 results in release of arachidonic acid and the production of PDF_2 (M.G. Thompson and Palmer 1998).

We also know from these studies that indomethacin, which blocks prostaglandin synthesis, also reduces (but does not abolish) the initial decrease and subsequent increase in protein synthesis. Thus, prostaglandins play a role but are not the whole story concerning the increased protein synthesis that occurs with stretch.

Increased Transcription of Protooncogenes *c-Fos, c-Jun,* and *c-Myc*

Evidence for a change in protein synthesis in stretched cells is given by a rapid increase after the initiation of the stimulus in the expression of several protooncogenes, the products of which are important transcription factors that influence the expression of several proteins involved in cell adaptation. Three of these have been investigated specifically with reference to muscle: *c-Fos, c-Jun,* and *c-Myc.* (Many others have been found in other tissues, but their responsiveness and importance have not been demonstrated in muscle.)

The protooncogene *c-Myc* codes for a phosphoprotein that can bind to DNA and that appears to play a role in mitosis (Hesketh and Whitelaw 1992). It is normally down-regulated in fully differentiated muscle fibers, reflecting a reduced cell proliferation. The mRNA for *c-Myc* increases within three hours following the induction of compensatory overload and remains elevated for 24 hours before returning to baseline (Whitelaw and Hesketh 1992). The increase in *c-Myc* expression has been proposed to be involved in the activation of ribosome synthesis early in the adaptation period.

The *c-Fos* and *c-Jun* oncogenes are components of the activator protein-1 (AP-1) transcription complex, which binds to promoters of several growth-associated genes. Both *c-Fos* and *c-Jun* are induced in response to mechanical stimulation. In stretched rabbit muscle, *c-Fos* and *c-Jun* expression increases transiently during the first three hours of stretch, then returns to baseline (Dawes et al. 1996; figure 5.4). Interestingly, these changes in *c-Fos* and *c-Jun* expression are different in amplitude when chronically stretched muscles are also electrically stimulated (Goldspink et al. 1995; figure 5.5). For example, rabbit muscles show a 20-fold and 15-fold increase in mRNAs for *c-Fos* and *c-Jun*, respectively, when passively stretched, but when electrical stimulation is added, their levels increase further to 80-fold and 60-fold, respectively, even though no increase occurs with the electrical stimulation alone (Goldspink et al. 1995). In addition, it is clear that *c-Fos* and *c-Jun* are not signals arising from the exact same stimulus. For example, when stretch is reapplied after a period without stretch, the responses of *c-Fos* and *c-Jun* are qualitatively as well as quantitatively different from the response to the initial stretch. The response of these two protooncogenes to stretch may therefore occur along two slightly different pathways.

Much still has to be learned about the importance of these early changes in expression of *c-Myc, c-Fos,* and *c-Jun* with stretch and in resistance training.

Increased Transcription of Muscle Regulatory Factor Genes

The products of the muscle regulatory factor (MRF) genes, which are active during muscle development and are present at negligible levels in normal adult muscles, include Myf-5, MyoD1, MRF4, and myogenin (Jacobs-El, Zhou, and Russell 1995). In rat tibialis anterior subjected to stretch and electrical stimulation for two hours, mRNAs for MRF4 and Myf-5 increased significantly (11-fold and 6-fold, respectively). Since binding sites for these factors are found upstream of many muscle genes, this may signify another pathway whereby stretch activates protein synthesis. In rat hindlimb muscles subjected to compensatory overload, myogenin and MyoD expression increased within 12 hours of the stimulus, possibly as a result of increased IGF-1 production (Adams, Haddad, and Baldwin 1999).

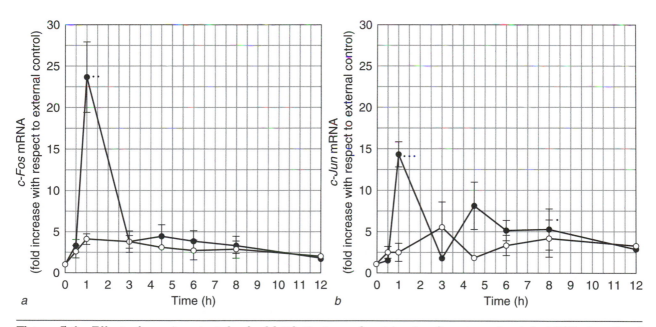

Figure 5.4   Effect of passive stretch of rabbit latissimus dorsi *in vivo* (to approximately 115% of optimal muscle length) on (*a*) *c-Fos* and (*b*) *c-Jun* mRNA. Filled circles are stretched; open circles are contralateral control muscles.

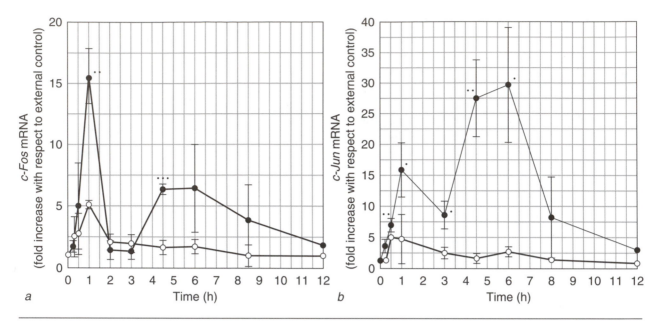

Figure 5.5 Effects of passive stretch (as in figure 5.4) combined with electrical stimulation (10 Hz) on (*a*) *c-Fos* and (*b*) *c-Jun* mRNA. Compare with figure 5.4.

Reprinted, by permission, from N.J. Osbaldeston et al., 1995, "The temporal and cellular expression of *c-fos* and *c-jun* in mechanically stimulated rabbit latissimus dorsi muscle," *Biochemical Journal 308*: 465-471. © 1995 the Biochemical Society.

Posttranscriptional Changes

We have already mentioned that increased RNA activity (in units of protein synthesis per unit time per unit RNA) occurs within hours of a bout of resistance exercise. Baar and Esser (1999) showed that six hours after electrical stimulation–induced lengthening contractions of rat hindlimb muscles, there was an increase in the size of the polysome pool, indicating increased initiation, elongation, or termination of protein synthesis. These investigators also found increased phosphorylation of the 70-kilodalton S6 protein kinase ($p70^{S6K}$) three and six hours after stimulation, which remained increased, but at lower levels, up to 36 hours postexercise. This kinase is involved in the phosphorylation of the small ribosomal subunit protein S6, which is involved in the regulation of protein synthesis and is mitogen stimulated (Sugden and Clerk 1998). In cardiac myocytes, the phosphorylation of this protein is required for stretch-induced increase in protein synthesis. In Baar and Esser's (1999) experiment, there was a relationship between the degree of phosphorylation of this protein at six hours and the increase in mass of the muscle after six weeks of the stimulation paradigm (figure 5.6).

Control of protein synthesis at the translation level also depends to a large extent on the activation levels of numerous eukaryotic initiation factors (eIF), which facilitate peptide initiation at the ribosome. Sixteen hours following an acute bout of strength training in rats, the activity of member of this family, eIF2B, is elevated, along with protein synthesis (Farrell et al. 1999).

Effects on Intracellular Proteolytic Systems

Intracellular proteolytic systems include the calcium-activated neutral proteases (calpains), lysosomal proteases, and the ATP-ubiquitin-dependent pathway (DeMartino and Ordway 1998; M.G. Thompson and Palmer 1998).

Calpains I and II are proteases activated by micromolar and millimolar concentrations of calcium, respectively (Melloni and Pontremoli 1989). There is some evidence that activation of calpain, which demonstrates a specificity toward myofibrillar and cytoskeletal proteins, is associated with the disruption of Z-lines in myofibrils from exhausted rat muscles (Belcastro, Albisser, and Littlejohn 1996; Belcastro 1993). Calpain activation may be secondary to the increased intracellular Ca^{2+} that accompa-

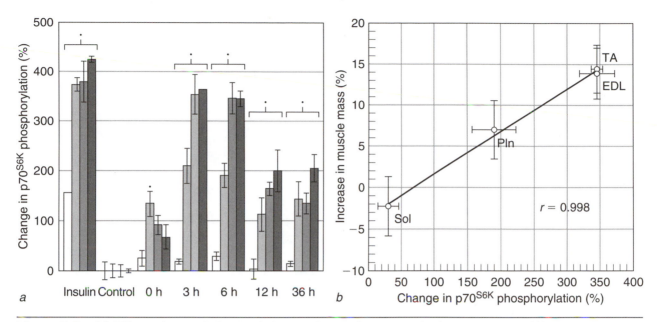

Figure 5.6   (*a*) Effect of a single bout of electrical stimulation–evoked contractions (60 contractions at 100 Hz, each lasting 3 s) on phosphorylation of 70-kDa S6 protein kinase (p70^{S6K}) in rat soleus (Sol; open bars), plantaris (Pln; lightly shaded bars), extensor digitorum longus (EDL; darkly shaded bars), and tibialis anterior (TA; filled bars). Dots indicate results different from control. (*b*) Correlation between change in phosphorylation of p70^{S6K} at six hours after stimulation and muscle wet mass measured after stimulation two days per week for six weeks.

Reprinted from Baar and Esser 1999.

nies exercise-associated muscle damage. After a period of 10 days of reduced load bearing, muscles exposed to two days of weight bearing demonstrate increased calpain II expression (Spencer and Tidball 1997).

Lysosomal proteases include cathepsins D, L, B, and H, as well as carboxypeptidases and aminopeptidases (DeMartino and Ordway 1998). During three days of muscle stretch, large increases (200% to 400%) occur in activities of cathepsins B and L and dipeptidyl aminopeptidase, with or without stimulation (Goldspink et al. 1995). Perhaps these enzymes are instrumental in the muscle remodeling that occurs with stretch. Vandenburgh et al. (1990) found an increase in cathepsin H activity within one hour after the imposition of cyclic stretch in myotube cultures.

The ubiquitin-proteosome system may be the major source of nonlysosomal protein degradation. Ubiquitin is a protein that is known to form complexes with abnormal proteins, thus tagging them for further degradation by covalently binding and modifying the abnormal protein. The tagged proteins are then degraded by a large protease complex. Elevated ubiquitin levels

(60%) have been found in the biceps brachii of subjects who had performed eccentric contractions 48 hours before (H.S. Thompson and Scordilis 1994). Much work remains to be done in the research of the ubiquitin-proteosome system as an important component of muscle remodeling during training.

Role of Muscle Damage

CONCEPT

4   Some tissue disruption is bound to occur as a result of the high forces generated during resistance training. This is particularly evident during the eccentric phase of dynamic contractions, where the force is generated by less muscle area than during the concentric phase (Gibala et al. 1995). Phenomena seen in muscles after eccentric contractile activity include cytoskeletal disruptions, Z-disk streaming, A-band disorganization, evidence of membrane damage, sarcoplasmic reticulum damage, hypercontracted regions, localized areas of increased intracellular calcium, and invasion of cells (Saxton and Donnelly 1996; Komulainen et al. 1998; Fridén and Lieber 1992,

1996, 1998; Fridén, Seger, and Ekblom 1998; Warren et al. 1995; Lowe et al. 1995; Yasuda et al. 1997; figure 5.7). Effects are more severe in type II fibers (Fridén and Lieber 1998). Muscle force capacity is reduced following eccentric damage and remains depressed for several days or up to several weeks (Lowe et al. 1995; J.N. Howell, Chleboun, and Conatser 1993; Brown et al. 1996). Interestingly, intrinsic proteolytic mechanisms possibly are not involved in the increased protein degradation that follows resistance exercise, since degradation does not increase during the six hours immediately following the insult (Lowe et al. 1995). Degradation of proteins is temporally related to phagocytic infiltration into fibers during the period 24 to 120 hours following the injury (Lowe et al. 1995).

During the period following eccentric contraction–induced damage, an inflammatory response ensues. This includes edema and the infiltration of the damaged fibers by neutrophils and macrophages. The attraction of inflammatory cells may occur via wound hormones, which are released from the injured muscle (Tidball 1995). Activation of calpain via elevated calcium levels as a consequence of muscle damage may also generate a chemoattractive signal (Raj, Booker, and Belcastro 1998). Fibroblasts, macrophages, and neutrophils, as well as the damaged fiber itself, produce a number of substances that take part in the degeneration–regeneration process, including platelet-derived growth factor (PDGF), basic fibroblast growth factor (bFGF), interleukins (IL), transforming growth factors (TGF), and tumor necrosis factor (TNF; Clarkson and Sayers 1999; Tidball 1995).

The importance of these events in the adaptive response of muscle to resistance training is not currently known. Phillips and colleagues (1999) have observed that the stimulatory effect of a single bout of eccentric knee extensions on muscle protein synthesis and breakdown is attenuated in trained subjects. The fact that the degree of muscle damage produced by acute eccentric exercise is also decreased in these subjects may suggest a link between muscle damage mechanisms and protein synthesis–degradation.

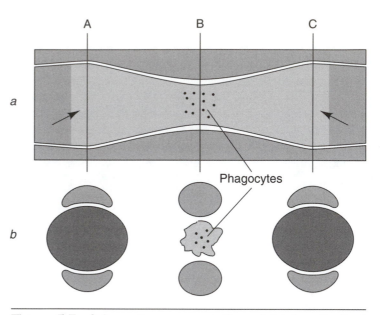

Figure 5.7   Schematic representation of (*a*) longitudinal and (*b*) cross sections of a muscle fiber (B) with segmental damage following eccentric contractions, surrounded by two normal fibers (A and C). Arrows show hypercontraction zone, bilateral to the necrotic zone. Phagocytes are shown in the necrotic zone.

Reprinted from Fridén and Lieber 1998.

The Chronic Effect of Resistance Overload on Muscle Phenotype

The results of strength training on muscle phenotype are presented in the sections that follow. For the most part, I have not made any attempt to separate the effects of dynamic and static types of training (or explosive and heavy resistance training). Although there are many similarities in the responses to these two types of resistance training, there are also some differences, which are highlighted when appropriate.

Fiber Cross-Sectional Area

Although the maximum limit to which fibers can hypertrophy in response to resistance training is not known, increases in cross-sectional area of 30% to 70% are the most commonly reported (Thorstensson, Sjodin, and Karlsson 1975; Sale et al. 1987; Alway et al. 1988, 1990; Alway, MacDougall, and Sale 1989; Staron et al. 1989;

MacDougall et al. 1991). In bodybuilders, fibers can range up to 2.5 times larger than in sedentary subjects (Alway et al. 1988). Generally speaking, type II fibers show larger hypertrophic responses than type I (MacDougall et al. 1980; Alway, MacDougall, and Sale 1989; Staron et al. 1989; Hather et al. 1991; Hortobágyi, Hill, et al. 1996). As one would expect, training programs involving heavier resistances or eccentric actions demonstrate more pronounced cell hypertrophy than programs involving faster or lighter-load contractions or without eccentric actions (Ewing et al. 1990; Hather et al. 1991; Hortobágyi, Hill, et al. 1996).

Fiber-Type Composition

The most common histochemical change in fiber-type proportions, when there is one, is a decrease in IIB fibers and a corresponding increase in IIAB/IIA proportion (Häkkinen, Newton, et al. 1998; Jürimäe et al. 1997; Hortobágyi, Hill, et al. 1996; Staron et al. 1989, 1994; Ploutz et al. 1994; Hather et al. 1991; Klitgaard, Zhou, and Richter 1990; Colliander and Tesch 1990). This has been shown to occur most frequently in knee extensors but also has been shown in

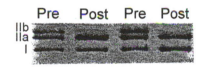

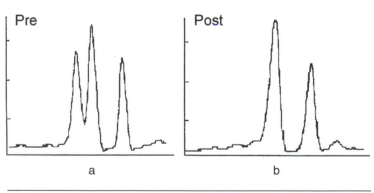

Figure 5.8 Heavy resistance training of the quadriceps for 19 weeks alters myosin heavy-chain composition in vastus lateralis. Myofibrils were separated using polyacrylamide gel electrophoresis (PAGE), and the gel was subsequently subjected to densitometric scan (*a* and *b*). Posttraining biopsies contained negligible MHC IIb.

Reprinted from Adams et al. 1993.

trained biceps and triceps brachii. This change has been substantiated using electrophoresis to separate the isoforms in biopsy samples from vastus lateralis (Adams et al. 1993; figure 5.8) and triceps brachii (Jürimäe et al. 1996). In women performing strength training of the trapezius, a decrease in MHC I and IIb complement with a corresponding increase in IIa has been reported (Kadi and Thornell 1999).

Fiber Number

Does hyperplasia occur with resistance training? One way that this particular problem is investigated is to examine the muscles of bodybuilders who have produced extreme muscle hypertrophy through many years of resistance training. In groups of untrained subjects, intermediate-level bodybuilders, and elite bodybuilders, estimates of fiber numbers, based on average muscle area measured from tomographic scans, divided by average fiber area taken from biopsy material showed that muscle (biceps brachii) cross-sectional area is more closely related to muscle fiber area ($r = 0.71$ to 0.81) than to estimated fiber number ($r = 0.35$ to 0.6; Sale et al. 1987; MacDougall et al. 1984).

Although the conclusion might be that hyperplasia does not play a role in extreme hypertrophic response, the wide variation in the numbers of muscle fibers among subjects, as well as the errors associated with the estimation techniques, leave enough possibility that small degrees of hyperplasia might take place with training. This wide variation among individuals may signify genetic variation. The authors of these studies pointed out that many of the subjects with the largest muscles also possessed the largest number of fibers (MacDougall et al. 1984; figure 5.9).

Other investigators (Larsson and Tesch 1986) reported that a larger muscle circumference of the biceps brachii in bodybuilders, with no difference in the average fiber area, signified that hyperplasia was involved. They also found evidence of a higher fiber density (using an intramuscular EMG technique by which the number

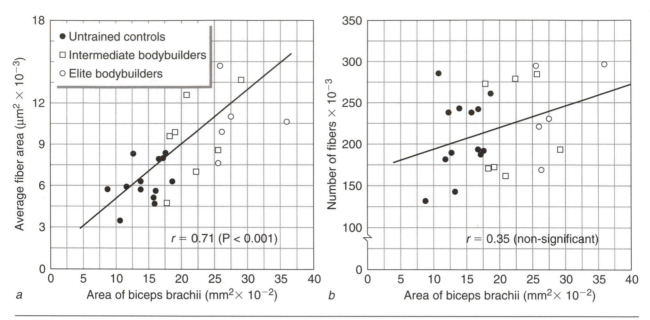

Figure 5.9 (*a*) Correlations between biceps brachii area and average muscle fiber area and (*b*) between biceps area and fiber number. Biceps area was determined using computerized tomographic scanning, and fiber areas were determined from biopsy material. Size of the biceps is better related to the size of individual fibers than to the number of fibers.

Reprinted from MacDougall et al. 1984.

of fibers belonging to the same motor unit within the recording range of the electrode is estimated), which would support their claim. However, the number of subjects in their bodybuilder group was limited ($n = 4$).

Hyperplasia is easier to investigate in cats than in humans for obvious technical reasons. In one experiment, cats were operantly conditioned to perform forearm flexion movements against a resistance for a food reward. Left–right comparisons of fiber numbers in the flexor carpi radialis muscles, performed by counting fibers after maceration of the muscle via nitric acid digestion, revealed a difference of approximately 9% (Gonyea et al. 1986). Further experimentation with cats has confirmed satellite cell activation by uptake of tritiated thymidine in exercised muscles, as well as the copresence of larger and smaller fibers than controls, which might indicate *de novo* fiber formation (Giddings, Neaves, and Gonyea 1985; Giddings and Gonyea 1992).

Hyperplasia occurs to a much greater extent in chickens and quail using a technique of stretching the muscles that normally support the wing by application of a weight. When the stretch is intermittent (24 hours on, 48 to 72 hours off), hyperplasia is not present, even after the same total time of stretch application (Antonio and Gonyea 1993). The applicability of these findings to the mammalian system is not known.

A metanalysis (Kelley 1996) regarding evidence for or against hyperplasia during muscle overload in animals came to the following conclusions: (1) Significant hyperplasia probably occurs (average increase in fiber number is 15%); (2) the avian subjects showed much more hyperplasia than mammalian subjects (21% vs. 8%); and (3) the degree of hyperplasia as a result of stretch exceeded that from compensatory overload, which in turn exceeded that from voluntary exercise (21% vs. 12% vs. 5%). We must consider, however, that these studies were published subsequent to the finding of a significant increase in fiber number. How many set out to find an increase but did not?

It has been shown that satellite cell activation appears to be necessary for the hypertrophic response and that nuclear proliferation occurs with resistance overload. Both of these phenomena are essential conditions for hyperplasia. Rosenblatt, Yong, and Parry (1994) showed that compensatory hypertrophy of the rat tibialis anterior in response to ablation of its synergists does not occur if the muscle is first exposed to radiation, which inhibits mitotic activity and thus satellite cell proliferation. Phelan and Gonyea (1997) substantiated this and showed that no small fibers were produced during the compensatory overload after muscle irradiation. Both compensatory overload (D.L. Allen et al. 1995) and resistance training (MacDougall et al. 1980; Kadi and Thornell 2000) result in an increase in myonuclei (figure 5.10). Whether or not increased mitotic activity in response to resistance overload leads eventually to the production of new muscle fibers and the ultimate fate of these fibers vis à vis their innervation and participation within existing motor units remain to be demonstrated.

Muscle Composition

We have already considered the adaptations in fiber-type composition, which, in combination with changes in fiber areas, have implications for muscle composition. There is some evidence that the percentage of collagenous and non-contractile proteins decreases in extremely hypertrophied fibers (Sale et al. 1987). D.A. Jones and Rutherford (1987) presented evidence from computerized tomographic scans

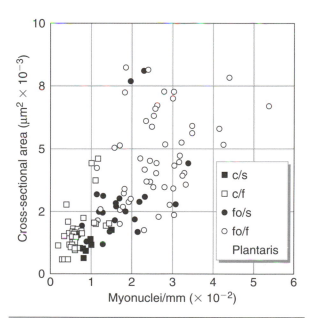

Figure 5.10   Relationship between muscle fiber area and number of myonuclei per millimeter in cat plantaris. Control slow (c/s) and fast (c/f) fibers and functionally overloaded (for 3 months) slow and fast fibers (fo/s and fo/f) are shown.

Reprinted from Allen et al. 1995.

that resistance-trained muscles were slightly more dense in myofibrillar packing. The activities of most energetic enzymes measured are unchanged after resistance training, including Mg^{2+}-stimulated ATPase, creatine kinase, phosphofructokinase, citrate synthase, lactic dehydrogenase, hexokinase, and succinate dehydrogenase (Thorstensson, Hulten, et al. 1976; Tesch, Thorsson, and Colliander 1990; Ploutz et al. 1994). Myokinase activity may increase slightly after resistance training (Thorstensson, Hulten, et al. 1976).

Fiber Ultrastructural Changes

Modest nuclei/fiber ratio increases have been found after resistance training that causes a 31% to 39% increase in fiber area (MacDougall et al. 1980). In compensatory overload in animals, where hypertrophy is more extreme (a 2.8-fold increase in area, for example), a more than threefold increase in myonuclei per fiber can occur (D.L. Allen et al. 1995). This myonucleus proliferation, which probably results from the activation and proliferation of satellite cells and their subsequent fusion with existing fibers, may be an adaptive attempt to maintain an adequate cytoplasmic volume per nucleus in the face of increased cytoplasmic volume (D.L. Allen et al. 1995).

Alway and colleagues (Alway et al. 1988; Alway, MacDougall, and Sale 1989) examined ultrastructural characteristics of muscle fibers before and after resistance training, as well as in the muscles of strength-trained athletes and controls. Their results suggest that significant hypertrophy can occur with very little change in the percentage of the fiber cross section occupied by sarcoplasmic reticulum–transverse tubular network, cytoplasm, lipids, and myofibrils. They did report, however, that gastrocnemius muscle fibers of strength-trained athletes had a mitochondrial fraction that was 30% less than controls. A possible decrease in mitochondrial enzyme (SDH) activity with intensive strength training has recently been substantiated (Chilibeck, Syrotiuk, and Bell 1999).

In addition to, or one could say in spite of, fiber hypertrophy, capillary-to-fiber ratios increase after both isometric and dynamic types of resistance training (Hather et al. 1991; Rube and Secher 1990).

Evoked Isometric Contractile Properties

Several reports have shown an increase in voluntary muscle strength after resistance training, without a corresponding increase in the isometric force response evoked via electrical stimulation of the nerve (Sale, Martin, and Moroz 1992; Davies et al. 1985). Part of this discrepancy may have to do with so-called neural effects that are called into play with voluntary contractions but that would not be seen when these contractions are evoked. This issue is dealt with in more depth in the section "Neural Effects of Resistance Training," page 161. Nonetheless, examination of the basic force- and speed-related properties of the muscle can be investigated independently of changes in neural mechanisms, using electrical stimulation–evoked contractions.

The most systematic study to date regarding this issue is that of Duchateau and Hainaut (1984), who trained adductor pollicis muscles of subjects using either dynamic (fast adductions at a load corresponding to 33% MVC) or static (5-second MVCs) types of overload for three months. They then measured evoked isometric contractile properties. Their findings showed unequivocally that the dynamic or isometric nature of the training stimulus has an impact on intrinsic muscle properties (figure 5.11). For example, isometric training resulted in larger increases in static strength–associated properties (twitch and tetanic forces, maximal muscle power), whereas dynamic training resulted in preferential enhancement of speed-associated properties (faster time course of the isometric twitch, higher peak rate of tetanic force development and maximum shortening velocity).

The intramuscular signals that determine whether the adaptations are dynamic-like or isometric-like are unknown. Behm and Sale (1993) showed that dynamic isokinetic training of the ankle flexors increased the maximal rate of twitch tension development and decreased the twitch time to peak tension of evoked contractions. Training also produced an increased

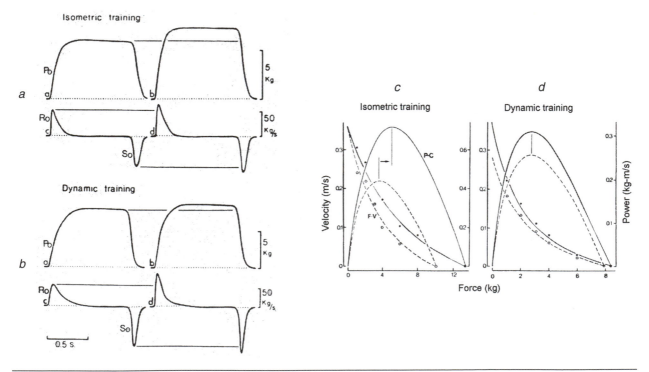

Figure 5.11   Effects of three months of moderate (*a*) isometric or (*b*) isotonic training on isometric contractile properties of adductor pollicis muscle. Isometric training resulted in greater increase in isometric force, while maximal rate of tension development increased more with dynamic training. P_o refers to maximum force, R_o to maximum rate of force development, and S_o to rate of force relaxation. Effects of (*c*) isometric and (*d*) dynamic training on force–velocity and power characteristics of adductor pollicis.

Reprinted from Duchateau and Hainaut 1984.

isokinetic torque response that was most marked at the higher velocities. Interestingly, these adaptations, which would be expected according to the previous experiments of Duchateau and Hainaut (1984), occurred whether or not the contraction was isometric or isokinetic at a high speed (5.23 radians per second). In this experiment, however, the subjects were asked to generate force as quickly as possible, regardless of whether the ensuing contraction was dynamic or static. Thus, the intramuscular signal specifying dynamic-like or static-like muscle adaptations is not related to the degree or speed of muscle fiber shortening or the load during the training. The signal may have to do with the rate of force development or the nature of the command from the nervous system.

Reports of the effects of resistance training on evoked contractile properties are mixed, some showing a more prolonged (Alway et al. 1988; Sale, Upton, et al. 1983) and others a faster (Alway, MacDougall, and Sale 1989; Sale et al. 1982) twitch response. Similarly, twitch force has been reported to decrease (Sale et al. 1982), increase (Sale, Upton, et al. 1983), or remain unchanged (Alway, MacDougall, and Sale 1989; Alway et al. 1988). These results vary because of differences in the subjects studied (some studies used trained subjects, while others compared sedentary subjects with strength-trained subjects), the nature of the training (dynamic or isometric), and the muscles examined.

The central message is that intrinsic muscle properties other than mass and strength change with resistance training. These changes are detectable using electrical stimulation–evoked contractions, and some of the properties that change (twitch tension and time course, rate of twitch tension development) have implications for muscle use during voluntary activation. These adaptations may be independent of the actual degree of shortening or lengthening performed by the muscle during the training but may depend on either the rate of activation of the contractile machinery or the nature of the signal reaching the muscle fiber from the nervous system.

Force, Velocity, and Power

Evidence about force–velocity characteristics and power comes from studies of voluntary contractions and thus might reflect the partial influence of neural factors, discussed in detail further on in this chapter. Nonetheless, certain general adaptive responses to resistance training are worth noting here, especially in light of our previous conclusion that some of the muscle changes are specific to the dynamic or static nature of the training.

Power lifters and bodybuilders had higher torques (plantar flexion, knee extension, elbow extension) than controls at both low and high velocities, with a fairly good correlation ($r = 0.84$) between torque at high and low velocities (Sale and MacDougall 1984). This suggests that a general training regimen involving a variety of loads and velocities increases strength throughout the entire range of muscle contractile speeds.

When speed of contraction during training is controlled, on the other hand, there is evidence that speed-specific adaptations occur. Caiozzo, Perrine, and Edgerton (1981) trained subjects at either slow (1.68 radians per second) or fast (4.19 radians per second) isokinetic speeds for four weeks, using maximal contractions. Slow training involved improvements in torque at all velocities except the fastest ones and caused the torque–velocity curve to level off at a lower level. High-velocity training resulted in improvements at the higher velocities. The leveling-off phenomenon and the effect of training on it suggested to the authors that motoneuron recruitment capacity was influenced by the training.

This velocity-specific training response has been subsequently found by other investigators. Narici et al. (1989) had subjects train knee extensors for 60 days at a relatively slow isokinetic speed (2.09 radians per second). Torque increased at the training velocity and at velocities below it but not above it. Similarly, Ewing and colleagues (1990) found that the effects of a training program of 10 weeks on isokinetic torque were specific to the speed of training. Although the mechanism for this speed-specific response is not known, consider again the findings of Behm and Sale (1993), who found a speed-specific adaptation of the torque–velocity relationship not only in subjects who performed the isokinetic movement during training, but also in subjects who intended to make this movement but ended up making an isometric contraction instead.

Explosive-type training enhances the performance of powerful movements. Häkkinen, Komi, and Alén (1985) showed that explosive-type training increased power-related measurements, such as rates of isometric force development during maximal, isometric, ballistic, voluntary contractions. Power measurements such as vertical jump height and drop-jump performance (countermovement contractions) are also enhanced preferentially when the resistance training includes dynamic movement (Häkkinen and Komi 1983a; Colliander and Tesch 1990).

Fatigue Resistance

A commonly held belief is that hypertrophy induces increased strength, but at the expense of fatigue resistance. This is not borne out in the literature.

M.J.N. McDonagh, Hayward, and Davies (1983) measured fatigue of elbow flexors before and after five weeks of isometric training. Fatigue was measured by stimulating the muscle nerve using the Burke protocol. They found that fatigue resistance was significantly increased after training. Similarly, Rube and Secher (1990) found that a program of five weeks of static training using maximal isometric knee extensions increased fatigue resistance, as tested by maximal extensions every fifth second. Interestingly, when the test was performed bilaterally, improvement was seen only in the groups that had trained bilaterally, while improvement on the unilateral fatigue test was evident only in the group that had trained unilaterally with that leg.

Muscle fatigability was also measured by Hortobágyi, Barrier, et al. (1996) in women before and after either concentric or eccentric training of the quadriceps. The fatigue test was 40 maximal isokinetic contractions. Training had no effect on fatigability, although it was demonstrated that concentric exercise was more fatiguing than eccentric.

Animal studies are rare in the strength-training literature, due to the difficulty of finding a model that mimics this type of training. Duncan, Williams, and Lynch (1998) recently studied the effects of a weight-training program in rats; the training consisted of conditioning rats to climb a vertical grid repeatedly with a weight, which became heavier as the animals adapted, attached to their tails. The extensor digitorum longus muscles, which were hypertrophied as a result of the 26-week training program, were slightly more fatigue resistant *in situ* than those of controls.

The Role of Eccentric Contractions

Not much has been said thus far regarding the role of eccentric (lengthening) contractions on the response to resistance training. Clearly, any resistance training using free weights or body weight as resistance and involving dynamic as opposed to isometric contractions includes an eccentric phase. The simple act of lifting and lowering a weight, even if we rigorously control the speed with which these two movements are made (which is not always the case in training studies), is complicated by the fact that recruitment of motor units is different for these two phases of contraction against the same resistance (see the discussion of this point in chapter 2). Thus, to determine the importance of the concentric versus the eccentric phase of the movement in the final training adaptations is no easy task. However, several investigators have conducted fairly controlled studies of the role of eccentric contractions as a training stimulus. Following is a summary of these findings.

Eccentric training results in bigger strength gains than concentric or isometric training with the same external load. Colliander and Tesch (1990) and Dudley and colleagues (1991) found this after comparing a concentric–eccentric training program with a program involving the same number of concentric-only contractions with the same load. Hortobágyi, Barrier, and colleagues (1996) demonstrated the superiority of eccentric to concentric training in optimal strength development by showing that eccentric-training strengths gains were bigger even when the load was maximal for the concentric training groups but submaximal for the eccentric groups. The difference in the efficacy of eccentric and concentric programs for maximal strength development is less evident when strength testing is not specific to the type of training (i.e., when eccentrically trained subjects are tested with concentric contractions), suggesting either muscle morphological or neural effects that are specific to the training mode. Neural effects are discussed in more detail later.

Eccentric training may result in specific muscle morphological changes. Responses to eccentric training are more evident when testing in eccentric mode than when testing in concentric or isometric modes (Seger, Arvidsson, and Thorstensson 1998; Higbie et al. 1996; Hortobágyi, Barrier, et al. 1996; Hortobágyi, Hill, et al. 1996). While this result has been used as evidence for a neural component to resistance training, several investigators have also suggested that the mode-specific training response may be due at least in part to mode-specific changes in muscle morphology.

An acute bout of eccentric contractions, for example, produces a length-dependent deficit in force production that is most pronounced at shorter lengths and that lasts several days. This may indicate a stretching of myofibrillar and myotendinous structures, resulting in a physiological muscle lengthening (Saxton and Donnelly 1996). Whitehead and colleagues (1998) found a similar shift in the force–length curve toward longer lengths for optimal force development after an acute bout of eccentric exercise. They found, in addition, that this effect and the degree of muscle swelling were exacerbated in a group that had trained concentrically for one week prior to the acute experiment. Their conclusion was that the concentric training resulted in shorter sarcomeres, which would render the muscle more sensitive to eccentric exercise–induced damage.

Fridén (1984) reported that after two months of eccentric training involving the knee extensors, there were myriad changes in sarcomere ultrastructure, including a large variation in sarcomere lengths that was more pronounced in the type II fibers. This author proposed that the described changes result in a better stretchability of the muscle fibers, thus reducing the risk of mechanical damage.

Hypertrophic patterns following concentric and eccentric training are different not only in extent, but also in proximodistal pattern. Seger, Arvidsson, and Thorstensson (1998), using magnetic resonance imaging (MRI) to examine hypertrophy of the knee extensors, found that eccentric and concentric training resulted in different hypertrophic patterns. For the eccentrically trained group, muscle girth increased in the distal portion of the muscle only, while in the concentrically trained group, there was a tendency for increased girth only at the muscle midbelly, not distally.

The structural damage caused by eccentric exercise may help explain some of these functional and structural effects that appear to be specific to eccentric exercise. Foley and colleagues (1999) used MRI to examine elbow flexors for up to 56 days following a single bout of eccentric exercise. They found that muscle compartment volume was decreased by 10% two weeks after the exercise and attributed this to a complete and irreversible necrosis of a population of fibers that were most susceptible. According to these investigators, such an occurrence would explain the repeated-bout effect, whereby a second eccentric exercise resulted in less evidence of muscle damage and pain. Thus, fibers less susceptible to damage would be recruited to replace the necrosed fibers during the second bout, with reduced damage. This interesting hypothesis of the repeated-bout effect remains to be substantiated.

Muscle hypertrophy may be more marked with eccentric than with concentric training (Dudley et al. 1991; Hather et al. 1991). In addition, the force–velocity curve changes after eccentric training of the elbow flexors: Its curvature becomes more pronounced, and V_o increases. The biceps becomes faster contracting. These changes should be considered in light of the morphological changes described earlier.

Neural Effects of Resistance Training

CONCEPT

6 Much emphasis has been placed on the so-called neural basis for increased performance following resistance-type training. Some of this evidence is somewhat indirect in that the neural component of the strength increase is arrived at by eliminating other possibilities (increased strength with no change in muscle girth, for example). Some of this evidence, however, is direct in that it is supported by neurophysiological data. In the following sections, I discuss the evidence for a neural component, from the weakest to the strongest evidence.

Difference in Strength vs. Muscle Girth

Strength increases faster than muscle girth during the initial stages of strength training, suggesting that factors other than morphological (i.e., neural) are involved (Ploutz et al. 1994; Higbie et al. 1996; Häkkinen, Kallinen, et al. 1998; Seger, Arvidsson, and Thorstensson 1998; Van Cutsem, Duchateau, and Hainaut 1998). Even in untrained subjects, only about 50% of the variation in quadriceps MVC can be explained by quadriceps cross-sectional area (D.A. Jones, Rutherford, and Parker 1989), and it is difficult to believe that the remainder is due entirely to neural factors. Several investigators feel that even short-term resistance training might influence muscle specific tension, angle of pennation, connective tissue content, and muscle torque–length relationships (J.N. Howell, Chleboun, and Conatser 1993; D.A. Jones and Rutherford 1987; D.A. Jones, Rutherford, and Parker 1989). All these factors, singly or in combination, might produce this apparent discrepancy between increased maximal strength and increased muscle girth.

D.A. Jones, Rutherford, and Parker (1989) proposed that connective tissue attachments might be altered by resistance training, with the result that force expression per fiber increases. Such an adaptation has yet to be demonstrated experimentally. In addition, D.A. Jones and Rutherford (1987) demonstrated evidence of increased quadriceps muscle density after resistance training and proposed that this might be due to either decreased intramuscular fat content or increased myofibrillar packing density, both of which would increase specific muscle tension.

Narici and colleagues (1989) measured cross-sectional area changes in the major knee extensors vastus lateralis, vastus intermedius, vastus medialis, and rectus femoris resulting from a 60-day isokinetic resistance-training program of the knee extensors. Their nuclear magnetic resonance data, which they obtained by examining the cross-sectional area of these muscles at seven places along the femur, demonstrate that the increase in quadriceps cross-sectional area with resistance training is not uniform along its length and that the pattern of increase in girth for the different muscles is not identical (figure 5.12). Thus, we must consider changes in the performance of this muscle group (which is by far the one most studied for resistance-training effects) in the light of possible heterogeneous changes in the length–tension curves, cross-sectional areas, and angles of fiber pull on distal tendon that might occur among the different muscles of this complex group.

Recently, another variable has been added to this issue. Ploutz and colleagues (1994), using contrast shift in MRI, estimated recruitment of fibers during contractions of quadriceps muscles against various submaximal loads. Their results suggest that less absolute muscle cross-sectional area is used to accomplish the same submaximal load after resistance training (for nine weeks, two times per week; figure 5.13). Such a change might imply an increase in muscle specific tension, as opposed to the decreased specific tension proposed by Kawakami and colleagues (1995). This change might also imply a difference in recruitment strategy, such that the same force is accomplished by recruiting fewer motor units at higher firing frequencies.

There are other, more compelling reasons to believe in a neural component to resistance training than comparison of increases in muscle girth and maximal strength.

Specificity of Task

Increases in performance following resistance training are generally most evident when performing the task used during the training. This task specificity has been used as evidence for a neural component in resistance training. For example, eight weeks of dynamic knee-extension training resulted in a 67% increase in 1-RM performance for squats, while isometric knee-extension MVC increased only 13% (Thorstensson and Karlsson 1976). This specificity extends to the velocity at which the training is performed. Caiozzo, Perrine, and Edgerton (1981) and Ewing and colleagues (1990) found that training of knee extensors on an isokinetic device increased torque more at or near the training velocities. Similarly, Narici and colleagues (1989) found that knee-extensor torque was improved at isokinetic speeds at or below, but not above, the speed used for training.

Häkkinen, Komi, and Alén (1985) demonstrated task specificity of response by comparing the adaptations to heavy resistance train-

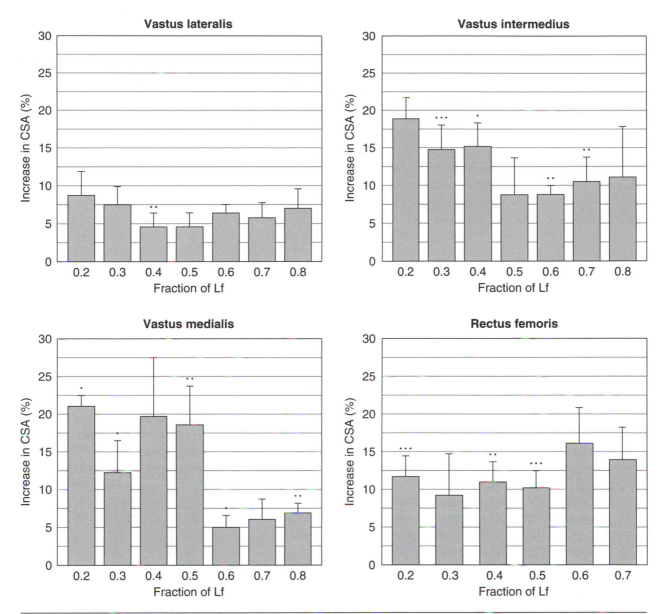

Figure 5.12 Percentage increase in cross-sectional area of four quadriceps muscles resulting from a 60-day training program of isokinetic (2.09 rad/s) knee extensions at various points along the femur's length (Lf), extending from proximal to the greater trochanter (0.2) to near the knee (0.8). Cross-sectional areas were measured using nuclear magnetic resonance imaging. Note the spatial differences in hypertrophic responses among muscles.

Reprinted from Narici et al. 1989.

ing and to explosive training. Their results demonstrated that explosive training (jumping without a load or with a light load) increased not only maximal force and integrated EMG, but also the maximal rates of force development and integrated EMG. While the value of absolute EMG measurements as an index of response to strength training is open to dispute (discussed later), changes in the rate of EMG development,

corrected for the maximal measured value, seem intuitively more meaningful.

The task-specificity phenomenon extends to unilateral versus bilateral tasks. Bilateral training increases bilateral performance more than unilateral performance, while unilateral training increases unilateral performance more than bilateral (Häkkinen et al. 1996). This is another fairly convincing argument for resistance

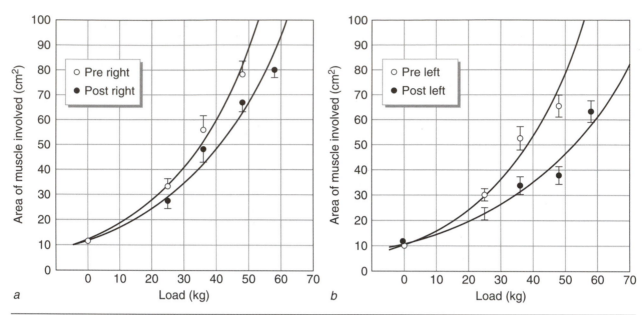

Figure 5.13 Average cross-sectional area of (*a*) right untrained and (*b*) left trained quadriceps recruited during concentric contractions with various loads. Activated muscle area was estimated from elevated spin-spin relaxation times in magnetic resonance images following five sets of 10 repetitions with the load. Symbols indicate measurements before (open circles) and after (filled circles) nine weeks of training of the left quadriceps using knee extensions with heavy weights.

Reprinted from Ploutz et al. 1994.

training–induced changes in the nervous system.

The model of eccentric training provides still another example of possible neural effects during resistance training. We have already seen that eccentric training is superior to concentric training in development of maximal strength, but only when strength is tested using the training mode. For example, eccentric training increases eccentric performance more than concentric or isometric performance, while cross-mode effects (i.e., effects of eccentric training on concentric strength or of concentric training on eccentric strength) are not different between the two training modes (Hortobágyi, Hill, et al. 1996; figure 5.14).

Surface EMG Response During MVC

Many reports have shown an increase in the integrated muscle EMG during an MVC after a resistance-training program. The assumptions are that less than maximal recruitment occurs before training and that reproducible surface EMG measurements can be taken from the same subjects on different occasions separated by several months. The evidence seems quite strong that

recruitment is maximal for most muscles tested in untrained subjects (see chapter 2), raising questions as to the source of this change in maximal EMG with resistance training. However, the frequency of reports of this response renders it difficult to ignore as a physiological adaptation. Several observations seem to support this increased EMG response as a real phenomenon. For instance, this change can occur without a concomitant change in the amplitude of the M wave (Yue and Cole 1992; Van Cutsem, Duchateau, and Hainaut 1998). Also, integrated EMG (iEMG) increases in the contralateral untrained muscle performing an MVC (Narici et al. 1989; Moritani and De Vries 1979). In addition, in subjects who trained with both knee extensors, maximal iEMG is higher after training, but only when tested during the bilateral test (i.e., not when tested unilaterally). Similar specificity of response was found with subjects who trained with only one leg; the increase in iEMG was not seen when the test was with both legs (Häkkinen et al. 1996).

Imaginary Strength Training

In 1992, Yue and Cole published a somewhat controversial paper that demonstrated that

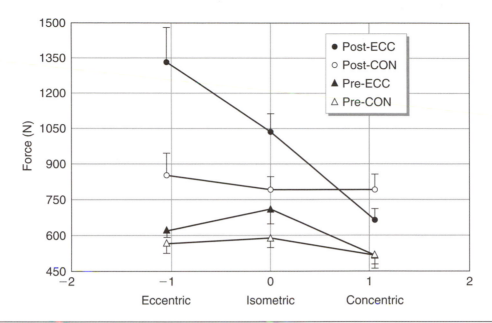

Figure 5.14 Changes in knee-extensor muscle force following eccentric (ECC) and concentric (CON) training. Training comprised 36 sessions over 12 weeks using maximal concentric or eccentric contractions of one leg at a speed of 1.05 rad/s.

Adapted from Hortobágyi et al. 1996.

strength increases could be produced by imagining the training sessions. In this study of the abductor muscles of the fifth digit (abductor digiti minimi), subjects were asked during each session either to perform 15 isometric MVCs or to imagine doing this while keeping the muscle totally relaxed. After four weeks (20 sessions), the group training with the MVCs increased force by 30%, while those who imagined performing the MVCs increased almost as much (22%). Other previously demonstrated effects of strength training were in evidence in this study: increased maximal EMG and an improvement in the contralateral muscle. The authors suggested an effect of training on the motor program, which resulted in more complete recruitment of the muscle (figure 5.15).

This result may be specific to muscles that are normally not used very much in everyday activities. For example, maximal isometric abduction of the little finger is not a normal action, nor is it easy to perform. R.D. Herbert, Dean, and Gandevia (1998) recently attempted to duplicate the result of Yue and Cole (1992), but using the elbow flexors. After eight weeks of imagined isometric training, no effects similar to those reported by Yue and Cole were observed. Furthermore, training, either imaginary

or real, had no influence on the extent of muscle activation during an MVC, which was always near 100% in all groups. These results demonstrate that if there is indeed an effect of imagined training, it does not occur with all muscle groups. This phenomenon will surely be revisited many more times before we can reach definite conclusions regarding its generality and the mechanisms involved.

Contralateral Effects

Contralateral effects, also known as cross-education, refer to the evidence of altered performance of the muscle contralateral to the trained one. A classic and often-quoted demonstration of this phenomenon is the results of Moritani and De Vries (1979), who found increased MVC and maximal iEMG in the biceps brachii contralateral to one subjected to resistance (dynamic weight) training, which became evident by two weeks. There are several other fascinating observations concerning this phenomenon:

1. Resistance training has a contralateral effect on force sensation. This was demonstrated by Cannon and Cafarelli (1987), who found an effect of an isometric training program for one

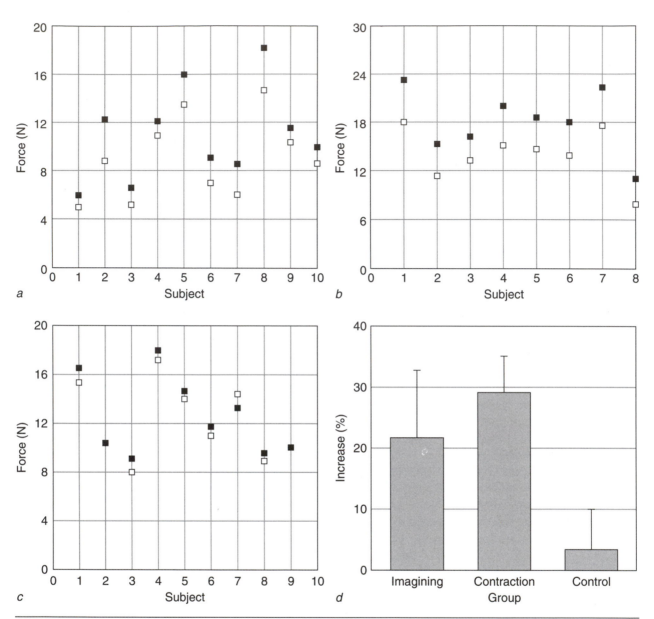

Figure 5.15   Imagining the contractions during a strength-training program can result in increased strength, results from Yue and Cole's (1992) experiment. One group (contraction) trained abductors of the fifth digit for four weeks with maximal isometric contractions, while another group (imagining) imagined performing these contractions during the training period. Individual pretraining (open squares) and posttraining (filled squares) values for maximal isometric force are shown for (*a*) the imagining group, (*b*) the contraction group, and (*c*) a control group. Means values are summarized in *d*. Both imagining and contraction group means were statistically significant.

Reprinted from Yue and Cole 1992.

adductor pollicis on the MVC of the contralateral muscle. MVC of the contralateral muscle increased 9.5% after voluntary training, but not after training using electrical stimulation (both muscles' MVC increased by 15%). They also found a difference between voluntary and electrical-stimulation training in the ability of the contralateral muscle to match trained muscle force, indicating a contralateral effect of voluntary resistance training on force sensation.

2. Häkkinen and colleagues (1996) examined the effects of unilateral and bilateral heavy resistance training of the knee extensors. They found that increases in MVC and maximal iEMG

were higher when the test used was the same as that used during the training. Similarly, Rube and Secher (1990) examined fatigability after five weeks of static strength training with either one-legged or two-legged leg extension that resulted in a significant increase (36% to 59%) in maximal extension torque; they found that fatigability was specific to the task used during the training. For example, if training was with both legs, fatigability was increased with a bilateral, but not a unilateral, test.

3. Contralateral effects extend to the coactivation of agonists and antagonists. Carolan and Cafarelli (1992) found a 32% increase in MVC with the trained leg and a 16% increase with the untrained leg after eight weeks of unilateral isometric training involving 30 MVCs per session. They also found a significant decrease in coactivation of biceps femoris during knee extension in both trained and untrained legs.

4. Ploutz et al. (1994) used MRI to estimate the muscle cross-sectional area used during contractions of the knee extensors (see figure 5.12). Eccentric training decreased the absolute muscle area recruited per unit force in both the trained and the contralateral leg.

5. Seger, Arvidsson, and Thorstensson (1998) found a mode-specific effect of training on the contralateral limb: Significant increases in torque were found only at the isokinetic velocity at which training was performed and only using the training mode (eccentric or concentric).

6. An imagined-resistance training program produced an increased MVC in the abductor muscles of the fifth digit but also increased the MVC of the contralateral muscle (Yue and Cole 1992).

Demonstrations of this contralateral effect will surely continue to appear in the literature, and there is little doubt that it is a real phenomenon, although the underlying mechanisms involved remain elusive.

Decrease in Activation of Antagonists

Carolan and Cafarelli (1992) had subjects train for eight weeks on a task requiring 30 MVCs of knee extensors per session. Effects of the training were evident as early as one week into the training program. Coactivation of biceps femoris decreased in both trained and untrained

muscles during the first week of training. It was estimated that about 33% of the increase in MVC during the first week could be attributed to the decreased coactivation, and this contribution decreased to 10% by the end of the eight-week training period. Thus, MVC continued to increase while coactivation did not change significantly after one week (figure 5.16).

Changes in Motor Unit Recruitment

Milner-Brown, Stein, and Lee (1975) were the first to demonstrate experimentally that strength training results in directly measurable changes in the way motor units are recruited during effort. In their experiment, they determined the degree of synchronization of motor unit recruitment during effort by comparing the pattern of firing of single motor units, measured with intramuscular electrodes, with the pattern of the surface EMG. Generally, their experimental paradigm asked the question: When the single motor unit fires, does the surface EMG show that a lot of other units are firing at the same time? They compared weight lifters with untrained control subjects, and they also studied a group of subjects before and after six weeks of weight training. Their study revealed a higher degree of synchronization of motor units in the first dorsal interosseus muscle in trained subjects than in untrained subjects. They also found that weight training resulted in increased amplitudes of V2 and V3 responses (reflexes evoked during voluntary activation) measured at the muscle in response to stimulation of the peripheral nerve. The V2 and V3 responses—with latencies of about 56 and 83 milliseconds, respectively, following nerve stimulation—represent transcortical reflexes. Thus, increased synchronization of motor units during effort in strength-trained individuals is accompanied by enhanced transcortical reflexes, suggesting that firing synchronization may be linked to a supraspinal mechanism.

A word on synchronization of motor units is warranted at this point. The time course of synchronization that I refer to here is of little consequence to the rapidity of force development: The relatively small differences in the degree of motor unit synchronization ratios between trained and untrained individuals would not translate into differences in power, as one might be inclined to conclude. For instance, a certain

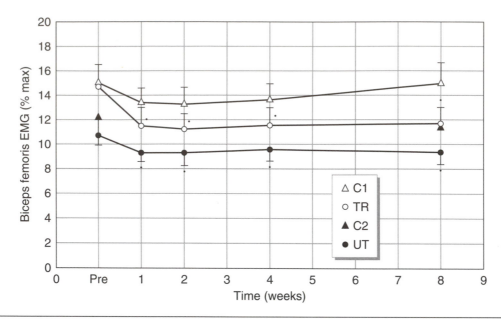

Figure 5.16   Strength training of the knee extensors results in a decrease in coactivation of the knee flexors. Decreased EMG of flexors (expressed as a percentage of maximal flexor EMG) during a maximal extension occurred one week after beginning the training program in the trained leg (TR). A smaller but still significant decreased coactivation was apparent in the nontrained contralateral leg (UT). The training consisted of 30 maximal isometric knee extensions per day, three times per week, for eight weeks.

Reprinted from Carolan and Cafarelli 1992.

degree of asynchrony would be expected in peak force generation among motor units, even if the appearance of their action potential in the EMG were perfectly synchronized, because of differences in contractile properties. Rather, synchronization is seen as an index of the strength of presynaptic influences from common sources on motoneurons, which determine the degree of coincident generation of action potentials in motoneurons within the pool (Semmler and Nordstrom 1998).

Thus, the results of Milner-Brown, Stein, and Lee (1975) suggest an effect of strength training on supraspinal mechanisms that are responsible for the recruitment of motor units, possibly by an enhancement of the efficacy of synapses on motoneurons from supraspinal sources. These original findings have been supported by more recent studies.

Sale, MacDougall, et al. (1983) substantiated the findings of enhanced reflex responses after strength training. They found increased V1 and V2 reflex responses in all muscles trained during the study (which included hypothenar muscles, extensor digitorum brevis, brachioradialis, and soleus) after 9 to 22 weeks of training. Since the

reflexes were evoked while the subjects were performing an MVC (unlike the 1975 study of Milner-Brown, Stein, and Lee, where the effort was submaximal), they suggested that the elevated reflex responses indicated increased excitation of motoneurons during MVC. This again suggests that activation is not maximal in untrained subjects. However one interprets the source of this training-induced effect, it does support the notion that the nervous system responds to strength training.

Semmler and Nordstrom (1998) recently confirmed an increased synchrony of motor unit activation in the first dorsal interosseus muscles of strength-trained subjects and interpreted their findings as an increased corticospinal activity accompanying this task as a result of training. An interesting observation in their study was that skilled individuals (experienced musicians who use their fingers extensively to play their instruments) demonstrated levels of motor unit synchronization, common drive to all motor units, and tremor amplitude during maintained submaximal contractions that were all less than those of controls and strength-trained individuals. For the skilled subjects, the task involved the trained muscles but not the task used in the daily training. For the strength-trained subjects, on the other hand, the task included neither trained muscles nor training task. Research must continue in this area to attempt to ascertain the muscle group specificity versus task-related specificity of these adaptations in the way motor units are recruited.

Recent evidence from Van Cutsem, Duchateau, and Hainaut (1998) provides substantial evidence for a neural effect of training. In their study, they asked subjects to train their ankle dorsiflexors for 12 weeks (five times per week) by moving a load representing 30% to 40% of 1-RM as quickly as possible. At the end of the study, recruitment of motor units during ballistic contractions was examined using intramuscular electrodes. The researchers found that ballistic contractions after the training period were faster, with a more rapid onset of EMG (figure 5.17). They also found that maximal instantaneous firing rates of motor units during ballistic contractions were higher and showed less decrease in frequency after training. In addition, the percentage of motor units showing incidents of doublets (two spikes of the same motor unit separated by five milliseconds or less) increased from 5.2% of the control units to 32.7% of the trained units. They suggested that ballistic-type training causes increased motoneuron excitability, which leads to the previously described changes during voluntary excitation.

Summary

The stimuli for altered muscle protein synthesis during strength training are related to the mechanical stress and strain to which the muscle is exposed during the activity. A number of stimuli are known to be involved, including activation of mitogen-activated protein kinases, protein kinase C, focal adhesion kinase, and phospholipases A and C and the production of IGF-1, prostaglandins, and mitogenic substances from damaged fibers and their infiltrating cells. It is the combination of the effects of these stimuli on transcriptional and translational processes and on activation of degradative enzyme systems that determines the ultimate phenotypic response to strength training. The result is a muscle that is larger and stronger, has more myonuclei, and may have fewer type IIB fibers. Functional changes occur that to a certain extent show specificity to the type of contraction used during the training (eccentric, concentric, or isometric; ballistic or ramp). The extent to

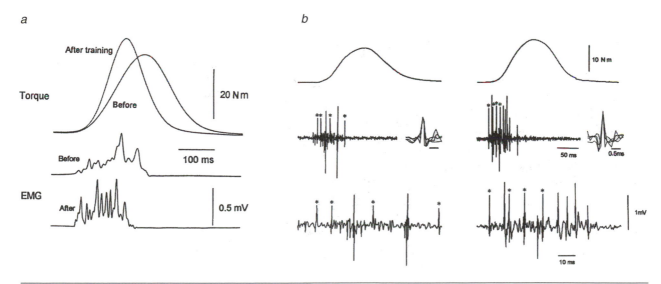

Figure 5.17   (*a*) After 12 weeks of dynamic training of the ankle dorsiflexors (dorsiflexions as fast as possible against a load of 30–40% of 1-RM), ballistic contractions at the same relative torque show a higher rate of tension and rectified EMG rise. (*b*) After training, motor unit firing was different during a ballistic contraction: Initial high firing rates were maintained longer during the contraction.

Reprinted from Van Cutsem, Duchateau, Hainaut 1998.

which these specific changes can be explained by muscle morphological factors as opposed to changes in neural mechanisms of recruitment is currently the subject of much debate. There is some evidence, nonetheless, that central nervous system changes that would be expected to improve performance occur with strength training.

chapter 6

© CORBIS

Neuromuscular Responses to Decrease in Normal Activity

"**B**ind one arm in a sling, and keep it utterly idle for a month, and meanwhile ply the other busily with heavy work, such as swinging a hammer, axe, or dumb-bell, and is it hard to say which will be the healthier, the plumper, the stronger—the live arm, at the end of the month?"

William Blaikie, *How to Get Strong and How to Stay So,* 1879

Most Important Concepts From This Chapter

1 Many of the mechanisms that are altered by increased activity in the form of exercise training are also involved in the phenotypic changes that result from decreased neuromuscular usage. These mechanisms include changes in protein synthesis and degradation and may in some cases involve muscle damage when activity is resumed.

2 In general, muscles that are most affected by disuse are those that are normally used most often. These muscles demonstrate a clear change in phenotype toward faster-contracting muscles with higher proportions of type II fibers.

3 Changes in the nervous system with decreased usage are evident from measurements of decreased function of the neuromuscular junction, decreased maximal motoneuron firing rates after immobilization, and decreased motoneuron excitability following spinal cord transection. The capacity to voluntarily generate a maximal contraction decreases following immobilization.

4 Models of relative disuse reported in the literature range from those of complete muscle inactivation (spinal cord isolation, pharmacological nerve blockade) to those where weight bearing decreases more than activation (hindlimb suspension in rats, spaceflight in rats and humans). Models that vary in extent of change in activation, load bearing, and length-change activity are discussed as to their effects on mass, fiber-type proportions, metabolic enzyme levels, and functional properties.

5 The efficacy of countermeasures to disuse atrophic responses is summarized.

In the previous two chapters, we considered the responses of the neuromuscular system to increases above normal levels in activation patterns that favor the development of either increased endurance or strength. Here we consider the responses of the neuromuscular apparatus to reductions in normal activity.

In a certain sense, the title of this chapter is a misnomer. Many of the models that we consider a reduction in activity per se are not really, if we measure the total integrated electromyographic signal during a 24-hour period and use this as an index of activity. This is the case for certain models of animal immobilization. The EMG of several hindlimb muscles that are fixed in length as a result of immobilization of their joints decreases early in the immobilization period, then returns to a level in the still immobilized muscle that is not different from controls (Fournier et al. 1983), in spite of the atrophic response. Activity involving load bearing by the muscle, on the other hand, is more likely the missing component in many forms of neuromuscular deficit resulting from "reduced" activity, although in joint immobilization, EMG activity would be accompanied by "load bearing" against the resistance of the immobilized joint. Thus, the decrease in normal activity referred to in the title includes decreased activation of the muscle fibers (as we see in spinal isolation or pharmacological nerve blockade), as well as decreased load bearing of the type seen during normal activation (such as in spaceflight and hindlimb suspension in animals). Although some of these models may arguably constitute more of a *change* in normal activity than a *decrease,* the atrophic response that most often results suggests that the change involves a lack of something seen during normal activity. It will become clear in this chapter that we do not know completely what these missing components are. Often, administration of countermeasures that should, in principle, attenuate the atrophic response fails to do so, demonstrating our lack of complete understanding in this area.

General Principles Underlying Neuromuscular Responses to Reduced Activity

CONCEPT

1 I examine several conditions of reduced normal activity in later sections, beginning with those that are most complete (spinal cord isolation and transection and nerve paralysis via pharmacological blockade) and ending with those in which activity may still be present in some form or other (bed rest, hindlimb suspension, and joint immobilization). In each section the effects on the neuromuscular apparatus are presented, and where appropriate, attempts at attenuating the deficit using countermeasures are discussed, primarily where these attempts are designed to provide insight into the missing component responsible for the atrophic responses. Before looking at these conditions, however, several general principles that are applicable to all models are presented in the following sections.

There Are Species Differences in Response to Altered Activity

Before surveying the literature in the area of reduced activity, we must first take into consideration species differences in response to reduced activity as we try to arrive at general conclusions. For example, in cats that have undergone spinal isolation, changes in fiber proportions of hindlimb muscles are less marked than those seen in humans with spinal cord injuries (Round et al. 1993; T.P. Martin et al. 1992; Roy et al. 1992; Pierotti et al. 1994). Similarly, there may be species-related differences in the changes in protein synthesis and degradation

when rats and humans are exposed to reduced activity (Ferrando et al. 1996). There are many more examples, which are mentioned throughout this chapter as they become relevant. Such differences are to be expected, given the variations among the various species used as experimental subjects in normal activity patterns, locomotion patterns (bipedal vs. quadripedal), fiber-type compositions of the corresponding muscles (e.g., human leg muscles appear to be less extreme in fiber-type composition than the corresponding muscles in rats and cats), metabolic rates, relative life spans, and so on.

Atrophic Response Is Most Marked in Muscles Most Normally Used

CONCEPT

2 It makes sense that the most marked atrophic response is in muscles most used, since these are the muscles that experience the most change in their normal activity patterns. Thus, in humans, most of the protein loss is from the leg musculature, as opposed to arm musculature (Ferrando et al. 1996; Leblanc et al. 1992). The most vulnerable muscles appear to be postural muscles that contain a relatively high proportion of type I fibers and that cross a single joint. The least prone to inactivity-related deficits are muscles that are not used in posture, cross more than one joint, and have a relatively high proportion of type IIB fibers (Roy and Acosta 1986; Lieber et al. 1988). Susceptibility of muscles to reduced activity appears to be more related to their function than to their fiber-type composition. For example, in cats with a spinal cord transection, flexors with a high percentage of type I fibers (capsularis, 76% type I fibers) and those with a high percentage of type II fibers (tibialis anterior and semitendinosus, both with more than 90% type II fibers) both atrophy very little (Roy and Acosta 1986).

Passive Stretch During the Reduced Activity Modifies the Atrophic Response

We have seen in the previous chapter the importance of stretch during the hypertrophic response. It is possible that this is the primary event that is lacking in muscles subjected to various models of decreased use. Atrophic responses are almost invariably more severe in situations where the muscle is kept in a shortened position relative to its normal physiological length, such as in limb or joint immobilization (Jokl and Konstadt 1983; Spector et al. 1982; Simard, Spector, and Edgerton 1982). Leterme and colleagues (1994) found that many of the atrophic changes in rat soleus in response to hindlimb unweighting, including losses in muscle mass, decreased twitch and tetanic force, decreased specific tension, shortening of the twitch response, and decrease in type I fiber proportion, were attenuated or even offset completely when the ankle was immobilized in dorsiflexion, thus stretching the soleus. Goldspink (1977b) showed that in denervated muscles, the decrease in protein synthesis and increase in protein degradation are less pronounced when the muscle is stretched. Attempts have been made to attenuate decreased-use atrophy using such interventions as periodic eccentric contractions (Kirby, Ryan, and Booth 1992) and passive cyclical stretching (Dupont-Versteegden et al. 1998; Roy et al. 1992), based on the hypothesis that stretch should stimulate protein synthesis. I discuss these examples further on.

The Muscle Atrophic Response Involves Changes in Protein Synthesis and Degradation

Changes in protein synthesis and degradation in response to reduced activity have been established by measuring these rates *in vivo* (Ferrando et al. 1996; Morrison et al. 1987) or *in vitro* (Goldspink 1977b). The increase in protein degradation with decreased use may be more marked in animals than in humans (Ferrando et al. 1996). The degradation response includes increases in proteolytic enzymes, such as cathepsin D (Loughna, Goldspink, and Goldspink 1987), acid phosphatases and acid proteases (Witzmann, Troup, and Fitts 1982), and ubiquitin (Riley et al. 1992).

Based on previous studies of the time course of change in mRNAs and their associated proteins, Booth and Kirby (1992) proposed that the atrophic response to decreased activity occurs in three major phases (see figure 6.1). Phase I is a rapid decrease in protein synthesis due to changes in trans-

lation. Phase II is characterized by a dramatic increase in protein degradation. Phase III includes a decrease in protein degradation and the attainment of a new muscle steady state, in which synthesis and degradation are decreased from normal values and muscle mass stabilizes in an atrophied state.

The Protein Synthesis Response Is Very Rapid

Decreased protein synthesis is evident in soleus within six hours of hindlimb suspension (Thomason, Biggs, and Booth 1989) and hindlimb immobilization (Watson, Stein, and Booth 1984; Booth and Seider 1979) in the rat. The mRNA coding for alpha-actin is reduced in soleus after one day of hindlimb suspension (G. Howard, Steffen, and Geoghegan 1989).

Protein Transcriptional Changes Are Apparent

The mRNAs coding for alpha-actin and for several unidentified small-molecular-weight proteins are substantially reduced in soleus

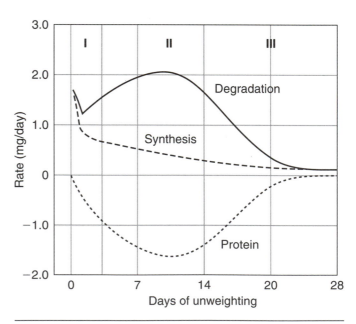

Figure 6.1 Time course of changes in rates of protein synthesis, degradation, and accumulation/loss during absence of weight bearing. Note the achievement in phase III of an equilibrium in synthesis and degradation in a state of atrophy.

Reprinted from Booth and Kirby 1992.

muscles of rats during hindlimb suspension (G. Howard, Steffen, and Geoghegan 1989). In addition, with this same species, model, and muscle, after four weeks, changes in mRNAs for MHC type I (decrease) and types IIx and IIb (increase) mirror the changes in their corresponding proteins (Haddad et al. 1998). Following five days of spaceflight, alpha-actin mRNA is decreased in several extensors (Thomason et al. 1992). In bilateral hindlimb immobilization in rats, the mRNAs for MHCs change toward faster isoforms in soleus, plantaris, and gastrocnemius (Jänkälä et al. 1997), and there is a significant decrease in mRNA for cytochrome c (Booth et al. 1996).

Posttranscriptional Changes Contribute to the Atrophic Responses

After seven days of hindlimb suspension in the rat, soleus contains significantly less RNA, indicating a decreased capacity for translation (G. Howard, Steffen, and Geoghegan 1989). This decreased RNA also occurs after 10 days of unilateral leg unloading in vastus lateralis in humans (Gamrin et al. 1998). In addition, mRNA for MHC I is unchanged from control levels after seven days, in spite of a significant decrease in protein synthesis, thus implicating posttranscriptional mechanisms (Thomason, Biggs, and Booth 1989). Immobilization of the rat soleus in a shortened position results in a reduction in RNA per milligram of muscle (Goldspink 1977b). Ribosomal efficiency, expressed as protein synthesis per microgram of RNA, decreases in this situation. The effect of protein synthesis by posttranscriptional mechanisms is very evident during the first six hours following immobilization, at which time protein synthesis decreases with no corresponding change in mRNA levels (Morrison et al. 1987; Watson, Stein, and Booth 1984).

Protein Response to Decreased Use Is Heterogeneous

It is safe to say that there is no typical phenotypic response to a period of decreased muscle usage. Even though there are overall changes in protein synthesis and degradation, it is also clear that not all proteins respond the same way. This is particularly clear from the work of Morri-son and colleagues (1987), who looked at protein synthesis in the gastrocnemius–plantaris muscle group of the rat following bilateral immobilization of the legs with muscles in a shortened position. They used constant infusion of labeled leucine to allow measurement of the synthesis rates of several proteins. After seven days of immobilization, they found that the synthesis rates of actin, cytochrome c, and mixed proteins were 33%, 81%, and 67%, respectively, of control values. This result probably demonstrates the variety of signals or various sensitivities to these signals that are responsible for controlling synthesis of different proteins. Supporting this idea are the results of Booth and colleagues (1996), who measured mRNA coding for cytochrome c in muscles of rats with both hindlimbs immobilized. They found that after seven days of immobilization, the decreases in cytochrome c mRNA in both tibialis anterior and soleus were similar when these muscles were immobilized in shortened and stretched positions, in spite of the much less marked atrophy occurring in the stretched condition.

Motoneuron Innervation May Influence the Atrophic Response to Decreased Use

CONCEPT

3 The influence of motoneuron innervation might seem obvious, but in fact the motoneuron most likely exerts a subtle effect on the properties of the muscle, known classically as the trophic influence, which is separate from the influence imparted via activation of the muscle fibers. The most obvious example of the trophic influence is the increase in extrajunctional acetylcholine receptors that occurs within days of muscle denervation, which is less marked when complete activity is abolished, but the motor nerve is left intact, via pharmacological nerve block with tetrodotoxin (TTX; Pestronk, Drachman, and Griffin 1976). The implication from this type of experiment is that substances are secreted by nerve terminals that influence certain properties of muscle fibers, since pharmacological blockade of action potential propagation does not interfere with axonal transport or basal secretion of these substances. The identity of the substances involved in this trophic control is not yet known, nor has their role in inactivity-related atrophy been demonstrated unequivocally. This trophic influence, if indeed present, may influence only a few muscle properties. In fact, many properties of TTX-inactivated and denervated muscles are very similar (Gardiner, Favron, and Corriveau 1992; R.N. Michel, Parry, and Dunn 1996).

H.L. Davis and Kiernan (1981) discovered that an extract from rat sciatic nerve, when injected into denervated rat hindlimb muscles, significantly attenuated the atrophic response. In general, their calculations determined that, following denervation, 60% of the loss in muscle mass was due to disuse and 40% to the loss of motoneuron-derived trophic support. Furthermore, the effect appeared to be fiber-type specific, in that the atrophic response of IIA fibers was similar with or without injection of the extract, while IIB fibers showed less atrophy in the presence of the extract. The results of this rather rudimentary experiment are still intriguing. Whether or not such a trophic influence that affects muscle mass will ever be found is unknown.

Reduced Activity Influences Motoneurons and Their Afferents

Neuromuscular junctions demonstrate a variety of degenerative changes with reduced activity (Fahim 1989; Fahim and Robbins 1986; Pachter and Eberstein 1984; Malathi and Batmanabane 1983). Neuromuscular junction safety factor may be compromised during immobilization, as evidenced by an increased incidence of "jitter" (i.e., increased variability in stimulus–response latency across the junctions) after four weeks of leg immobilization in humans (Grana, Chiou-Tan, and Jaweed 1996). Maximal firing frequencies of motoneurons during MVC decrease immediately after six weeks of immobilization (Duchateau and Hainaut 1990), perhaps due to decreased excitability (Sale et al. 1982). The result is difficulty in performing a maximal contraction (Vandenborne et al. 1998; White, Davies, and Brooksby 1984; Koryak 1998, 1999; Duchateau 1995). Decreased intrinsic excitability (Hochman and McCrea 1994b) and a transient increase in efficacy of Ia synapses on motoneurons (Baker and Chandler 1987b; Mendell, Cope, and Nelson 1982) have been demonstrated distal to a lesion in the spinal cord of the cat.

Models of Decreased Neuromuscular Usage

CONCEPT

4 The following sections consider several of the major models of reduced neuromuscular activity and the major adaptations that occur in each case. In all but one model, the amount of available pertinent information warrants subheadings. The discussion includes atrophic responses, concerning the relative degree of atrophy among muscles; metabolic enzyme changes; fiber-type changes, concerning primarily adaptations in myosin expression; muscle functional changes; and other noteworthy observations. When there is sufficient literature available on the subject and that information sheds some light on the components that are absent from the affected muscle, a section on attempts to attenuate the atrophic responses is included.

Pharmacological Blockade of the Motoneuron

The pharmacological blockade model allows the estimation of at least short-term neuromuscular responses to decreased activity that would occur in other models, since this model represents zero activity, as confirmed by EMG recordings before and after the blockade (St.-Pierre and Gardiner 1985). The procedure that has been used most frequently is blockade of the motor nerve with tetrodotoxin (TTX), a sodium conductance channel blocker. The muscles distal to the blockade are thus inactivated, since motor impulses from the spinal cord motoneurons cannot pass the blockade. TTX is usually delivered to the nerve via an osmotic pump and Silastic cuff system, and periods of paralysis for up to four weeks have been studied.

There are limitations to this model of decreased activity. Although it appears that bulk anterograde axon transport is not influenced by TTX (Lavoie, Collier, and Tenenhouse 1976; Pestronk, Drachman, and Griffin 1976), this has not been studied in very much detail. In addition, there is no control over muscle length or the excursions in passive tension in the inactive muscles that would occur as the animal moves its paralyzed limb about the cage. Goldspink (1977a) showed quite clearly the effect of passive stretch *in vivo* by showing that denervated muscles atrophy less if they are maintained in a stretched state. Finally, the dimensions of the delivery system limit paralysis to a relatively short (four-week) period of time.

Atrophic Responses

The responses to TTX-induced inactivation are consistent in the literature. Unlike most other models of decreased usage, the slow soleus does not preferentially atrophy in this model (R.N. Michel et al. 1994; St.-Pierre and Gardiner 1985). Within mixed fast muscles, such as the medial gastrocnemius or plantaris, fiber atrophic responses are more pronounced in type II than in type I fibers, which is once again not consistent with the atrophic pattern found in other models. In comparing the model of hindlimb unweighting with the TTX model, it appears that the response of the soleus in the two conditions is more similar than is that of mixed fast muscles, which are more atrophied in the TTX condition (R.N. Michel et al. 1994; Gardiner, Favron, and Corriveau 1992). Comparing the atrophic responses of TTX blockade with those of other models invites the conclusion that the small amount of residual activity remaining in models such as limb immobilization, unweighting, or spaceflight might be sufficient to attenuate atrophy in fast muscles but not enough for the slow soleus.

Metabolic Enzyme Changes

The major enzyme change noted, besides the obvious loss in total muscle protein, is a decrease in concentrations of several energetic enzymes. Enzymes studied to date that demonstrate a decrease with pharmacological nerve blockade include hexokinase, phosphorylase, alpha-glycerophosphate dehydrogenase, pyruvate kinase, lactate dehydrogenase, beta-hydroxyCoA-dehydrogenase, citrate synthase, malate dehydrogenase, succinate dehydrogenase, and myofibrillar ATPase (St.-Pierre et al. 1988; Seburn, Coicou, and Gardiner 1994; R.N. Michel et al. 1994; Turcotte, Panenic, and Gardiner 1991; figure 6.2).

Fiber-Type Changes

In soleus and mixed fast ankle extensors, the proportions of fibers containing only one myosin heavy chain decrease, and proportions of hybrid fibers increase (R.N. Michel, Parry, and Dunn 1996). There is also *de novo* appearance of embryonic/neonatal myosin heavy chain in a large proportion of fibers (Cormery et al. 1999; R.N. Michel, Parry, and Dunn 1996; figure 6.3).

Muscle Functional Changes

The fiber atrophic response is reflected in decreased tetanic force in fast muscles, whereas in soleus a decrease in specific tension also occurs (Spector 1985). The twitch becomes potentiated, nonetheless, with a corresponding increase in the ratio of twitch to tetanic force (P_t/P_o ratio), and also demonstrates a longer time course (Gardiner, Favron, and Corriveau 1992; St.-Pierre and Gardiner 1985; table 6.1 and figure 6.4). These changes in twitch time course and P_t/P_o appear to result from changes in calcium release and uptake mechanisms (S. Howell, Zhan, and Sieck 1997). At the same time, fast and slow muscles become faster contracting: The rate of rise of tetanic force increases, as does the velocity of unloaded shortening in the case of soleus (St.-Pierre and Gardiner 1985; Gardiner, Favron, and Corriveau 1992). These whole-muscle changes are reflected at the level of the muscle unit; units of all sizes and types experience the same changes (Gardiner and Seburn 1997).

Fatigue resistance appears to be either unchanged or slightly increased following TTX-induced reduction in activity (St.-Pierre et al. 1988; Gardiner and Seburn 1997). Certainly, the full range of variety of motor unit types, based on fatigue resistance using the standard Burke protocol, is not significantly altered (Gardiner and Seburn 1997). This property is difficult to measure in this model, due to the changes in time course of the twitch, which significantly alter the force–frequency response. As a result of this change, the measured fatigue resistance at the same subtetanic frequency for control and TTX-paralyzed muscles implies that TTX-paralyzed muscles contract at a higher proportion of their maximal tetanic force during the protocol. Thus, these paralyzed muscles are more than likely highly fatigue resistant compared with their control counterparts, considering the absolute amount of contractile work that the muscle is capable of during the fatigue protocol.

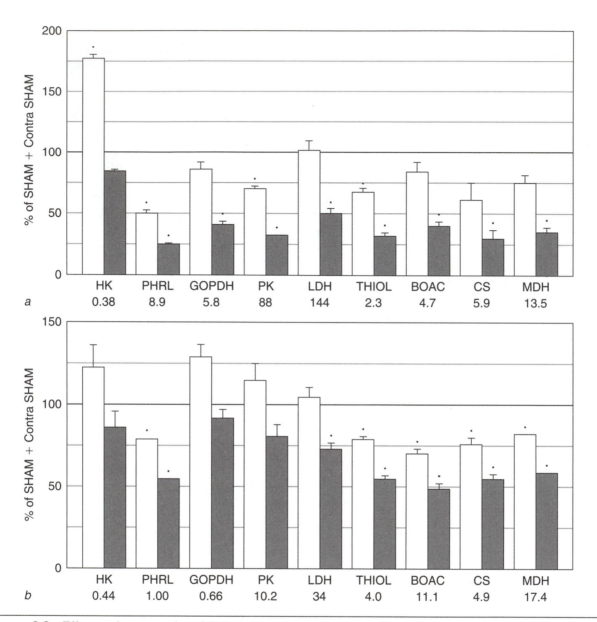

Figure 6.2 Effects of two weeks of TTX-induced muscle paralysis on enzyme activities (*a*) in muscle homogenates of rat plantaris and (*b*) from fiber bundles of soleus. Enzyme activities are expressed per unit of dry weight (filled bars) and per whole muscle (open bars). Dots denote a significant difference from control mean. Control values are normalized to 100%: Thus, values beneath this line indicate a decrease, while those extending above the line indicate an increase, with TTX.

Reprinted from Michel et al. 1994.

Other Noteworthy Observations

It should be noted that most of the effects of TTX-induced disuse also occur in cats subjected to prolonged (three weeks) barbiturate anesthesia and a similar duration of spinal isolation (C.J.F. Davis and Montgomery 1977).

TTX was used in one of the first experimental demonstrations of a possible retrograde effect of muscle status on properties of the innervating motoneuron. Czeh and colleagues (1978) found that TTX application to the soleus nerve of the cat for eight days produced a decrease in the afterhyperpolarization (AHP) duration of the motoneurons innervating soleus. More important, they found that when they electrically stimulated the nerve distal to the cuff with a chronic, low-frequency pattern, they prevented this decrease in motoneuron AHP

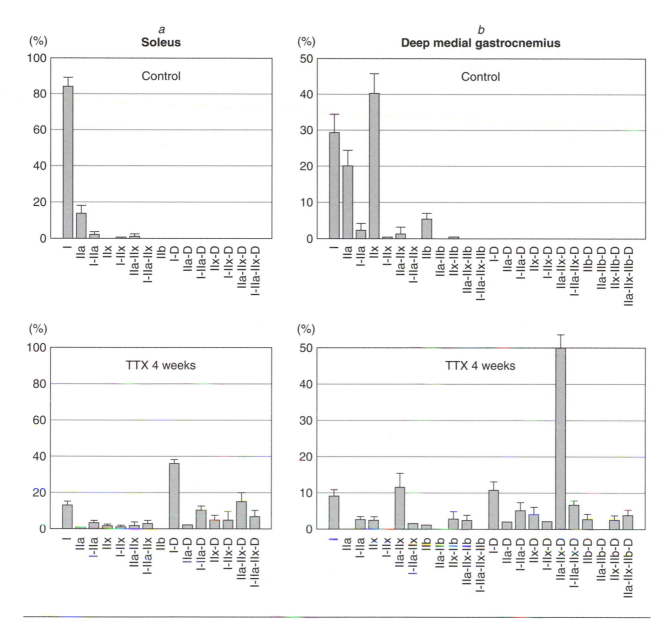

Figure 6.3 Effects of four weeks of muscle paralysis using chronic nerve application of TTX on MHC proportions of fibers (identified using immunohistochemistry) in (*a*) soleus and (*b*) deep medial gastrocnemius. Both muscles show more hybrid fibers and the expression of developmental (D) MHC.

Reprinted from Cormery et al. 2000.

duration. They were not able to duplicate this effect when stimulation was proximal to the TTX cuff (i.e., between the TTX cuff and the spinal cord). With distal but not proximal stimulation, muscle atrophy was attenuated. These results suggest that the motoneurons are sensitive, via a retrograde signaling effect, to the innervated muscle's atrophic or activity state. This effect is presumably not coming from the muscle via afferents, since afferents were blocked by TTX in the case of stimulation dis-

tal to the cuff and were stimulated antidromically (with no effect) when stimulation was proximal to the cuff and when there were no evoked contractions of the soleus. This mechanism may be more important for slow postural muscles, since the results of Czeh and colleagues (1978) concerning the change in AHP duration have not been duplicated when examining muscles that are more heterogeneous in muscle fiber-type composition (Gardiner and Seburn 1997).

Table 6.1
**Effects of Two Weeks of TTX-Induced Muscle Inactivation
on Contractile Properties of Rat Medial Gastrocnemius**

	Control ($n = 11$)	TTX-treated ($n = 9$)
Muscle wet weight (g)	599 (22)	**316 (19)**
Twitch tension (g)	143 (11)	138 (11)
Tetanic tension (g)	889 (63)	**227 (17)**
Twitch/tetanus ratio	0.16 (0.01)	**0.50 (0.02)**
Twitch time to peak (ms)	15.0 (0.5)	**17.3 (0.8)**
Twitch half-relaxation time (ms)	10.6 (0.7)	**17.1 (1.5)**
Max. tetanic dP/dt (%P_o/ms)	3.2 (0.1)	**5.5 (0.3)**
Fatigue index	34 (5)	42 (5)

Significant differences from control mean are shown in boldface. Numbers in parentheses represent SE.

Reprinted from Gardiner, Favron, and Corriveau 1992.

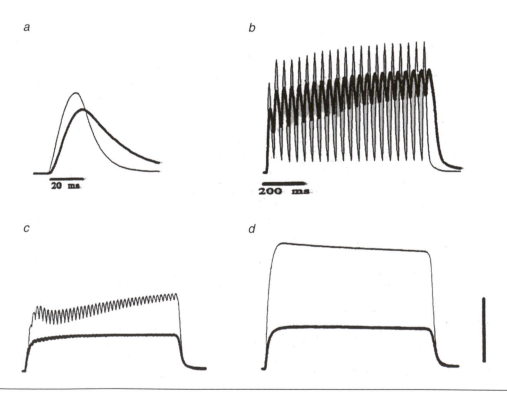

Figure 6.4 Effect of two weeks of TTX-induced paralysis on isometric *in situ* contractile properties of rat gastrocnemius. The thicker line in each recording shows results after TTX paralysis. Vertical calibration bar (bottom right) represents 150 g for (*a*) twitch and (*b*) 25-Hz contractions, and 450 g for (*c*) 50-Hz and (*d*) 200-Hz contractions. Note that twitch is slightly longer and contractions are more fused at subtetanic frequencies in the TTX condition. Note also that twitch force is less affected than tetanic force.

Reprinted from Gardiner, Favron, and Corriveau 1992.

The end plate is also influenced by disuse. After eight days of TTX-induced paralysis, activity of the enzyme acetylcholinesterase (AChE) is reduced by 52% in both soleus and extensor digitorum longus of the rat. Interestingly enough, AChE activity is further reduced in extensor digitorum longus when its nerve is sectioned, but not in soleus (Butler, Drachman, and Goldberg 1978). This shows that, as with relative degree of muscle atrophy, inactivation is more like denervation in soleus than in other muscles, probably because of the soleus' higher dependence on activity for the maintenance of its structure and function.

Paralysis with TTX for two weeks also results in an increased homonomous, but not heteronomous, EPSP amplitude (Gallego et al. 1979b). This change is due to alterations in afferent connections and not to intrinsic properties such as motoneuron excitability, which does not appear to change in this model.

Spinal Isolation

Spinal isolation (SI) causes neuromuscular inactivation as complete as does TTX nerve blockade (Roy et al. 1992; Steinbach, Schubert, and Eldridge 1980; Pierotti et al. 1991). There are fundamental differences between the two models. In spinal isolation, in which the spinal cord is transected above and below the motoneurons of interest and the associated dorsal roots are cut, motoneurons are deprived of supraspinal and afferent input, as well as some interneuron influence, depending on the level of the lesion (Mendell, Cope, and Nelson 1982). In TTX blockade, supraspinal influences are intact. The extent to which this difference influences neuromuscular responses is unknown. In fact, the responses of the same species to the two conditions are quite similar.

Atrophic Responses

It is probable that the degree of atrophy seen following short periods of SI is similar to that seen with TTX. SI gives us the opportunity to consider more long-term effects of neuromuscular inactivity, and some of them are not exactly what one might expect. Atrophic responses are more pronounced in SI than after spinal cord transection, probably because in spinal cord transection, afferents are intact and

hyperreflexia results in some muscle activation (Alaimo, Smith, and Edgerton 1984; Hochman and McCrea 1994b; Grossman et al. 1998).

Generally, muscles with more slow fibers atrophy more than muscles with more fast fibers. Interestingly, this does not seem to be the case for TTX paralysis, which suggests either a short-term versus long-term difference or a difference due to an intact versus interrupted supraspinal influence. The former suggestion is favored, since the time course of atrophy of all muscles in rats with SI is similar at two weeks, but slow muscles continue to atrophy after this, while fast muscles stop atrophying and may even increase slightly in weight (Grossman et al. 1998). Extensors atrophy more than flexors, and (by extrapolation) slow extensors atrophy more than fast extensors. At the fiber level, it appears that the predominant fiber in the muscle atrophies the most. Fiber sizes become more similar after SI, suggesting that fibers have a minimal size that they can attain as a consequence of complete inactivity (Roy et al. 1992).

Metabolic Enzyme Changes

After long-term (six months) SI, succinate dehydrogenase (SDH), as measured using quantitative histochemistry, decreases in fibers of medial gastrocnemius (Jiang et al. 1991; Jiang, Roy, and Edgerton 1990; figure 6.5), with no change evident in soleus fibers. It is worth recalling that SDH also decreases in gastrocnemius fibers with TTX paralysis. On the other hand, the activity of the enzyme glycerol-3-phosphate dehydrogenase (GPDH) increases in fibers of medial gastrocnemius (MG) and soleus (Roy et al. 1996; Graham et al. 1992; Jiang et al. 1991) and tibialis anterior (Pierotti et al. 1994). This increase accompanies the change in phenotype of fibers to become more like type II. In soleus, myofibrillar ATPase, measured biochemically, increases (Roy, Talmadge, et al. 1998).

Fiber-Type Changes

All cat hindlimb muscles examined so far have increased proportions of type II fibers after six months of SI (Roy et al. 1992; figure 6.6). The ankle extensors medial gastrocnemius and soleus show a dramatic increase in the expression of MHC II; by six months of SI, all fibers in medial gastrocnemius contain MHC II. In soleus,

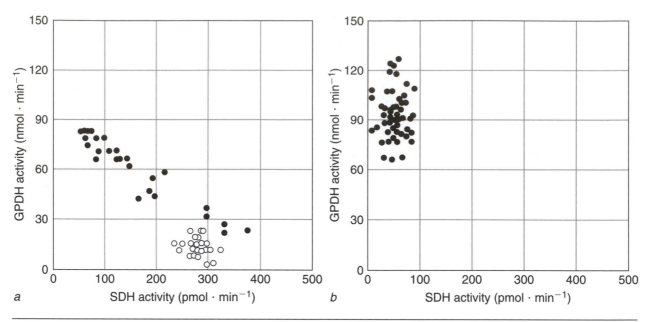

Figure 6.5 Plot of enzyme activities for glycerol-3-phosphate dehydrogenase (GPDH) and succinate dehydrogenase (SDH) in single fibers from medial gastrocnemius muscles of (*a*) a control cat and (*b*) a cat subjected to spinal isolation six months earlier. Filled circles and open circles represent type II and I fibers, respectively.

Reprinted from Jiang et al. 1991.

conversion is somewhat less complete, with only 30% to 45% of the fibers expressing MHC II after six months. Both muscles demonstrate an increased incidence of hybrid fibers. The changes seen in cat soleus are similar to those seen in rats that have undergone SI (Grossman et al. 1998). In the ankle flexor tibialis anterior, on the other hand, the loss of activity has less marked consequences: Only slightly fewer type I fibers are present six months after SI, with few hybrid fibers and very little atrophy present (Pierotti et al. 1994).

Muscle Functional Changes

Knowledge is limited about the changes in muscle functional properties of animals that have undergone SI. In soleus, contractile property changes appear to reflect the histochemical changes of the muscle: Twitches become shorter in duration, V_{max} is elevated, and the muscle is less fatigue resistant (Roy, Pierotti, et al. 1998; Roy et al. 1996). The potentiated twitch and elevated twitch/tetanic ratio that are seen with TTX-induced paralysis are not present in the SI condition, suggesting that the biochemi-

cal basis for this change (which is most likely related to altered sarcoplasmic reticulum function) is regularized with longer periods of inactivity.

Many of the functional changes in fast muscles caused by SI and spinal cord transection are probably similar. More data are available regarding spinal cord transection, which is considered in a later section.

Attempts to Attenuate the Atrophic Response

Roy and colleagues (Roy, Pierotti, et al. 1998; Roy et al. 1992) administered 30 minutes of passive stretch per day to cat hindlimb muscles during the six months after SI. Their technique involved positioning the cat on a treadmill and manipulating the leg to simulate motion of the knee and ankle joints normally seen during voluntary locomotion, at a rate of one step per second. This procedure reduced soleus weight loss by 12%, twitch shortening by 9%, P_o decrease by 21%, and V_{max} increase by 21% (figure 6.7). The weights of other hindlimb muscles examined were not significantly attenuated by this procedure.

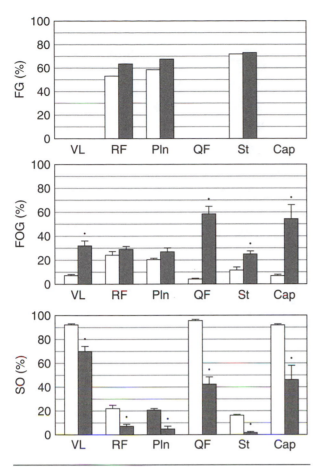

Figure 6.6 Relative contribution of each fiber type to the whole-muscle cross section in control cats (open bars) and cats with spinal isolation (filled bars). Dots indicate a significant difference from controls. FG, FOG, and SO represent type II fibers with low (FG) or high (FOG) oxidative enzyme activities, and type I (SO) fibers. Muscles are vastus intermedius (VI), rectus femoris (RF), plantaris (Pln), quadriceps femoris (QF), semitendinosus (St), and capsularis (Cap).

Reprinted from Roy et al. 1992.

Other Noteworthy Observations

It would be intuitively satisfying to find that all fibers in a muscle of mixed histochemical profile became the same after SI. This might tell us that activity plays a major role in creating the heterogeneity of muscle fiber and motor unit properties that is evident within mixed muscles. As it turns out, a significant degree of heterogeneity remains among muscle fibers following six months of complete inactivity resulting from SI (see figure 6.6), suggesting that a significant proportion

of this heterogeneity is not activity dependent (Pierotti et al. 1994).

Spinal Cord Transection

Since animals become hyperreflexic following spinal cord transection (ST), there is a small amount of residual activity in hindlimb muscles (Alaimo, Smith, and Edgerton 1984). Nonetheless, the atrophic responses are, for the most part, quite comparable with those seen in SI, although slightly less severe (Grossman et al. 1998; Talmadge, Roy, and Edgerton 1996b). There is currently more literature available from ST experiments, including effects on single motor units, properties of motoneurons, and afferent and motoneuron efficacy, than there is from SI experiments. ST is particularly interesting because (1) it more closely mimics the human condition of spinal cord injury than does SI and (2) the intact reflexes allow the study of the role of reflex-generated locomotor activity in attenuating the atrophic responses. As we saw in chapter 4, this model has been instrumental in demonstrating how the spinal cord, even when deprived of supraspinal influences, can "learn" rudimentary motor tasks.

Atrophic Responses

Extensors (triceps surae, plantaris, flexor hallucis longus, quadriceps femoris, and vasti), but not flexors (tibialis anterior, extensor digitorum longus, semitendinosus, capsularis), are atrophied six months following ST in cats (Roy et al. 1999; Roy and Acosta 1986). Similarly, in rats after short (Dupont-Versteegden et al. 1998) and long (Lieber, Fridén, et al. 1986) periods after ST, soleus atrophies markedly, while extensor digitorum longus is relatively unaffected in mass. Atrophy of up to 56% of fiber area occurs within six months of spinal cord injury in humans (Castro et al. 1999). A general observation is that atrophy is related more to the function of the muscle than to its fiber composition (Roy and Acosta 1986).

Metabolic Enzyme Changes

In vastus lateralis, gastrocnemius, and soleus muscles of patients with complete transection of the spinal cord 10 months to 10 years previously, phosphofructokinase tended to be higher and SDH lower than control levels. Note that this

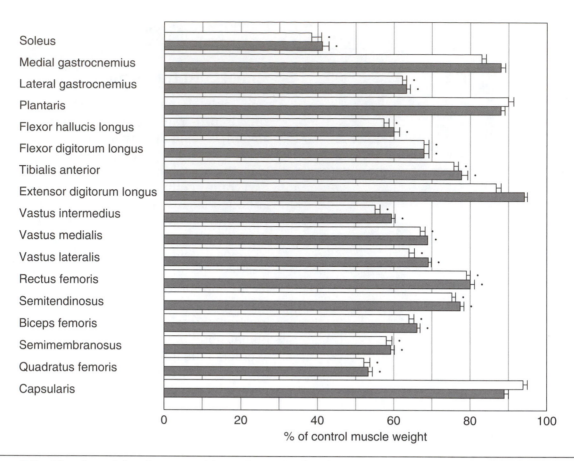

Figure 6.7 Muscle weights in cats subjected to spinal isolation six months previously (open bars) and spinal isolation plus passive limb manipulation (filled bars). Dots indicate significant differences from control values.

Reprinted from Roy et al. 1992.

trend is similar to that seen in SI (increased glycolytic, decreased oxidative enzymes). However, for shorter periods after injury in humans (up to six months), glycolytic and oxidative enzyme activities may actually increase (Castro et al. 1999). In humans, myofibrillar ATPase, measured using quantitative histochemical techniques, is reduced in tibialis anterior 2 to 11 years postinjury, as is fiber SDH (to 48% and 67% of controls in type I and II fibers, respectively; T.P. Martin et al. 1992).

Fiber-Type Changes

At longer periods following ST, all muscles tend to revert to a pure type II fiber composition, leading to the idea that type II is the default fiber type in the absence of activity. For example, in humans up to 10 years following transection and in rats after 1 year, fiber composition of the muscles, including those (like soleus) that normally have a high percentage of type I, is converted to exclusively or almost exclusively type II (G. Grimby et al. 1976; Lieber, Fridén, et al. 1986; T.P. Martin et al. 1992; Round et al. 1993; Shields 1995; Talmadge, Roy, and Edgerton 1999). Before conversion is complete—for example, in cats six months after ST—muscles show a decrease in type I and increases in type IIA (for slow muscles) or IIB (for fast muscles; Roy et al. 1996; figure 6.8). During conversion from type I to II, fibers pass through a period during which multiple forms of MHC are expressed, as in the case of SI (Talmadge, Roy, and Edgerton 1996b).

Muscle Functional Changes

Generally, the contractile properties of fast muscles, either extensors or flexors, appear to be influenced somewhat less by ST than slow muscles, although both tend to speed up (Roy

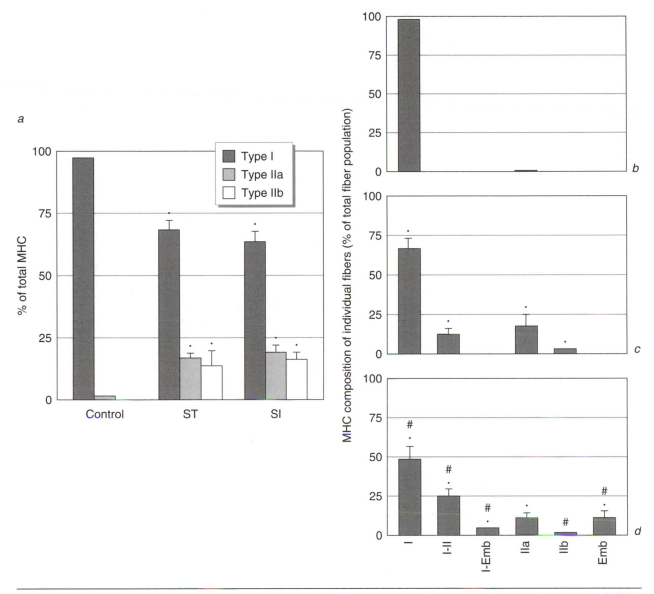

Figure 6.8 (*a*) Myosin heavy-chain composition in soleus muscles from control, spinal transected (ST), and spinal isolated (SI) cats, based on densitometric scans of SDS-PAGE MHC gels. Asterisks indicate values that are different from control values. Fiber composition in soleus of (*b*) control, (*c*) ST, and (*d*) SI rats based on MHC immunohistochemistry. Dots indicate values different from control values, and # indicates SI values different from those for ST soleus.

Reprinted, by permission of Wiley-Liss, Inc., a subsidiary of John Wiley & Sons, Inc. from R.R. Roy et al., 1996, "Neural influence on slow muscle properties: Inactivity with and without cross-reinnervation," *Muscle & Nerve 19*: 704-714.

et al. 1984; Lieber, Johansson, et al. 1986; L.A. Smith, Eldred, and Edgerton 1993; Roy, Talmadge, et al. 1998; Gerrits et al. 1999). Before histochemical conversion to type II fibers is complete, fatigue resistance either remains unchanged or decreases slightly, with a greater tendency for decrease in soleus, which is in keeping with histochemical changes (L.A. Smith, Eldred, and Edgerton 1993; Roy, Talmadge, et

al. 1998). In people with long-term spinal cord injury, muscle fatigability decreases markedly in soleus (Shields 1995) and even in fast muscles such as quadriceps (Gerrits et al. 1999) and tibialis anterior (Stein et al. 1992).

The study of motor unit adaptations gives us more detailed information regarding responses to ST. As one might expect, twitches become faster both in soleus and medial gastrocnemius

(L.A. Smith, Eldred, and Edgerton 1993; Cope et al. 1986). There is evidence of an overall decrease in average muscle unit fatigue resistance in medial gastrocnemius, with a decrease in type FR and a corresponding increase in FF muscle units (Munson et al. 1986). In soleus, on the other hand, muscle unit twitches do not show significant decreases in their fatigue resistance, in spite of changes toward a fast phenotype (Cope et al. 1986).

ST has been used to demonstrate how the relationships between the properties of muscle fibers within a motor unit and the motoneuron that innervates it are maintained when whole-muscle properties change. For example, as outlined earlier, the distribution of muscle units changes in cat medial gastrocnemius, with the appearance of more fast-fatiguing units at the expense of those that are more fatigue resistant. The properties of the motoneurons innervating these units change correspondingly, such that the normal motoneuron–muscle unit property relationships are maintained (Munson et al. 1986). One motoneuron property that changes with ST is a shortening of the afterhyperpolarization (AHP) (Hochman and McCrea 1994b; Czeh et al. 1978; Cope et al. 1986; Munson et al. 1986), which seems intuitively appropriate for a motoneuron innervating muscle fibers that are generally becoming faster contracting. Gallego and colleagues (1979a) showed that immobilization of the soleus of ST cats in a lengthened position prevented the shortening of the AHP that occurred with ST alone and suggested that information regarding the atrophic state of the muscle was somehow communicated to the innervating motoneuron, possibly via retrograde mechanisms. Other models have also presented evidence of such a retrograde mechanism, operating through mechanisms other than via afferents (Foehring and Munson 1990; Czeh et al. 1978). Thus, substances related to muscle functional state are probably taken up by endocytosis at nerve terminals and transported to motoneuron cell bodies, where the appropriate motoneuron changes are orchestrated. The current thought is that, after ST, the decrease in nerve activity causes muscle changes but also that changes in muscle properties probably influence the properties of nerve through a retrograde mechanism. A feedback system like this has been proposed, involving neurotrophins produced by active muscles, taken up by specific receptors in the motoneuron, and influencing motoneuronal gene expression (Funakoshi et al. 1995).

Attempts to Attenuate the Atrophic Response

The nature of the lesion in the cat allows investigators to administer regular activity, in the form of weight support or treadmill walking, in an attempt to determine the factors that promote atrophic responses in this model. Treadmill exercise for 30 minutes per day, five times per week, for a period of six months following ST attenuates the loss of mass and tension-generating capacity and the increase in MHC II of slow soleus, with little effect on fast extensor gastrocnemius or flexor tibialis anterior (Roy, Pierotti, et al. 1998; Roy et al. 1999). Some scattered attenuating effects of the exercise were also noted on fiber-type changes in several hindlimb muscles (Roy and Acosta 1986). Weight-support training for the same period of time as the treadmill walking appears, however, to be more beneficial for medial gastrocnemius, but also causes an increased V_{max} and decreased fatigue resistance compared with ST alone. Clearly, the factors involved in producing the muscle responses to ST are multidimensional and are not restricted to merely a quantitative decrease in muscle activation (figure 6.9).

Electrical stimulation of paralyzed muscles for 24 weeks attenuates some of the atrophic effects, as one might expect. Stimulation of one paralyzed tibialis anterior for up to eight hours per day in human subjects normalized the decreased muscle fiber SDH, increased type I fiber proportions, and increased fatigue resistance (T.P. Martin et al. 1992).

Other Noteworthy Observations

The spinal cord transection of cats has also supplied us with information concerning the electrophysiological responses of alpha-motoneurons to this procedure. It appears that changes are modest and are not always consistent, due to the varying times at which measurements have been taken after the spinal insult and to the various or unidentified muscles that the motoneurons innervate. It has been reported, in fact, that motoneuron electrophysiological properties are not altered following ST (Baker and Chandler 1987a). In general, where

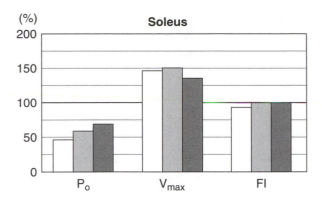

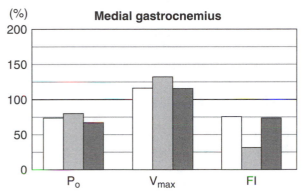

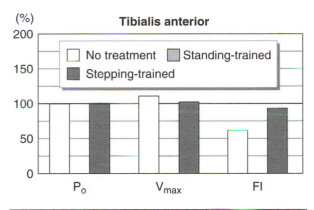

Figure 6.9 Summary of changes in three cat hindlimb muscles after spinal cord transection six months earlier, with no supplemental treatment, daily standing training, and step training.

Reprinted, by permission of Wiley-Liss, Inc., a subsidiary of John Wiley & Sons, Inc., R.R. Roy et al., 1999, "Differential response of fast hindlimb extensor and flexor muscles to exercise in adult spinal cats," *Muscle & Nerve* 22: 240.

AHP duration tends to decrease. In addition, passive membrane properties show that motoneurons exhibit decreased excitability, as evidenced by a number of indices: increased rheobase current (Cope et al. 1986); an increase in the measured threshold voltage V_{th} (Hochman and McCrea 1994b); an increase in the product of rheobase and input resistance (rheobasic voltage; Cope et al. 1986), which is an estimate of V_{th}; and decreased cell capacitance (estimate of total membrane area) with no change in cell input resistance (Hochman and McCrea 1994b). Interestingly, changes not consistent with this slow-to-fast conversion are motoneuron size, which decreases, according to electrophysiological measurements (Hochman and McCrea 1994b; Gustafsson, Katz, and Malmsten 1982), and axon conduction velocity, which decreases slightly in fast muscle motoneurons (Hochman and McCrea 1994b) but increases slightly for soleus motoneurons (Cope et al. 1986).

ST also results in a transient increase in the amplitudes of homonymous and heteronymous composite EPSPs (Baker and Chandler 1987b; Hochman and McCrea 1994a; Mendell, Cope, and Nelson 1982). This may be due to the sprouting of existing inputs, activation of latent ineffective synaptic contacts, or both. Mendell, Cope, and Nelson (1982) demonstrated that the increased EPSP amplitude was more marked and longer lasting when the transection was closer to the motoneurons of interest and proposed that there was sprouting of afferents onto synaptic sites on the motoneuron surface that had been vacated by interneurons injured by the transection. The hyperreflexia seen in ST animals and in humans with spinal injuries is related to this increased EPSP amplitude and is apparently not due to increased motoneuron excitability, based on the preceding information.

Hindlimb Unweighting and Spaceflight

The rat hindlimb suspension model (or hindlimb unweighting, HU) reduces weight bearing, with a significant amount of residual EMG activity present in the muscle after an initial decrease (Blewett and Elder 1993; Alford et al. 1987). The procedure involves preventing weight bearing by the hindlimbs by suspending them using either a tail cast or a body harness. The response of the soleus is more similar

changes are reported, they appear consistent with an increase in the proportion of FF and decreases in FI and FR units in fast muscles such as the medial gastrocnemius, and a decrease in S and corresponding increase in F units in soleus. For example, as pointed out previously,

to that in the complete inactivity models of ST, SI, and TTX blockade than are the responses of the fast ankle extensors (R.N. Michel and Gardiner 1990), suggesting that phasic movements in the unweighted condition supply sufficient stimulus to attenuate or prevent atrophy in fast muscles but not in slow muscles. HU has been used as a model for spaceflight and appears to evoke very similar atrophic responses, based on studies that have compared the two models directly (Ohira et al. 1992; Chi et al. 1992). The experiments with HU are more numerous than those involving rats in space, for obvious practical reasons. The results from the space experiments are limited by the fact that there is usually a period of several hours between the return of the animals to weight bearing and the collection of muscle samples, which might influence the results.

The research in this area is often contradictory and inconsistent and includes various durations of HU and spaceflight and, in the spaceflight, various periods of subsequent weight support before the harvesting of samples, all of which tend to be somewhat overwhelming to anyone trying to arrive at general conclusions regarding the effects of these conditions. I therefore often arrive at global interpretations that may not always agree with all the literature on the subject, a regrettable oversimplification. I include together the principal results from HU and spaceflight experiments in rats. In the section "Other Noteworthy Observations" on page 192, I present, among other things, the limited information currently available on neuromuscular responses in rhesus monkeys and humans to spaceflight.

Atrophic Responses

Atrophic responses are most evident in slow muscles such as the soleus and vastus intermedius and are less evident in fast extensors. At the other extreme, atrophy is much less, and may in fact be absent, in ankle flexors and in the IIB regions of the ankle extensors (T.P. Martin, Edgerton, and Grindeland 1988; R.N. Michel and Gardiner 1990; Gardetto, Schluter, and Fitts 1989; Winiarski et al. 1987; Roy et al. 1987). Decreased concentration of RNA is found in both soleus and gastrocnemius after seven days of HU (Babij and Booth 1988). The extensor digitorum longus, on the other hand, has been re-ported even to hypertrophy within the first week of HU, most likely due to the muscle's abnormal action against the resistance of the hanging foot (G. Howard, Steffen, and Geoghegan 1989). Because of the severe and consistent atrophic responses that occur in the soleus during HU and its importance as a postural muscle, it has received by far the most attention in the research literature.

Metabolic Enzyme Changes

In soleus, biochemical changes include increases in both SDH and in the glycolytic enzymes GPDH, pyruvate kinase, and phosphofructokinase (T.P. Martin, Edgerton, and Grindeland 1988; Chi et al. 1992; Graham et al. 1989; Hauschka, Roy, and Edgerton 1987), although citrate synthase (CS) has been reported to decrease after seven days of HU (Fell, Steffen, and Musacchia 1985). In fast ankle extensors, SDH, CS, and cytochrome c decrease, while GPDH either remains unchanged or increases (Fell, Steffen, and Musacchia 1985; Savard et al. 1989; Jiang et al. 1992). These changes seem to be consistent with a histochemical change in soleus from type I toward type IIA (which has higher oxidative and glycolytic enzyme activities) and in fast muscles from I/IIA toward IIB. Myofibrillar ATPase, measured biochemically, is increased in soleus after seven days in space (T.P. Martin, Edgerton, and Grindeland 1988). Mitochondrial changes, if they do occur, may be pathway specific. For example, the capacity to oxidize long-chain fatty acids was significantly reduced (by 37%) in both red and white vastus after nine days of exposure to zero gravity, while the capacity to oxidize pyruvate and the activities of the mitochondrial enzymes CS and 3-hydroxyacyl-CoA dehydrogenase were unchanged in the same muscles (Baldwin, Herrick, and McCue 1993). Mitochondria in the subsarcolemmal region preferentially decrease in content, with evidence of autolytic degeneration (Riley et al. 1992).

Fiber-Type Changes

Responses of soleus and red regions of antigravity muscles are characterized by a decrease in the proportions of pure type I fibers and an increase in proportions of fibers containing MHC II and hybrids (McDonald, Blaser, and Fitts 1994; Talmadge, Roy, and Edgerton 1996a;

Caiozzo et al. 1996; Ishihara et al. 1997; figure 6.10). These changes are reflected in quantitative changes in the MHC proteins and their mRNAs (Haddad et al. 1993; Stevens et al. 1999). In addition, some fibers in the soleus of rats that have undergone HU synthesize MHC IIx *de novo,* since there is virtually none in control soleus muscles (Talmadge, Roy, and Edgerton 1995). The slow-to-fast transitions appear to occur in

the order MHC Iβ → MHC IIa → MHC IIx → MHC IIb. In rabbits, and to a lesser degree in rats, the alpha form of MHC is transiently expressed between MHC Iβ and MHC IIa (Stevens, Sultan, et al. 1999; Stevens, Gohlsch, et al. 1999). One stimulus for the *de novo* expression of fast MHCs in soleus may be the production of the myogenic regulatory factor MyoD, which appears to promote MHC IIb transcription, during the first few days of HU (Wheeler et al. 1999). No notable changes occur in the fiber proportions of the fast muscles medial gastrocnemius or tibialis anterior (Roy et al. 1987).

Muscle Functional Changes

As expected from the histochemical changes, soleus demonstrates a faster twitch response, a potentiated twitch (as measured by an increased P_t/P_o), and faster V_{max} and V_o (Winiarski et al. 1987; M.E. Herbert, Roy, and Edgerton 1988; Gardetto, Schluter, and Fitts 1989; Pierotti et al. 1990; Diffee et al. 1991; McDonald and Fitts 1993; table 6.2). Interestingly, only a portion of type I fibers in soleus demonstrate the increased V_{max}, indicating that type I fibers do not all have the same sensitivity to this condition and giving more support to the idea that there are more than one species of type I fibers (Gardetto, Schluter, and Fitts 1989). Another consistent finding is a reduced specific tension in the soleus (Winiarski et al. 1987; Pierotti et al. 1990; Widrick and Fitts 1997; Toursel, Stevens, and Mounier 1999), which has been proposed to occur due to a decrease in cross-bridge number per unit of cross-sectional area (McDonald and Fitts 1995). A decrease in fatigue resistance of the soleus after 15 days of HU has been reported (McDonald, Delp, and Fitts 1992). As for fast muscle responses, gastrocnemius shows an elevated P_t/P_o (Pierotti et al. 1990) and a slightly decreased fatigue resistance (Winiarski et al. 1987).

Attempts to Attenuate the Atrophic Response

CONCEPT

5 Table 6.3 presents a summary of the effects of various countermeasures on the weight loss seen in the soleus as a result of HU. The results tend to suggest that weight support for two hours per day is sufficient to reduce the atrophy by about half. Increasing the intensity of contractions during this period by

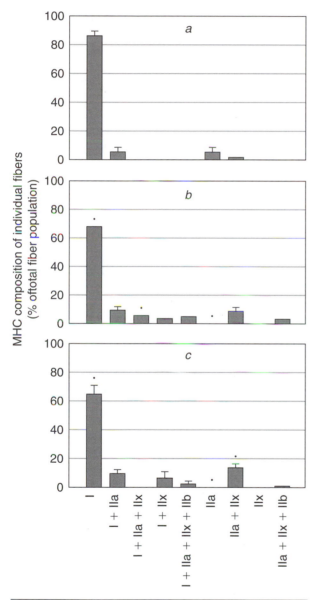

Figure 6.10 Fiber-type percentages, determined using immunohistochemistry, in rat soleus (*a*) from control, (*b*) after 14 days of hindlimb suspension, and (*c*) after 14 days in space. Dots indicate differences from corresponding control value.

Reprinted from Talmadge, Roy, and Edgerton 1996.

Table 6.2
Effects of 28 Days of Hindlimb Suspension on Rat Soleus and Plantaris Muscle Fibers

	Soleus	Plantaris
Muscle mass, mg	−45%	−12.1%
Twitch tension, g	−49.1%	−42.9%
Tetanic tension, g	−59.3%	−23.6%
Specific tetanic tension, g/cm^2	−28.8%	−15.3% (ND)
Twitch time to peak, ms	−34.7%	−19.1%
V_{max}, fiber lengths/s	+36.2%	−0.2% (ND)
V_o, fiber lengths/s	+34.6%	(not measured)

ND indicates not significantly different from control value.

Reprinted from Diffee et al. 1991.

Table 6.3
Effects of Countermeasures During Hindlimb Unweighting on Degree of Soleus Muscle Atrophy

Source, duration of HU	Countermeasures	Atrophy attenuation
Hauschka, Roy, and Edgerton 1987, 4 wk	Rats dropped from a height of 58 cm, 10 times/d	0%
Thomason, Herrick, and Baldwin 1987, 4 wk	Ground support, 2 h/d	37%
	Ground support, 4 h/d	37%
	Uphill running, 1.5 h/d, 20 m/min, 30% grade	48%
Shaw et al. 1987, 4 wk	Treadmill, up to 1.5 h/d, 33.5 cm/s, 30% grade	50%
Hauschka et al. 1988, 1 wk	Treadmill, 5 m/min, 4 × 10 min/d, 19% grade	46%
M.E. Herbert, Roy, and Edgerton 1988, 1 wk	Grid climbing, 32 times/d, load of 75% body weight	43%
Graham et al. 1989, 4 wk	Treadmill, 1.5 h/d, 20 m/min, 30% grade	57%
Pierotti et al. 1990, 1 wk	Treadmill, 0.2 m/s, 19% grade, 10 min every 6 h	26%
Kirby, Ryan, and Booth 1992, 10 d	Eccentric resistance training via electrical stimulation, 4 × 6 reps every 48 h	76%

Source, duration of HU	Countermeasures	Atrophy attenuation
D'Aunno et al. 1992, 1 wk	Ground support, 4 × 15 min/d	61%
	Acceleration to 1.2 g, 60 min/d	39%
Linderman et al. 1994, 5 d	Grid climbing, 3 × 10 reps, 50% body weight attached to tail	7%
	Above exercise plus two daily injections of growth hormone	14%
Diffee et al. 1993, 4 wk	Isovelocity contractions, 4 × 10 reps every 48 h	33%
	Isometric contractions, 5 × 10 reps every 48 h	44%
Bangart, Widrick, and Fitts 1997, 2 wk	Weight bearing, 4 × 10 min/d	16%
Widrick and Fitts 1997, 2 wk	Grid climbing, 40 times/d with load of 500 g	40%
D.L. Allen, Linderman, Roy, Grindeland, et al. 1997, 2 wk	Grid climbing, 10 reps with 40% attached body weight, 3 times/d	65%
	Daily injections of GH/IGF-1	0%
	Both of the above combined	100%
Canon et al. 1998, 3 wk	Chronic stimulation, 10 Hz, 8 h/d	29%
Naito et al. 2000, 8 d	60 min of hyperthermia immediately before HU	38%

Attenuation of atrophy was calculated from muscle wet weights and is expressed as [(HU-T − HU)/(C−HU)] × 100, where HU-T indicates the value for the HU-and-treatment group and HU is that of the HU-only group. Thus, 20% attenuation indicates that the soleus of the HU-and-treatment group lost only 80% of the weight lost in the HU group.

adding grid climbing, treadmill exercise, or eccentric exercise seems generally to have added beneficial effect. Muscle mass loss can be partially attenuated by 40 to 50 evoked slow-velocity or isometric contractions every two days during HU. Treadmill exercise seems more effective than grid climbing, suggesting that moderate, rhythmic tension excursions spread over a long period of time are what is lacking in the soleus in rats undergoing HU. Chronic stimulation for eight hours per day has a modest effect, suggesting that metabolic activity alone is not the missing component.

Interestingly, a 60-minute period of heat stress immediately before the HU period has a significant attenuating effect on the degree of soleus atrophy, presumably by increasing the muscle levels of heat-shock proteins, which subse-quently reduce the decrement in protein synthesis that occurs during the first few days (Naito et al. 2000). The effects of a combination of regular periods of mild heat stress and exercise on muscle atrophic responses during longer periods of decreased use have yet to be investigated, but the results so far appear promising.

The data in table 6.3 concern only soleus mass. Other changes in soleus that are attenuated with countermeasures include the decreased stiffness (Widrick and Fitts 1997), the decreased myofibrillar protein concentration and specific tetanic tension (Pierotti et al. 1990; Widrick and Fitts 1997; Kirby, Ryan, and Booth 1992), the increase in apoptotic myonuclei (Allen, Linderman, Roy, Grindeland, et al. 1997), the decreased content of type I myosin (Diffee et al. 1993; Thomason, Herrick, and Baldwin

1987), the decrease in twitch time course (Pierotti et al. 1990; Leterme and Falempin 1994), the decreased calcium sensitivity (Bangart, Widrick, and Fitts 1997), and the decrease in myonuclei (Allen, Linderman, Roy, Grindeland, et al. 1997).

And what about fast muscles? It appears that 30 daily ladder climbs against a load can have a slight but significant attenuating influence on the loss of mass in gastrocnemius and that this effect is amplified when rats are also treated daily with growth hormone (Linderman et al. 1994, 1995). The combination of exercise and growth hormone treatment (GH alone or in combination with IGF-1) may prove a powerful recipe for attenuation of atrophy due to decreased weight support.

Similarly, during 10 days of HU, the 16% loss in muscle mass of plantaris is virtually eliminated when rats are given 30 minutes of daily treadmill walking up an incline (Wehring, Cal, and Tidball 2000).

Other Noteworthy Observations

Researchers examined human vastus lateralis responses to 5 and 11 days of spaceflight (Zhou et al. 1995; Edgerton et al. 1995). Their principal finding was that the proportion of type I fibers slightly but significantly decreased (from 49% to 39%), with a corresponding increase in type II fibers. In addition, the magnitude of the atrophic response was in the order IIB > IIA > I, with atrophy in the range of 16% to 36%. They also found an increased GPDH/SDH ratio in type I fibers. These results are consistent with rat experiments as far as the tendency for type I to II conversion, the corresponding increase in the glycolytic enzyme GPDH, and the rapidity of the atrophic response. The atrophic responses among the fiber types are not consistent with data from rat experiments, most probably reflecting a species difference. Interestingly, changes in fiber sizes and SDH activity of the medial gastrocnemius, tibialis anterior, and soleus of a rhesus monkey after a spaceflight of 14 days were minor (Bodine-Fowler et al. 1992).

Allen, Linderman, Roy, Bigbee, and colleagues (1997) have demonstrated that apoptosis, or programmed cell death, may be up-regulated in muscles after two weeks of HU. In soleus, they found that the severe atrophy was accompanied by a reduction in the number of myonuclei per millimeter of fiber length (and thus per muscle)

of 17%. They also estimated apoptotic activity by measuring the number of myonuclei demonstrating double-stranded DNA fragmentation, using deoxynucleotidyl transferase (TDT) histochemical staining. After two weeks of HU, the number of TDT-positive myonuclei increased 15-fold in soleus. A significant increase was seen even at three days, indicating that increase in apoptosis, like loss in mass, is an early event.

Spaceflight and HU appear to render muscles more susceptible to damage when weight support is resumed. One reason for this may be a disproportionate loss of thin filaments that apparently occurs under microgravity conditions (Riley et al. 2000). This has been demonstrated in the soleus muscles of astronauts following 17 days in space. In general, slow fibers are more susceptible than fast fibers during the period after reloading subsequent to HU (Vijayan, Thompson, and Riley 1998) and exhibit eccentric contraction–like sarcomere lesions, such as breakage and loss of myofilaments, Z-line streaming, connective tissue tearing, and thrombosis of the microcirculation (Riley et al. 1992; Krippendorf and Riley 1994; Riley et al. 1996).

In spite of muscle changes with HU and spaceflight, there is apparently no effect on the size or SDH activity of the innervating motoneurons (Ishihara et al. 1996).

Unilateral Lower-Limb Suspension and Bed Rest in Humans

Unilateral lower-limb suspension (ULLS) is an attempt to eliminate weight bearing of one limb whereby one ankle is supported by a shoulder strap and the subject walks with crutches (figure 6.11). The results of this procedure on neuromuscular structure and function have been studied for periods of ULLS ranging from six days to six weeks. The general pattern of atrophy, as measured using MRI techniques, is knee extensors 16%, knee flexors 7%, soleus 17%, and gastrocnemius 26% (Hather et al. 1992; table 6.4). Note the difference in the relative responses of soleus versus gastrocnemius between humans and rats, which is probably a reflection of the greater similarity in fiber composition of these two muscles in humans.

Changes in fiber composition (in vastus lateralis) have not been found for periods of up to four weeks. While the activity of the mitochondrial enzymes has been shown to decrease slightly but significantly (18%) after four weeks, no detectable effect on muscle fatigue resistance has been found (Berg et al. 1993). There is some evidence that muscle specific tension may decrease. For example, the cross-sectional area of quadriceps activated during the same absolute load is increased after ULLS (Ploutz-Snyder et al. 1995; figure 6.12). Similarly, the maximal knee extensor torque at all isokinetic velocities decreases more than the decrease in cross-sectional area of the quadriceps (Dudley et al. 1992; Ploutz-Snyder et al. 1995). While these findings are consistent with the decreased specific tension interpretation, they may also indicate an altered pattern of recruitment as a result of ULLS. For example, maximal EMG may decrease (Dudley et al. 1992), and the deficit in torque shows a very rapid recovery phase following resumption of weight bearing (Berg and Tesch 1996).

The strong evidence of contralateral influences of one limb on the other (see chapter 5) raises the question of whether or not this phenomenon might serve to attenuate the atrophic responses to ULLS.

Bed rest is another particularly interesting human model with which to study the effects of prolonged inactivity. Periods ranging from 1

Figure 6.11 Unilateral lower-limb unloading model of Berg and colleagues.

Reprinted from Berg et al. 1991.

Table 6.4
Changes in Cross-Sectional Areas in the Thigh Before and After Six Weeks of Unilateral Lower-Limb Suspension

Region	Area presuspension, cm²	Area postsuspension, cm²	Percentage change
Total thigh	175 (14)	173 (15)	–2
Total muscle	93.7 (9.7)	83.1 (9.1)	–12*
Knee extensors	38.3 (3.9)	32.3 (3.6)	–16*
Knee flexors	27.3 (7.3)	25.4 (7.4)	–7*
Vastus lateralis	12.2 (0.8)	10.2 (1.0)	–16*
Vastus intermedius	10.4 (0.9)	9.0 (0.9)	–14*
Vastus medialis	12.9 (0.2)	11.0 (1.1)	–15*
Rectus femoris	5.4 (0.4)	5.2 (0.3)	–2

Data are means, with SE in parentheses.

* Significant differences ($p < 0.05$)

Reprinted from Hather et al. 1992.

to 17 weeks have been studied. Many of the findings confirm what has been found in other models. For example, most-used muscles (leg and back muscles) atrophy more than least-used muscles (arm muscles), and extensors more than flexors (Ferrando et al. 1996; Leblanc

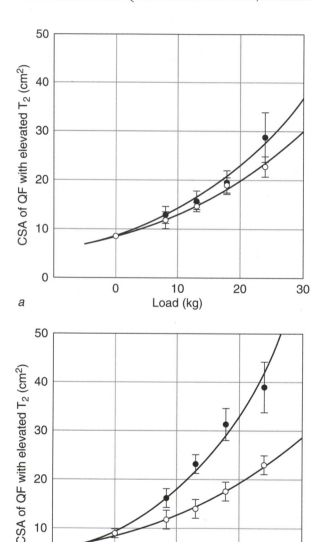

Figure 6.12 Average cross-sectional area (CSA) of (*a*) right weight-bearing and (*b*) left unweighted quadriceps femoris (QF) recruited during concentric contractions with various loads. Activated CSA was estimated from elevated spin-spin relaxation times in MRIs following five sets of 10 repetitions with the load. Symbols indicate before (open circles) and after (filled circles) five weeks of unweighting of left quadriceps.

Reprinted from Ploutz-Snyder et al. 1995.

et al. 1992). As with the ULLS model, there is no evidence for a major difference in susceptibility of soleus versus other ankle extensors to atrophy during bed rest (Hikida et al. 1989). Soleus type I fibers show an increased V_o, even with no change in MHC complement (Widrick, Romatowski, Bain, et al. 1997). There is some evidence of a loss in specific tension in vastus lateralis (Berg, Larsson, and Tesch 1997; Larsson et al. 1996). A major component of the torque decrease after bed rest may be due to neural mechanisms, by which the subject experiences a decreased ability to maximally recruit the muscle (Berg, Larsson, and Tesch 1997; Duchateau 1995; Koryak 1999; figures 6.13 and 6.14).

One particular study (Duchateau 1995) of the effects of bed rest on the triceps surae demonstrated quite clearly the involvement of a neural component (figures 6.15 and 6.16). This particular study described the responses of a single subject and examined maximal isometric torque and activation (by looking at maximal EMG and twitch interpolation) of the triceps surae (TS). The 46% decrease in voluntary isometric torque was attributed to a 19% decrease in muscle force-generating capacity and a 33% decrease in central activation. This was substantiated by

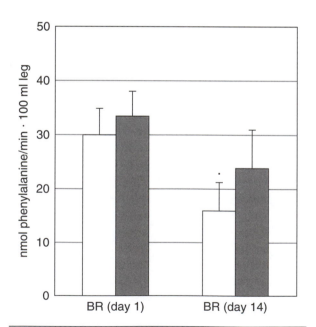

Figure 6.13 Bed rest (BR) for 14 days results in decreased skeletal muscle protein synthesis (filled bars) and slightly decreased breakdown (open bars).

Reprinted from Ferrando et al. 1996.

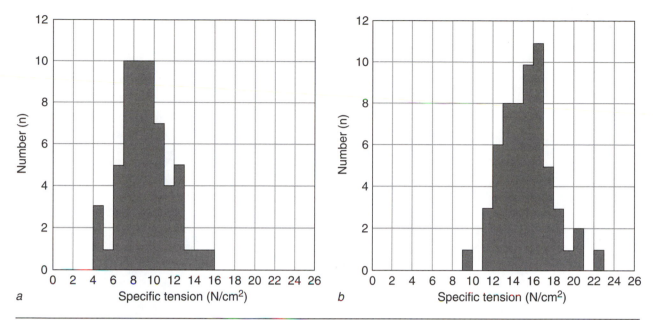

Figure 6.14   Bed rest for six weeks results in a decreased specific tension in quadriceps muscle fibers; *a* and *b* represent samples taken before and after bed rest, respectively.

Reprinted from Larsson et al. 1996.

a rapid recovery of voluntary torque during the two weeks following bed rest. As we shall see in the next section on joint immobilization, reduced ability to maximally recruit muscles may be a general feature of neuromuscular systems deprived of a minimal level of activity.

Joint Immobilization

Major differences between joint immobilization and the other models already considered have implications for the adaptations that are seen. First of all, not unlike the model of hindlimb suspension, there is a considerable amount of recorded EMG activity in the muscles surrounding an immobilized joint, depending on the muscle length at immobilization, with more activity in muscles immobilized in a lengthened position (Fournier et al. 1983; Edgerton et al. 1975). Second, the factor that probably most singles out this model is that muscles are fixed in length during the period of immobilization. This allowed Tabary et al. (1972) to demonstrate that muscles adapt in the immobilized state by adjusting the number of sarcomeres in series in an attempt to retain the normal sarcomere spacing. Furthermore, it allows us to see the effect of initial stretch, or lack of stretch, on protein synthesis and degradation profiles. Finally, even if muscle activation as measured by

residual EMG is similar in immobilization and hindlimb suspension, the fixation of the joint during immobilization means that forces generated during the activation are different: Forces are isometric and may also be higher, since resistance to the contraction is high, being limited only by the strength of the immobilizing material. Because of these considerations and the variability of angles at which joints can be immobilized, the responses recorded in the literature are quite variable. However, it is quite clear that the responses to immobilization seem to be less severe than to the other models of reduced activity already discussed, unless the muscle is immobilized in a shortened position.

Atrophic Responses

As in other models of decreased use, in joint immobilization the most vulnerable muscles appear to be postural muscles with an abundance of type I fibers that cross a single joint (Lieber et al. 1988). Generally speaking, type I fibers atrophy more than type II (Edgerton et al. 1975; Sargeant et al. 1977; St.-Pierre and Gardiner 1985; Gibson et al. 1987). The worst atrophic responses occur when the muscle is immobilized in a shortened position (figure 6.17). Under these conditions, decreases in mixed protein synthesis and specific protein synthesis are evident within six hours (Goldspink 1977b; Watson, Stein, and

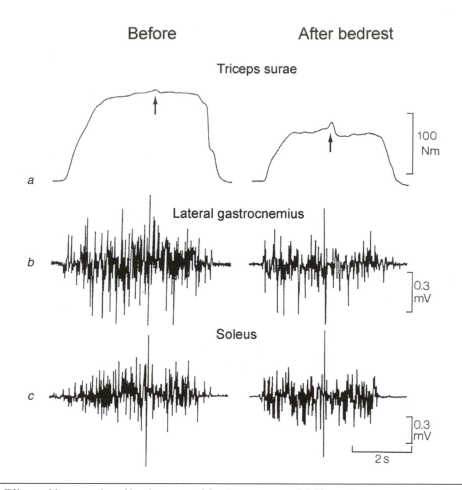

Figure 6.15 Effect of five weeks of bed rest on (*a*) triceps surae MVC, and corresponding EMG of (*b*) lateral gastrocnemius and (*c*) soleus. Arrows indicate supramaximal stimulation of the tibial nerve.

Reprinted from Duchateau 1995.

Booth 1984; Morrison et al. 1987). An important proportion of this initial decrease is caused by decreased RNA activity (protein synthesis per microgram of RNA; Goldspink 1977b) and possibly a reduced concentration of total RNA (Babij and Booth 1988; Goldspink 1977b). Increased protein degradation is seen in animal studies but not in human immobilization studies (Gibson et al. 1987). This may have to do with the relative length at which measured muscles were immobilized (most studies demonstrating significantly increased protein degradation involved muscles immobilized in shortened positions). Later, changes in transcription also contribute to the decreased protein synthesis (Watson, Stein, and Booth 1984; Morrison et al. 1987).

The heterogeneity of protein response during immobilization is quite marked. For example, after seven days of immobilization of the rat gastrocnemius and plantaris, the decreases in synthesis of actin, cytochrome c, and mixed proteins are 33%, 81%, and 67% of control levels, respectively (Morrison et al. 1987). In addition, the decreases in mRNA for cytochrome c in both rat soleus and tibialis anterior with immobilization are about the same (62–72% for soleus, 32–36% for tibialis anterior), whether or not the muscles are fixed in a lengthened or shortened position. Since loss of muscle mass is significantly more in the latter, this demonstrates the differences in the signals that control the transcription of mitochondrial and contractile proteins (Booth et al. 1996).

Increased protein degradation and decreased protein synthesis are accompanied by increases in the lysosomal enzymes cathepsins B, D, E, and L (Edes et al. 1982) and elevated acid phosphatases and proteases (Witzmann, Troup, and Fitts 1982). In addition, after 90 days of immobilization of rat gastrocnemius in a shortened position, calcium concentration more than doubled (Booth and Seider 1979a). This re-

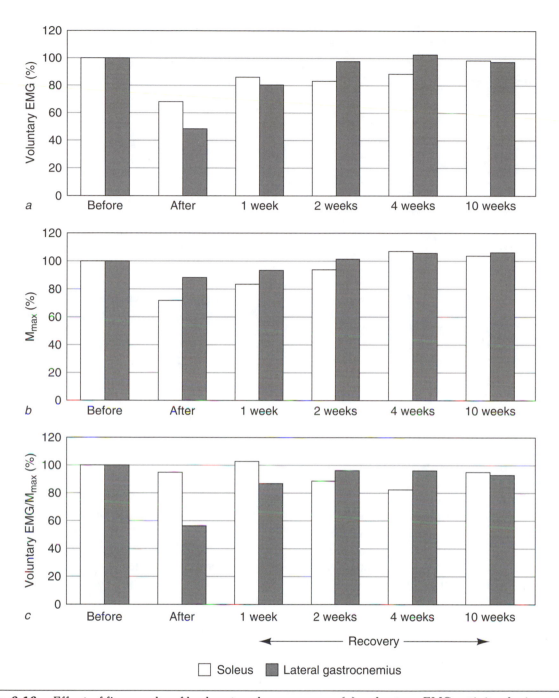

Figure 6.16   Effect of five weeks of bed rest and recovery on (*a*) voluntary EMG activity during an MVC of triceps surae, (*b*) maximal M wave, and (*c*) voluntary EMG normalized to the M wave. Note how decreased EMG in soleus, but not in lateral gastrocnemius, is accompanied by a corresponding decrease in M-wave amplitude.

Reprinted from Duchateau 1995.

sponse might play a role in activating calcium-dependent proteases.

The noncollagenous versus collagenous protein composition of the muscle changes with joint immobilization. In immobilized rat soleus, collagen content increases during the first few days of immobilization, mostly in the perimy-sium rather than the endomysium compartment, although there is a thickening of the latter after longer periods (Jozsa et al. 1990; P.E. Williams and Goldspink 1984). In the study by Williams and Goldspink, collagen fibers were found to be aligned at a more acute angle to the muscle fibers than in normal muscles, which

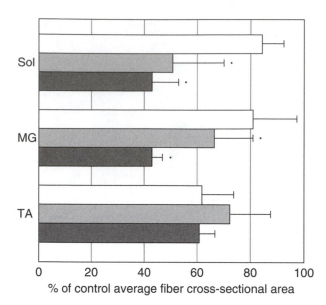

Figure 6.17   Effects of four weeks of rat hindlimb immobilization on fiber cross-sectional areas in three muscles. Open, shaded, and filled bars indicate that the muscle was immobilized in a lengthened, neutral, or shortened position, respectively.

Reprinted from Spector et al. 1982.

would be expected to play a role in the altered compliance that occurs. With chronic stretch, the increase in interstitial collagen results in decreased compliance and a resulting increase in muscle passive tension. Interestingly, previously immobilized muscles appear to be injured less readily by a series of eccentric contractions (Lapier et al. 1995). This was especially true for muscles immobilized in a lengthened position, in which the connective tissue area/fiber area ratio was the highest. This leads to speculation that a rapid adaptation in connective tissue may explain the reduction in damage seen following a few bouts of eccentric training (see chapter 5).

Metabolic Enzyme Changes

The literature on changes in metabolic enzyme levels with immobilization is inconclusive. Decreases have been reported in the concentrations of a variety of mitochondrial proteins in rats, rabbits, and humans (Blakemore et al. 1996; Jansson et al. 1988; Morrison et al. 1987; Edes et al. 1980; K.J. Jones 1993). In a subgroup of fibers of the immobilized rat soleus, activities of mitochondrial and glycolytic enzymes may actually increase, in keeping with their conversion from type I to IIA (Fitts, Brimmer, et al. 1989). In the same vein, several investigators have reported

no effect of immobilization on mitochondrial proteins. Similarly, the response of glycolytic enzymes is equivocal, with investigators showing decreases, increases, or no change. This is most likely a reflection of the variety of species, durations of immobilization, muscles examined, and degree of activity of the muscles used in the studies. When we restrict our view to human muscles immobilized in a neutral position, it appears that immobilization may result in a slight but significant decrease in mitochondrial as well as glycolytic enzyme proteins (Blakemore et al. 1996; Jansson et al. 1988).

Models where activity is truly decreased, such as a human joint immobilized due to an injury, might provide a more appropriate solution to this question. It is worth noting that the effect of short periods of complete disuse is decreased activity of both mitochondrial and glycolytic enzymes (St.-Pierre et al. 1988; R.N. Michel et al. 1994). There is some evidence that sarcoplasmic reticulum ATPase and rate of calcium uptake might increase with immobilization (Jakab et al. 1987; Kim, Witzmann, and Fitts 1982).

Fiber-Type Changes

The general trend in joint immobilization is a conversion of type I toward type II fibers, as we have seen with other models of decreased use. The change in fiber-type percentage is most evident in muscles containing the highest proportions of type I (Oishi, Ishihara, and Katsuta 1992; Booth and Kelso 1973; Edgerton et al. 1975), perhaps because the percentage changes in such muscles are easier to measure. Jänkälä and colleagues (1997) reported that after one week of hindlimb casting in rats, the myosin heavy-chain mRNA profiles of soleus, plantaris, and gastrocnemius were all changed toward the expression of faster isoforms (figure 6.18). A determining factor in this conversion from type I to II is the degree of stretch on the immobilized muscle. Pattullo and colleagues (1992) found that the proportion of type I fibers in rabbit tibialis anterior immobilized for six weeks in a lengthened position was five times greater than that seen in controls. They concluded that stretch alone produced this somewhat paradoxical result, since EMG was not any greater in the stretched position.

Muscle Functional Changes

Generally, predominantly slow muscles like the rat soleus develop shorter twitches with im-

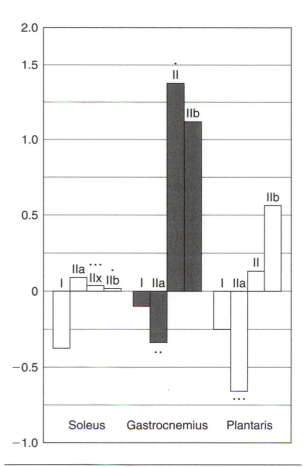

Figure 6.18 Effect of rat hindlimb immobilization for one week (cast, maximal plantar flexion) on expression of various MHCs in soleus, gastrocnemius, and plantaris. Values are expressed as the difference from control muscles in the number of molecules of the specific mRNA $\times 10^8$ per microgram of total RNA. Dots indicate significantly different from control values ($\cdot$ $p < 0.05$, $\cdot\cdot$ $p < 0.01$, $\cdot\cdot\cdot$ $p < 0.005$).

Reprinted from Jänkälä et al. 1997.

mobilization, while the duration of twitches in predominantly fast muscles becomes longer (Witzmann, Kim, and Fitts 1983; Witzmann, Kim, and Fitts 1982). In human mixed muscles like triceps surae and adductor pollicis, it appears that the fast motor unit responses may predominate, since twitches get longer (Davies, Rutherford, and Thomas 1987; Duchateau and Hainaut 1987; White, Davies, and Brooksby 1984). This increase is seen transiently in immobilized cat peroneus longus motor units of all types (Petit and Gioux 1993). This phenomenon most likely reflects the effects of inactivity on sarcoplasmic reticulum function, perhaps in combination with alterations in myofibrillar proteins, since V_{max}

appears to increase in both slow and fast muscles, in keeping with the tendency for all muscles to convert to faster MHC isoforms (Witzmann, Kim, and Fitts 1982). Twitch tension may also be potentiated, resulting in an elevated twitch/tetanic ratio, at both the whole-muscle and single-motor unit level (Nordstrom et al. 1995; Duchateau and Hainaut 1987). The maximal force capacity of the immobilized muscle generally decreases in proportion to the decreased cross-sectional area, although in muscles immobilized in a shortened position, specific tetanic tension may also be compromised (Jokl and Konstadt 1983; Witzmann, Kim, and Fitts 1983; Simard, Spector, and Edgerton 1982; Witzmann, Kim, and Fitts 1982).

Soleus immobilized in a shortened position loses fatigue resistance (Witzmann, Kim, and Fitts 1983). Interestingly, a decrease in fatigue resistance also occurs in immobilized adductor pollicis in humans, which, like the rat soleus, possesses a high proportion of type I fibers (figure 6.19). On the other hand, reports are also found in the literature that immobilization results in increased fatigue resistance (Robinson, Enoka, and Stuart 1991; Vandenborne et al. 1998; Semmler, Kutzscher, and Enoka 1999) or no change (Gardiner and Lapointe 1982; White, Davies, and Brooksby 1984; St.-Pierre et al. 1988; Miles et al. 1994; Nordstrom et al. 1995).

There is a clear transient decrease in the ability of subjects to activate their previously immobilized muscles (figures 6.20 and 6.21). This is evident from a greater decline in muscle cross-sectional area than in maximal isometric force (Miles et al. 1994), a greater decline in MVC than in the tetanic force evoked via supramaximal electrical stimulation (Hainaut and Duchateau 1989; Duchateau and Hainaut 1987; White, Davies, and Brooksby 1984), or an increased twitch response superimposed on an MVC (twitch interpolation technique; Vandenborne et al. 1998; Behm and St.-Pierre 1997).

Duchateau and Hainaut (1990) reported that after six to eight weeks of immobilization of the adductor pollicis and first dorsal interosseus muscles in human subjects, minimal motor unit firing rates (firing rates at recruitment) were unaffected, but maximal firing rates were markedly decreased, especially in the low-threshold units (table 6.5).

These results are consistent with the previous findings of Sale and colleagues (1982), who

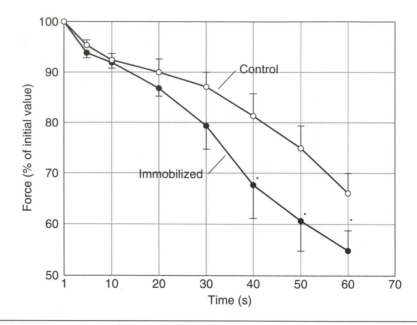

Figure 6.19 Fatigue response to electrical stimulation of human adductor pollicis muscle following six weeks of voluntary immobilization of the thumb in a plaster cast. Stimuli were 60 intermittent 1-s contractions at 30 Hz, separated by 1-s intervals.

Reprinted from Davies, Rutherford, and Thomas 1987.

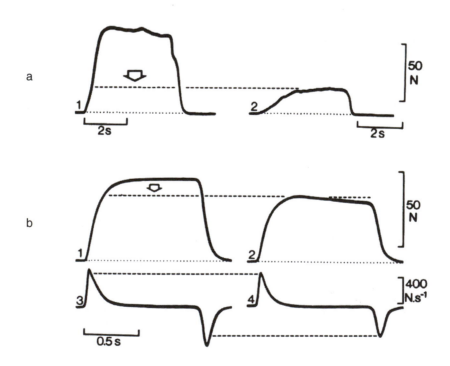

Figure 6.20 Effects of six weeks of immobilization of the thumb on (*a*) MVC and (*b*) response to supramaximal stimulation at 100 Hz. First derivative of the isometric force signals is also shown in *b*. Dotted lines and arrows indicate the extent of the effect of immobilization on these properties.

Reprinted from Duchateau and Hainaut 1987.

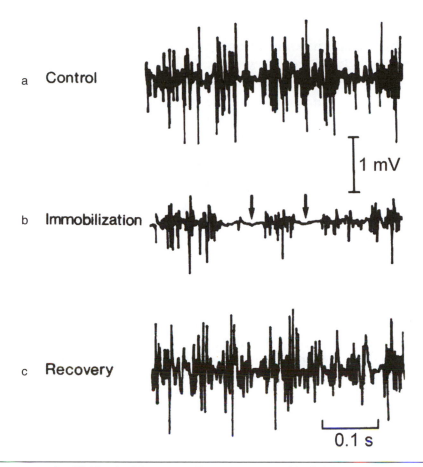

Figure 6.21   Intramuscular EMG of adductor pollicis during an MVC (*a*) in controls, (*b*) after six weeks of thumb immobilization, and (*c*) after one week of recovery. EMG silent periods are shown by arrows in *b*.

Reprinted from Duchateau and Hainaut 1987.

found a decreased reflex potentiation in immobilized human subjects performing an MVC. The standard interpretation of such a decrease is a decreased efficacy of the pathways involved in exciting motoneurons during an MVC.

Other Noteworthy Observations

The nervous system is undoubtedly influenced by immobilization. Neuromuscular junctions in immobilized muscles may increase in area, become more diffuse in appearance, and show denervation-like changes such as nerve terminal disruption, exposed junctional folds, and postsynaptic areas with few or no folds (Fahim and Robbins 1986; Pachter and Eberstein 1984; Malathi and Batmanabane 1983). Not much is known, however, regarding the functional state of neuromuscular junctions following immobilization-induced disuse. Robbins and Fischbach (1971) reported

an increased EPP amplitude at junctions of immobilized soleus muscles in rats. This increased efficacy is reminiscent of the transiently increased EPSP amplitude seen at Ia afferent–motoneuron connections following spinal cord transection and suggests an increased safety factor at such junctions. However, increased jitter has been found in immobilized leg muscles of human subjects (Grana, Chiou-Tan, and Jaweed 1996), which suggests a *reduced* junctional safety factor. Also, decreased miniature endplate potential (MEPP) frequency at soleus end plates after immobilization may indicate a functional deficit at this level (Zemková et al. 1990).

There are slight alterations in the responsiveness of spindle and tendon organ afferents, which are primarily reflections of the muscle atrophic and compliance changes (Nordstrom et al. 1995; Petit and Gioux 1993).

Table 6.5

Effects of 6–8 Weeks of Cast Immobilization on Firing Rates of Motor Units in Human Adductor Pollicis (AP) and First Dorsal Interosseus (FDI) Muscles

	Firing rate at recruitment (Hz)	Maximal firing rate (Hz)
Control AP	6.6 ± 16.1 (55 units)	22.6 ± 17.4 (55 units)
Immobilized AP	6.2 ± 15.3 (56 units)	**13.1 ± 3.7 (51 units)**
Control FDI	6.5 ± 13.6 (43 units)	31.0 ± 8.9 (39 units)
Immobilized FDI	6.1 ± 11.8 (41 units)	**19.0 ± 4.9 (38 units)**

Data are expressed as means ± standard deviation. Means significantly different from the corresponding control means are shown in boldface.

Adapted from Duchateau and Hainaut 1990.

Finally, are there trophic influences, released at nerve terminals, that affect muscle characteristics? Could some of the muscle changes seen with decreased usage be due to alterations in the secretion of these substances? Although this question is relevant for all models of reduced activity, it was with the joint immobilization model that H.L. Davis and Kiernan (1981) were able to demonstrate that an extract of nerve injected into muscles had an attenuating effect on atrophic responses in rats. These investigators found that a crude extract from sciatic nerve injected into a denervated extensor digitorum longus reduced the atrophy to the same level as in the contralateral immobilized muscle. They were able to estimate the effects of decreased activity versus loss of trophic substances on the extent of atrophy. They also determined that the trophic effect was different for different fiber types (Davis and Kiernan 1981), that the effect was also present when the extract was injected intraperitoneally (H.L. Davis 1985), and that the influence of the nerve extract on various muscle properties varied (e.g., posttetanic twitch potentiation, which was lost with denervation, was not affected by the extract injections, while the decrease in parvalbumin was completely prevented; H.L. Davis, Bressler, and Jasch 1988). Since that time, investigators have found several substances that are released from nerve terminals and affect postsynaptic structure and function. The extent to which the levels of these substances and their release at nerve terminals are altered by immobilization (or any other reduced-activity condition, for that matter) and their role in the atrophic responses remain to be revealed.

Summary

The research shows us that both activation by the nervous system and weight bearing are important components in determining muscles' stable protein content. Models that totally reduce these two components to zero, exemplified by spinal isolation and pharmacological blockade of the motor nerve, demonstrate the near-maximal extent of changes in muscle phenotype. The extent of atrophy can be to less than half of the original weight, and a near-complete conversion of fiber-type proportions to type II can occur. The lesser extent of the changes seen in spaceflight, hindlimb suspension, and immobilization attests to the residual effects of the small amounts of activation or weight bearing that remain in these models. Research into countermeasures to these atrophic changes, which may include electrical stimulation, exercise of various types, administration of hormones and growth factors, and even heat stress, is helping reveal the molecular pathways by which atrophic changes occur. In addition, this research is valuable for preventive and rehabilitative therapy for both individuals who suffer from pathological forms of disuse atrophy and astronauts who travel for long periods in a microgravity environment.

References

Abbruzzese, G., Morena, M., Spadavecchia, L., and Schieppati, M. 1994. Response of arm flexor muscles to magnetic and electrical brain stimulation during shortening and lengthening tasks in man. *J. Physiol.* 481:499–507.

Abu-Shakra, S.R., Cole, A.J., Adams, R.N., and Drachman, D.B. 1994. Cholinergic stimulation of skeletal muscle cells induces rapid immediate early gene expression: Role of intracellular calcium. *Mol. Brain Res.* 26:55–60.

Adam, A., De Luca, C.J., and Erim, Z. 1998. Hand dominance and motor unit firing behavior. *J. Neurophysiol.* 80:1373–1382.

Adams, G.R., and Haddad, F. 1996. The relationships among IGF-1, DNA content, and protein accumulation during skeletal muscle hypertrophy. *J. Appl. Physiol.* 81:2509–2516.

Adams, G.R., Haddad, F., and Baldwin, K.M. 1999. Time course of changes in markers of myogenesis in overloaded rat skeletal muscles. *J. Appl. Physiol.* 87:1705–1712.

Adams, G.R., Hather, B.M., Baldwin, K.M., and Dudley, G.A. 1993. Skeletal muscle myosin heavy chain composition and resistance training. *J. Appl. Physiol.* 74:911–915.

Adams, G.R., and McCue, S.A. 1998. Localized infusion of IGF-I results in skeletal muscle hypertrophy in rats. *J. Appl. Physiol.* 84:1716–1722.

Alaimo, M.A., Smith, J.L., and Edgerton, V.R. 1984. EMG activity of slow and fast ankle extensors following spinal cord transection. *J. Appl. Physiol. Respir. Environ. Exerc. Physiol.* 56:1608–1613.

Alberts, B., Bray, D., Lewis, J., Raff, M., Roberts, K., and Watson, J. 1994. *Molecular biology of the cell.* New York: Garland.

Aldrich, T.K. 1987. Transmission fatigue of the rabbit diaphragm. *Respir. Physiol.* 69:307–319.

Aldrich, T.K., Shander, A., Chaudhry, I., and Nagashima, H. 1986. Fatigue of isolated rat diaphragm: Role of impaired neuromuscular transmission. *J. Appl. Physiol.* 61:1077–1083.

Alford, E.K., Roy, R.R., Hodgson, J.A., and Edgerton, V.R. 1987. Electromyography of rat soleus, medial gastrocnemius, and tibialis anterior during hind limb suspension. *Exp. Neurol.* 96:635–649.

Allen, D.L., Linderman, J.K., Roy, R.R., Bigbee, A.J., Grindeland, R.E., Mukku, V., and Edgerton, V.R. 1997. Apoptosis: A mechanism contributing to remodeling of skeletal muscle in response to hindlimb unweighting. *Am. J. Physiol. Cell Physiol.* 273:C579–C587.

Allen, D.L., Linderman, J.K., Roy, R.R., Grindeland, R.E., Mukku, V., and Edgerton, V.R. 1997. Growth hormone IGF-I and/or resistive exercise maintains myonuclear number in hindlimb unweighted muscles. *J. Appl. Physiol.* 83:1857–1861.

Allen, D.L., Monke, S.R., Talmadge, R.J., Roy, R.R., and Edgerton, V.R. 1995. Plasticity of myonuclear number in hypertrophied and atrophied mammalian skeletal muscle fibers. *J. Appl. Physiol.* 78:1969–1976.

Allen, G.M., Gandevia, S.C., and McKenzie, D.K. 1995. Reliability of measurements of muscle strength and voluntary activation using twitch interpolation. *Muscle & Nerve* 18:593–600.

Alway, S.E., MacDougall, J.D., and Sale, D.G. 1989. Contractile adaptations in the human triceps surae after isometric exercise. *J. Appl. Physiol.* 66:2725–2732.

Alway, S.E., MacDougall, J.D., Sale, D.G., Sutton, J.R., and McComas, A.J. 1988. Functional and structural adaptations in skeletal muscle of trained athletes. *J. Appl. Physiol.* 64:1114–1120.

Alway, S.E., Stray-Gundersen, J., Grumbt, W.H., and Gonyea, W.J. 1990. Muscle cross-sectional area and torque in resistance-trained subjects. *Eur. J. Appl. Physiol.* 60:86–90.

Andersen, J.L., and Schiaffino, S. 1997. Mismatch between myosin heavy chain mRNA and protein distribution in human skeletal muscle fibers. *Am. J. Physiol. Cell Physiol.* 272:C1881–C1889.

Annex, B.H., Kraus, W.E., Dohm, G.L., and Williams, R.S. 1991. Mitochondrial biogenesis in striated muscles: Rapid induction of citrate synthase mRNA by nerve stimulation. *Am. J. Physiol. Cell Physiol.* 260:C266–C270.

Antonio, J., and Gonyea, W.J. 1993. Role of muscle fiber hypertrophy and hyperplasia in intermittently stretched avian muscle. *J. Appl. Physiol.* 74:1893–1898.

Appelberg, B., Hulliger, M., Johansson, H., and Sojka, P. 1983. Actions on gamma-motoneurones elicited by electrical stimulation of group III muscle afferent fibres in the hind limb of the cat. *J. Physiol.* (Lond.) 335:275–292.

Aronson, D., Boppart, M.D., Dufresne, S.D., Fielding, R.A., and Goodyear, L.J. 1998. Exercise stimulates c-Jun NH_2 kinase activity and c-Jun transcriptional activity in human skeletal muscle. *Biochem. Biophys. Res. Commun.* 251:106–110.

Aronson, D., Dufresne, S.D., and Goodyear, L.J. 1997. Contractile activity stimulates the c-Jun NH_2-terminal kinase pathway in rat skeletal muscle. *J. Biol. Chem.* 272:25636–25640.

Avela, J., and Komi, P.V. 1998. Reduced stretch reflex sensitivity and muscle stiffness after long-lasting stretch-shortening cycle exercise in humans. *Eur. J. Appl. Physiol. Occup. Physiol.* 78:403–410.

Aymard, C., Katz, R., Lafitte, C., Le Bozec, S., and Pénicaud, A. 1995. Changes in reciprocal and transjoint inhibition induced by muscle fatigue in man. *Exp. Brain Res.* 106:418–424.

Baar, K., Blough, E., Dineen, B., and Esser, K. 1999. Transcriptional regulation in response to exercise. *Exerc. Sport Sci. Rev.* 27:333–379.

Baar, K., and Esser, K. 1999. Phosphorylation of p70^{S6k} correlates with increased skeletal muscle mass following resistance exercise. *Am. J. Physiol. Cell Physiol.* 276:C120–C127.

Babij, P., and Booth, F.W. 1988. Alpha-actin and cytochrome c mRNAs in atrophied adult rat skeletal muscle. *Am. J. Physiol.* 254:C651–C656.

Bakels, R., and Kernell, D. 1993. Matching between motoneurone and muscle unit properties in rat medial gastrocnemius. *J. Physiol.* 463:307–324.

Bakels, R., and Kernell, D. 1994. Threshold-spacing in motoneurone pools of rat and cat: Possible relevance for manner of force gradation. *Exp. Brain Res.* 102:69–74.

Baker, L.L., and Chandler, S.H. 1987a. Characterization of hindlimb motoneuron membrane properties in acute and chronic spinal cats. *Brain Res.* 420:333–339.

Baker, L.L., and Chandler, S.H. 1987b. Characterization of postsynaptic potentials evoked by sural nerve stimulation in hindlimb motoneurons from acute and chronic spinal cats. *Brain Res.* 420:340–350.

Baldwin, K.M., Herrick, R.E., and McCue, S.A. 1993. Substrate oxidation capacity in rodent skeletal muscle: Effects of exposure to zero gravity. *J. Appl. Physiol.* 75:2466–2470.

Balestra, C., Duchateau, J., and Hainaut, K. 1992. Effects of fatigue on the stretch reflex in a human muscle. *Electroencephalogr. Clin. Neurophysiol. Electromyogr. Motor Control* 85:46–52.

Bangart, J.J., Widrick, J.J., and Fitts, R.H. 1997. Effect of intermittent weight bearing on soleus fiber force–velocity–power and force–pCa relationships. *J. Appl. Physiol.* 82:1905–1910.

Barany, M. 1967. ATPase activity of myosin correlated with speed of muscle shortening. *J. Gen. Physiol.* 50:197–218.

Barbeau, H., Norman, K., Fung, J., Visintin, M., and Ladouceur, M. 1998. Does neurorehabilitation play a role in the recovery of walking in neurological populations? *Ann. NY Acad. Sci.* 860:377–392.

Barbeau, H., and Rossignol, S. 1987. Recovery of locomotion after chronic spinalization in the adult cat. *Brain Res.* 412:84–95.

Bar-Or, O., Dotan, R., Inbar, O., Rothstein, A., Karlsson, J., and Tesch, P. 1980. Anaerobic capacity and muscle fiber type distribution in man. *Int. J. Sports Med.* 1:82–85.

Barstow, T.J., Jones, A.M., Nguyen, P.H., and Casaburi, R. 2000. Influence of muscle fibre type and fitness on the oxygen uptake/power output slope during incremental exercise in humans. *Exp. Physiol.* 85:109–116.

Barton-Davis, E.R., LaFramboise, W.A., and Kushmerick, M.J. 1996. Activity-dependent induction of slow myosin gene expression in isolated fast-twitch mouse muscle. *Am. J. Physiol. Cell Physiol.* 271:C1409–C1414.

Barton-Davis, E.R., Shoturma, D., and Sweeney, L. 1999. Contribution of satellite cells to IGF-1 induced hypertrophy of skeletal muscle. *Acta Physiol. Scand.* 167:301–305.

Bawa, P., and Lemon, R.N. 1993. Recruitment of motor units in response to transcranial magnetic stimulation in man. *J. Physiol.* 471:445–464.

Bazzy, A.R., and Donnelly, D.F. 1993. Diaphragmatic failure during loaded breathing: Role of neuromuscular transmission. *J. Appl. Physiol.* 74:1679–1683.

Behm, D.G., and Sale, D.G. 1993. Intended rather than actual movement velocity determines velocity-specific training response. *J. Appl. Physiol.* 74:359–368.

Behm, D.G., and St.-Pierre, D.M.M. 1997. Fatigue characteristics following ankle fractures. *Med. Sci. Sports Exerc.* 29:1115–1123.

Bélanger, M., Drew, T., Provencher, J., and Rossignol, S. 1996. A comparison of treadmill locomotion in adult cats before and after spinal transection. *J. Neurophysiol.* 76:471–491.

Belcastro, A.N. 1993. Skeletal muscle calcium-activated neutral protease (calpain) with exercise. *J. Appl. Physiol.* 74:1381–1386.

Belcastro, A.N., Albisser, T.A., and Littlejohn, B. 1996. Role of calcium-activated neutral protease (calpain) with diet and exercise. *Can. J. Appl. Physiol.* 21:328–346.

Belhaj-Saïf, A., Fourment, A., and Maton, B. 1996. Adaptation of the precentral cortical command to elbow muscle fatigue. *Exp. Brain Res.* 111:405–416.

Bellemare, F., Woods, J.J., Johansson, R., and Bigland-Ritchie, B. 1983. Motor-unit discharge rates in maximal voluntary contractions of three human muscles. *J. Neurophysiol.* 50:1380–1392.

Bennett, D.J., Hultborn, H., Fedirchuk, B., and Gorassini, M. 1998. Short-term plasticity in hindlimb motoneurons of decerebrate cats. *J. Neurophysiol.* 80:2038–2045.

Bentley, D., Smith, P., Davie, A., and Zhou, S. 2000. Muscle activation of the knee extensors following high intensity endurance exercise in cyclists. *Eur. J. Appl. Physiol.* 81:297–302.

Berg, H.E., Dudley, G.A., Häggmark, T., Ohlsén, H., and Tesch, P.A. 1991. Effects of lower limb unloading on skeletal muscle mass and function in humans. *J. Appl. Physiol.* 70:1882–1885.

Berg, H.E., Dudley, G.A., Hather, B., and Tesch, P.A. 1993. Work capacity and metabolic and morphologic characteristics of the human quadriceps muscle in response to unloading. *Clin. Physiol.* 13:337–347.

Berg, H.E., Larsson, L., and Tesch, P.A. 1997. Lower limb skeletal muscle function after 6 wk of bed rest. *J. Appl. Physiol.* 82:182–188.

Berg, H.E., and Tesch, P.A. 1996. Changes in muscle function in response to 10 days of lower limb unloading in humans. *Acta Physiol. Scand.* 157:63–70.

Bernardi, M., Solomonow, M., Nguyen, G., Smith, A., and Baratta, R. 1996. Motor unit recruitment strategy changes with skill acquisition. *Eur. J. Appl. Physiol. Occup. Physiol.* 74:52–59.

Bevan, L., Laouris, Y., Reinking, R.M., and Stuart, D.G. 1992. The effect of the stimulation pattern on the fatigue of single motor units in adult cats. *J. Physiol.* 449:85–108.

Bigland-Ritchie, B., Dawson, N., Johansson, R., and Lippold, O.C.J. 1986. Reflex origin for the slowing of motoneurone firing rates in fatigue of human voluntary contractions. *J. Physiol.* (Lond.) 379:451–459.

Bigland-Ritchie, B., Fuglevand, A., and Thomas, C. 1998. Contractile properties of human motor units: Is man a cat? *Neuroscientist* 4:240–249.

Bigland-Ritchie, B., Furbush, F., and Woods, J.J. 1986. Fatigue of intermittent submaximal voluntary contractions: Central and peripheral factors. *J. Appl. Physiol.* 61:421–429.

Bigland-Ritchie, B., Johansson, R., Lippold, O.C.J., Smith, S., and Woods, J.J. 1983. Changes in motoneurone firing rates during sustained maximal voluntary contractions. *J. Physiol.* (Lond.) 340:335–346.

Bigland-Ritchie, B., Jones, D.A., and Woods, J.J. 1979. Excitation frequency and muscle fatigue: Electrical responses during human voluntary and stimulated contractions. *Exp. Neurol.* 64:414–427.

Bigland-Ritchie, B., and Woods, J.J. 1984. Changes in muscle contractile properties and neural control during human muscular contraction. *Muscle & Nerve* 7:691–699.

Billeter, R., Heizmann, C.W., Howald, H., and Jenny, E. 1981. Analysis of myosin light and heavy chain types in single human skeletal muscle fibers. *Eur. J. Biochem.* 116:389–395.

Binder, M.D., Bawa, P., Ruenzel, P., and Henneman, E. 1983. Does orderly recruitment of motoneurons depend on the existence of different types of motor units? *Neurosci. Lett.* 36:55–58.

Binder, M.D., Heckman, C.J., and Powers, R.K. 1996. The physiological control of motoneuron activity. In *Handbook of physiology: 12. Exercise, regulation and integration of multiple systems,* ed. L.B. Rowell and J.T. Shepherd, 3-53. New York: Oxford University Press.

Biolo, G., Maggi, S.P., Williams, B.D., Tipton, K.D., and Wolfe, R.R. 1995. Increased rates of muscle protein turnover and amino acid transport after resistance exercise in humans. *Am. J. Physiol. Endocrinol. Metab.* 268:E514–E520.

Blakemore, S.J., Rickhuss, P.K., Watt, P.W., Rennie, M.J., and Hundal, H.S. 1996. Effects of limb immobilization on cytochrome c oxidase activity and GLUT4 and GLUT5 protein expression in human skeletal muscle. *Clin. Sci.* 91:591–599.

Blewett, C., and Elder, G.C.B. 1993. Quantitative EMG analysis in soleus and plantaris during hindlimb suspension and recovery. *J. Appl. Physiol.* 74:2057–2066.

Bodine-Fowler, S.C., Roy, R.R., Rudolph, W., Haque, N., Kozlovskaya, I.B., and Edgerton, V.R. 1992. Spaceflight and growth effects on muscle fibers in the rhesus monkey. *J. Appl. Physiol.* 73 suppl.: 82S–89S.

Bonato, C., Zanette, G., Manganotti, P., Tinazzi, M., Bongiovanni, G., Polo, A., and Fiaschi, A. 1996. "Direct" and "crossed" modulation of human motor cortex excitability following exercise. *Neurosci. Lett.* 216:97–100.

Bongiovanni, L.G., and Hagbarth, K.-E. 1990. Tonic vibration reflexes elicited during fatigue from maximal voluntary contractions in man. *J. Physiol.* (Lond.) 423:1–14.

Bongiovanni, L.G., Hagbarth, K.-E., and Stjernberg, L. 1990. Prolonged muscle vibration reducing motor output in maximal voluntary contractions in man. *J. Physiol.* (Lond.) 423:15–26.

Booth, F.W., and Kelso, J.R. 1973. Effect of hind-limb immobilization on contractile and histochemical properties of skeletal muscle. *Pflugers Arch.* 342:231–238.

Booth, F.W., and Kirby, C.R. 1992. Changes in skeletal muscle gene expression consequent to altered weight bearing. *Am. J. Physiol. Regul. Integr. Comp. Physiol.* 262:R329–R332.

Booth, F.W., Lou, W., Hamilton, M.T., and Yan, Z. 1996. Cytochrome c mRNA in skeletal muscles of immobilized limbs. *J. Appl. Physiol.* 81:1941–1945.

Booth, F.W., and Seider, M.J. 1979. Early change in skeletal muscle protein synthesis after limb immobilization of rats. *J. Appl. Physiol.* 47:974–977.

Booth, F.W., Tseng, B.S., Flück, M., and Carson, J.A. 1998. Molecular and cellular adaptation of muscle in response to physical training. *Acta Physiol. Scand.* 162:343–350.

Boppart, M., Aronson, D., Gibson, L., Roubenoff, R., Abad, L., Bean, J., Goodyear, L., and Fielding, R. 1999. Eccentric exercise markedly increases c-Jun NH_2-terminal kinase activity in human skeletal muscle. *J. Appl. Physiol.* 87:1668–1673.

Bosco, C., Montanari, G., Ribacchi, R., Giovenali, P., Latteri, F., Iachelli, G., Faina, M., Colli, R., DalMonte, A., LaRosa, M., Cortili, G., and Saibene, F. 1987. Relationship between the efficiency of muscular work during jumping and the energetics of running. *Eur. J. Appl. Physiol.* 56:138–143.

Botterman, B.R., and Cope, T.C. 1988a. Maximum tension predicts relative endurance of fast-twitch motor units in the cat. *J. Neurophysiol.* 60:1215–1226.

Botterman, B.R., and Cope, T.C. 1988b. Motor-unit stimulation patterns during fatiguing contractions of constant tension. *J. Neurophysiol.* 60:1198–1214.

Bottinelli, R., Betto, R., Schiaffino, S., and Reggiani, C. 1994a. Maximum shortening velocity and coexistence of myosin heavy chain isoforms in single skinned fast fibres of rat skeletal muscle. *J. Muscle Res. Cell Motil.* 15:413–419.

Bottinelli, R., Betto, R., Schiaffino, S., and Reggiani, C. 1994b. Unloaded shortening velocity and myosin heavy chain and alkali light chain isoform composition in rat skeletal muscle fibres. *J. Physiol.* 478:341–349.

Bottinelli, R., Canepari, M., Pellegrino, M.A., and Reggiani, C. 1996. Force–velocity properties of human skeletal muscle fibres: Myosin heavy chain isoform and temperature dependence. *J. Physiol.* 495:573–586.

Bottinelli, R., Canepari, M., Reggiani, C., and Stienen, G.J.M. 1994. Myofibrillar ATPase activity during isometric contraction and isomyosin composition in rat single skinned muscle fibres. *J. Physiol.* (Lond.) 481:663–676.

Bottinelli, R., and Reggiani, C. 1995a. Essential myosin light chain isoforms and energy transduction in skeletal muscle fibers. *Biophys. J.* 68:227S.

Bottinelli, R., and Reggiani, C. 1995b. Force–velocity properties and myosin light chain isoform composition of an identified type of skinned fibres from rat skeletal muscle. *Pflugers Arch.* 429:592–594.

Brasil-Neto, J.P., Cohen, L.G., and Hallett, M. 1994. Central fatigue as revealed by postexercise decrement of motor evoked potentials. *Muscle & Nerve* 17:713–719.

Briggs, F.N., Poland, J.L., and Solaro, R.J. 1977. Relative capabilities of sarcoplasmic reticulum in fast and slow mammalian skeletal muscles. *J. Physiol.* 266:587–594.

Brooke, M.H., Williamson, E., and Kaiser, K.K. 1971. The behavior of four fiber types in developing and reinnervated muscle. *Arch. Neurol.* 25:360-366.

Brown, S.J., Child, R.B., Donnelly, A.E., Saxton, J.M., and Day, S.H. 1996. Changes in human skeletal muscle contractile function following stimulated eccentric exercise. *Eur. J. Appl. Physiol.* 72:515–521.

Brownson, C., Isenberg, H., Brown, W., Salmons, S., and Edwards, Y. 1988. Changes in skeletal muscle gene transcription induced by chronic stimulation. *Muscle & Nerve* 11:1183–1189.

Brownson, C., Little, P., Jarvis, J.C., and Salmons, S. 1992. Reciprocal changes in myosin isoform mRNAs of rabbit skeletal muscle in response to the initiation and cessation of chronic electrical stimulation. *Muscle & Nerve* 15:694–700.

Brownstone, R.M., Jordan, L.M., Kriellaars, D.J., Noga, B.R., and Shefchyk, S.J. 1992. On the regulation of repetitive firing in lumbar motoneurones during fictive locomotion in the cat. *Exp. Brain Res.* 90:441–455.

Burke, R.E. 1967. Motor unit types of cat triceps surae muscle. *J. Physiol.* (Lond.) 193:141–160.

Burke, R.E. 1990. Motor unit types: Some history and unsettled issues. In *The segmental motor system,* ed. M. Binder and L. Mendell, 207-221. New York: Oxford University Press.

Burke, R.E., Dum, R., Fleshman, J., Glenn, L., Lev-Tov, A., O'Donovan, M., and Pinter, M. 1982. An HRP study of the relation between cell size and motor unit type in cat ankle extensor motoneurons. *J. Comp. Neurol.* 209:17–28.

Burke, R.E., Levine, D.N., Tsairis, P., and Zajac, F.E. 1973. Physiological types and histochemical profiles in motor units of the cat gastrocnemius. *J. Physiol.* (Lond.) 234:723–748.

Burke, R.E., and Tsairis, P. 1974. The correlation of physiological properties with histochemical characteristics in single muscle units. *Ann. NY Acad. Sci.* 228:145–159.

Burridge, K., and Chrzanowska-Wodnicka, M. 1996. Focal adhesions, contractility, and signalling. *Ann. Rev. Cell Devel. Biol.* 12:463–519.

Butler, I., Drachman, D., and Goldberg, A. 1978. The effect of disuse on cholinergic enzymes. *J. Physiol.* (Lond.) 274:593–600.

Caiozzo, V.J., Baker, M.J., and Baldwin, K.M. 1998. Novel transitions in MHC isoforms: Separate and combined effects of thyroid hormone and mechanical unloading. *J. Appl. Physiol.* 85:2237–2248.

Caiozzo, V.J., Haddad, F., Baker, M.J., Herrick, R.E., Prietto, N., and Baldwin, K.M. 1996. Microgravity-induced transformations of myosin isoforms and contractile properties of skeletal muscle. *J. Appl. Physiol.* 81:123–132.

Caiozzo, V.J., Perrine, J.J., and Edgerton, V.R. 1981. Training-induced alterations of the *in vivo* force-velocity relationship of human muscle. *J. Appl. Physiol. Respir. Environ. Exerc. Physiol.* 51:750–754.

Calancie, B., and Bawa, P. 1985. Voluntary and reflexive recruitment of flexor carpi radialis motor units in humans. *J. Neurophysiol.* 53:1194–1200.

Cannon, R.J., and Cafarelli, E. 1987. Neuromuscular adaptations to training. *J. Appl. Physiol.* 63:2396–2402.

Canon, F., Goubel, F., and Guezennec, C.Y. 1998. Effects on contractile and elastic properties of hindlimb suspended rat soleus muscle. *Eur. J. Appl. Physiol. Occ. Physiol.* 77:118–124.

Capaday, C. 1997. Neurophysiological methods for studies of the motor system in freely moving human subjects. *J. Neurosci. Methods* 74:201–218.

Carolan, B., and Cafarelli, E. 1992. Adaptations in coactivation after isometric resistance training. *J. Appl. Physiol.* 73:911–917.

Carp, J.S., and Wolpaw, J.R. 1994. Motoneuron plasticity underlying operantly conditioned decrease in primate H-reflex. *J. Neurophysiol.* 72:431–442.

Carp, J.S., and Wolpaw, J.R. 1995. Motoneuron properties after operantly conditioned increase in primate H-reflex. *J. Neurophysiol.* 73:1365–1373.

Carraro, F., Stuart, C.A., Hartl, W.H., Rosenblatt, J., and Wolfe, R.R. 1990. Effect of exercise and recovery on muscle protein synthesis in human subjects. *Am. J. Physiol. Endocrinol. Metab.* 259:E470–E476.

Carrier, L., Brustein, E., and Rossignol, S. 1997. Locomotion of the hindlimbs after neurectomy of ankle flexors in intact and spinal cats: Model for the study of locomotor plasticity. *J. Neurophysiol.* 77:1979–1993.

Carroll, S.L., Klein, M.G., and Schneider, M.F. 1997. Decay of calcium transients after electrical stimulation in rat fast- and slow-twitch skeletal muscle fibres. *J. Physiol.* 501:573–588.

Carroll, S.L., Nicotera, P., and Pette, D. 1999. Calcium transients in single fibers of low-frequency stimulated fast-twitch muscle of rat. *Am. J. Physiol.* 277:C1122–C1129.

Carson, J., and Wei, L. 2000. Integrin signaling's potential for mediating gene expression in hypertrophying skeletal muscle. *J. Appl. Physiol.* 88:337–343.

Castro, M., Apple, D., Staron, R.S., Campos, G.E.R., and Dudley, G.A. 1999. Influence of complete spinal cord injury on skeletal muscle within 6 mo of injury. *J. Appl. Physiol.* 86:350–358.

Chan, K.M., Andres, L.P., Polykovskaya, Y., and Brown, W.F. 1998. Dissociation of the electrical and contractile properties in single human motor units during fatigue. *Muscle & Nerve* 21:1786–1789.

Chau, C., Barbeau, H., and Rossignol, S. 1998. Early locomotor training with clonidine in spinal cats. *J. Neurophysiol.* 79:392–409.

Chesley, A., MacDougall, J.D., Tarnopolsky, M.A., Atkinson, S.A., and Smith, K. 1992. Changes in human muscle protein synthesis after resistance exercise. *J. Appl. Physiol.* 73:1383–1388.

Chi, M.M.-Y., Choksi, R., Nemeth, P.M., Krasnov, I., Ilyina-Kakueva, E., Manchester, J.K., and Lowry, O.H. 1992. Effects of microgravity and tail suspension on enzymes of individual soleus and tibialis anterior fibers. *J. Appl. Physiol.* 73 suppl.: 66S–73S.

Chi, M.M.-Y., Hintz, C.S., Coyle, E.F., Martin, W.H., III, Ivy, J.L., Nemeth, P.M., Holloszy, J.O., and Lowry, O.H. 1983. Effects of detraining on enzymes of energy metabolism in individual human muscle fibers. *Am. J. Physiol. Cell Physiol.* 244 (13): C276–C287.

Chi, M.M.-Y., Hintz, C.S., McKee, D., Felder, S., Grant, N., Kaiser, K.K., and Lowry, O.H. 1987. Effect of Duchenne muscular dystrophy on enzymes of energy metabolism in individual muscle fibers. *Metabolism* 36:761–767.

Chilibeck, P.D., Syrotiuk, D.G., and Bell, G.J. 1999. The effect of strength training on estimates of mitochondrial density and distribution throughout muscle fibers. *Eur. J. Appl. Physiol.* 80:604–609.

Chin, E., Olson, E., Richardson, J., Yang, A., Humphries, C., Shelton, J., Wu, H., Zhu, W., Bassel-Duby, R., and Williams, R. 1998. A calcineurin-dependent transcriptional pathway controls skeletal muscle fiber type. *Genes Development* 12:2499–2509.

Christakos, C.N., and Windhorst, U. 1986. Spindle gain increase during muscle unit fatigue. *Brain Res.* 365:388–392.

Christova, P., and Kossev, A. 1998. Motor unit activity during long-lasting intermittent muscle contractions in humans. *Eur. J. Appl. Physiol. Occup. Physiol.* 77:379–387.

Clamann, H.P., and Robinson, A.J. 1985. A comparison of electromyographic and mechanical fatigue properties in motor units of the cat hindlimb. *Brain Res.* 327:203–219.

Clark, B.D., Dacko, S.M., and Cope, T.C. 1993. Cutaneous stimulation fails to alter motor unit recruitment in the decerebrate cat. *J. Neurophysiol.* 70:1433–1439.

Clark, E., and Brugge, J. 1995. Integrins and signal transduction pathways: The road taken. *Science* 268:233–239.

Clarkson, P.M., and Sayers, S.P. 1999. Etiology of exercise-induced muscle damage. *Can. J. Appl. Physiol.* 24:234–248.

Colliander, E.B., and Tesch, P.A. 1990. Effects of eccentric and concentric muscle actions in resistance training. *Acta Physiol. Scand.* 140:31–39.

Conjard, A., Peuker, H., and Pette, D. 1998. Energy state and myosin heavy chain isoforms in single fibres of normal and transforming rabbit muscles. *Pflugers Arch.* 436:962–969.

Cope, T.C., Bodine, S.C., Fournier, M., and Edgerton, V.R. 1986. Soleus motor units in chronic spinal transected cats: Physiological and morphological alterations. *J. Neurophysiol.* 55:1202–1220.

Cope, T.C., Sokoloff, A.J., Dacko, S.M., Huot, R., and Feingold, E. 1997. Stability of motor-unit force thresholds in the decerebrate cat. *J. Neurophysiol.* 78:3077–3082.

Cope, T.C., Webb, C.B., Yee, A.K., and Botterman, B.R. 1991. Nonuniform fatigue characteristics of slow-twitch motor units activated at a fixed percentage of their maximum tetanic tension. *J. Neurophysiol.* 66:1483–1492.

Cormery, B., Pons, F., Marini, J.-F., and Gardiner, P.F. 1999. Myosin heavy chains in fibers of TTX-paralyzed rat soleus and medial gastrocnemius muscles. *J. Appl. Physiol.* 88:66–76.

Coyle, E.F., Coggan, A.R., Hopper, M.K., and Walters, T.J. 1988. Determinants of endurance in well-trained cyclists. *J. Appl. Physiol.* 64:2622–2630.

Coyle, E.F., Costill, D.L., and Lesmes, G.R. 1979. Leg extension power and muscle fiber composition. *Med. Sci. Sports Exerc.* 11:12–15.

Coyle, E.F., Sidossis, L.S., Horowitz, J.F., and Beltz, J.D. 1992. Cycling efficiency is related to the percentage of type I muscle fibers. *Med. Sci. Sports Exerc.* 24:782–788.

Crockett, J.L., Edgerton, V.R., Max, S.R., and Barnard, R.J. 1976. The neuromuscular junction in response to endurance training. *Exp. Neurol.* 51:207–215.

Czeh, G., Gallego, R., Kudo, N., and Kuno, M. 1978. Evidence for the maintenance of motoneurone properties by muscle activity. *J. Physiol.* (Lond.) 281:239–252.

Damiani, E., and Margreth, A. 1994. Characterization study of the ryanodine receptor and of calsequestrin isoforms of mammalian skeletal muscles in relation to fibre types. *J. Muscle Res. Cell Motil.* 15:86–101.

D'Aunno, D.S., Robinson, R.R., Smith, G.S., Thomason, D.B., and Booth, F.W. 1992. Intermittent acceleration as a countermeasure to soleus muscle atrophy. *J. Appl. Physiol.* 72:428–433.

Davies, C.T.M., Dooley, P., McDonagh, M.J.N., and White, M.J. 1985. Adaptation of mechanical properties of muscle to high force training in man. *J. Physiol.* 365:277–284.

Davies, C.T.M., Rutherford, L.C., and Thomas, D.O. 1987. Electrically evoked contractions of the triceps surae during and following 21 days of voluntary leg immobilization. *Eur. J. Appl. Physiol.* 56:306–312.

Davis, C.J.F., and Montgomery, A. 1977. The effect of prolonged inactivity upon the contraction characteristics of fast and slow mammalian twitch muscle. *J. Physiol.* (Lond.) 270:581–594.

Davis, H.L. 1985. Myotrophic effects on denervation atrophy of hindlimb muscles of mice with systemic administration of nerve extract. *Brain Res.* 343:176–179.

Davis, H.L., Bressler, B.H., and Jasch, L.G. 1988. Myotrophic effects on denervated fast-twitch muscles of mice: Correlation of physiologic, biochemical, and morphologic findings. *Exp. Neurol.* 99:474–489.

Davis, H.L., and Kiernan, J.A. 1981. Effect of nerve extract on atrophy of denervated or immobilized muscles. *Exp. Neurol.* 72:582–591.

Dawes, N.J., Cox, V.M., Park, K.S., Nga, H., and Goldspink, D.F. 1996. The induction of *c-fos* and *c-jun* in the stretched latissimus dorsi muscle of the rabbit: Responses to duration, degree and re-application of the stretch stimulus. *Exp. Physiol.* 81:329–339.

De Leon, R.D., Hodgson, J.A., Roy, R.R., and Edgerton, V.R. 1998. Locomotor capacity attributable to step training versus spontaneous recovery after spinalization in adult cats. *J. Neurophysiol.* 79:1329–1340.

De Luca, C.J., and Erim, Z. 1994. Common drive of motor units in regulation of muscle force. *TINS* 17:299-305.

De Luca, C.J., Foley, P.J., and Erim, Z. 1996. Motor unit control properties in constant-force isometric contractions. *J. Neurophysiol.* 76:1503–1516.

DeMartino, G.N., and Ordway, G.A. 1998. Ubiquitin-protease pathway of intracellular protein degradation: Implications for muscle atrophy during unloading. *Exerc. Sports Sci. Rev.* 26:219–252.

Dengler, R., Stein, R.B., and Thomas, C.K. 1988. Axonal conduction velocity and force of single human motor units. *Muscle & Nerve* 11:136–145.

Desaulniers, P., Lavoie, P., and Gardiner, P.F. 1998. Endurance training increases acetylcholine receptor quantity at neuromuscular junctions of adult rat skeletal muscle. *Neuroreport* 9:3549–3552.

Deschenes, M.R., Covault, J., Kraemer, W.J., and Maresh, C.M. 1994. The neuromuscular junction: Muscle fibre type differences, plasticity and adaptability to increased and decreased activity. *Sports Med.* 17:358–372.

Deschenes, M.R., Maresh, C.M., Crivello, J.F., Armstrong, L.E., Kraemer, W.J., and Covault, J. 1993. The effects of exercise training of different intensities on neuromuscular junction morphology. *J. Neurocytol.* 22:603–615.

Desmedt, J.E., and Godaux, E. 1977. Ballistic contractions in man: Characteristic recruitment pattern of single motor units of the tibialis anterior muscle. *J. Physiol.* 264:673–693.

Desmedt, J.E., and Godaux, E. 1978. Ballistic contractions in fast or slow human muscles: Discharge patterns of single motor units. *J. Physiol.* 285:185–196.

Desypris, G., and Parry, D.J. 1990. Relative efficacy of slow and fast motoneurons to reinnervate mouse soleus muscle. *Am. J. Physiol.* 258:C62–C70.

Devasahayam, S.R., and Sandercock, T.G. 1992. Velocity of shortening of single motor units from rat soleus. *J. Neurophysiol.* 67:1133–1145.

DeVol, D.L., Rotwein, P., Sadow, J.L., Novakofski, J., and Bechtel, P.J. 1990. Activation of insulin-like growth factor gene expression during work-induced skeletal muscle growth. *Am. J. Physiol. Endocrinol. Metab.* 259:E89–E95.

Diffee, G.M., Caiozzo, V.J., Herrick, R.E., and Baldwin, K.M. 1991. Contractile and biochemical properties of rat soleus and plantaris after hindlimb suspension. *Am. J. Physiol. Cell Physiol.* 260:C528–C534.

Diffee, G.M., Caiozzo, V.J., McCue, S.A., Herrick, R.E., and Baldwin, K.M. 1993. Activity-induced regulation of myosin isoform distribution: Comparison of two contractile activity programs. *J. Appl. Physiol.* 74:2509–2516.

Dix, D.J., and Eisenberg, B.R. 1990. Myosin mRNA accumulation and myofibrillogenesis at the myotendinous junction of stretched muscle fibers. *J. Cell Biol.* 111:1885–1894.

Djupsjöbacka, M., Johansson, H., and Bergenheim, M. 1994. Influences on the gamma-muscle-spindle system from muscle afferents stimulated by increased intramuscular concentrations of arachadonic acid. *Brain Res.* 663:293–302.

Djupsjöbacka, M., Johansson, H., Bergenheim, M., and Sjölander, P. 1995. Influences on the gamma-muscle-spindle system from contralateral muscle afferents stimulated by KCl and lactic acid. *Neurosci. Res.* 21:301–309.

Djupsjöbacka, M., Johansson, H., Bergenheim, M., and Wenngren, B. 1995. Influences on the gamma-muscle spindle system from muscle afferents stimulated by increased intramuscular concentrations of bradykinin and 5-HT. *Neurosci. Res.* 22:325–333.

Dolmage, T., and Cafarelli, E. 1991. Rate of fatigue during repeated submaximal contractions of human quadriceps muscle. *Can. J. Physiol. Pharmacol.* 69:1410–1415.

Donselaar, Y., Eerbeek, D., Kernell, D., and Verhey, B.A. 1987. Fibre sizes and histochemical staining characteristics in normal and chronically stimulated fast muscle of cat. *J. Physiol.* 382:237–254.

Donselaar, Y., Kernell, D., and Eerbeek, O. 1986. Soma size and oxidative enzyme activity in normal and chronically stimulated motoneurones of the cat's spinal cord. *Brain Res.* 385:22–29.

Dorlöchter, M., Irintchev, A., Brinkers, M., and Wernig, A. 1991. Effects of enhanced activity on synaptic transmission in mouse extensor digitorum longus muscle. *J. Physiol.* (Lond.) 436:283–292.

Dowling, J.J., Konert, E., Ljucovic, P., and Andrews, D.M. 1994. Are humans able to voluntarily elicit maximum muscle force? *Neurosci. Lett.* 179:25–28.

Duchateau, J. 1995. Bed rest induces neural and contractile adaptations in triceps surae. *Med. Sci. Sports Exerc.* 27:1581–1589.

Duchateau, J., and Hainaut, K. 1984. Isometric or dynamic training: Differential effects on mechanical properties of a human muscle. *J. Appl. Physiol.* 56:296–301.

Duchateau, J., and Hainaut, K. 1987. Electrical and mechanical changes in immobilized human muscle. *J. Appl. Physiol.* 62 (6): 2168–2173.

Duchateau, J., and Hainaut, K. 1990. Effects of immobilization on contractile properties, recruitment and firing rates of human motor units. *J. Physiol.* (Lond.) 422:55–65.

Duchateau, J., and Hainaut, K. 1993. Behaviour of short and long latency reflexes in fatigued human muscles. *J. Physiol.* (Lond.) 471:787–799.

Dudley, G.A., Duvoisin, M.R., Adams, G.R., Meyer, R.A., Belew, A.H., and Buchanan, P. 1992. Adaptation to unilateral lower limb suspension in humans. *Aviat. Space Environ. Med.* 63:678–683.

Dudley, G.A., Tesch, P., Miller, B., and Buchanan, P. 1991. Importance of eccentric actions in performance adaptations to resistance training. *Aviat. Space Environ. Med.* (June):543–550.

Duncan, N.D., Williams, D.A., and Lynch, G.S. 1998. Adaptations in rat skeletal muscle following long-term resistance exercise training. *Eur. J. Appl. Physiol. Occup. Physiol.* 77:372–378.

Dunn, S.E., Burns, J.L., and Michel, R.N. 1999. Calcineurin is required for skeletal muscle hypertrophy. *J. Biol. Chem.* 274:21908–21912.

Dunn, S.E., and Michel, R.N. 1997. Coordinated expression of myosin heavy chain isoforms and metabolic enzymes within overloaded rat muscle fibers. *Am. J. Physiol. Cell Physiol.* 273:C371–C383.

Dupont-Versteegden, E.E., Houlé, J.D., Gurley, C.M., and Peterson, C.A. 1998. Early changes in muscle fiber size and gene expression in response to spinal cord transection and exercise. *Am. J. Physiol. Cell Physiol.* 275:C1124–C1133.

Dux, L. 1993. Muscle relaxation and sarcoplasmic reticulum function in different muscle types. *Rev. Physiol. Biochem. Pharmacol.* 122:69–147.

Edes, I., Dosa, E., Sohar, I., and Guba, F. 1982. Effect of plaster cast immobilization on the turnover rates of soluble proteins and lactate dehydrogenase isoenzymes of rabbit m. soleus. *Acta Biochim. Biophys. Acad. Sci. Hung.* 17:211–216.

Edes, I., Sohar, I., Mazarean, H., Takacs, O., and Guba, F. 1980. Changes in the aerobic and anaerobic metabolism of skeletal muscle subjected to plaster cast immobilization. *Acta Biochim. Biophys. Acad. Sci. Hung.* 15:305–311.

Edgerton, V.R., Barnard, R.J., Peter, J.B., Maier, A., and Simpson, D.R. 1975. Properties of immobilized hindlimb muscles of the *Galago senegalensis. Exp. Neurol.* 46:115–131.

Edgerton, V.R., Zhou, M.-Y., Ohira, Y., Klitgaard, H., Jiang, B., Bell, G., Harris, B., Saltin, B., Gollnick, P.D., Roy, R.R., Day, M.K., and Greenisen, M. 1995. Human fiber size and enzymatic properties after 5 and 11 days of spaceflight. *J. Appl. Physiol.* 78:1733–1739.

Edstrom, J.-E. 1957. Effects of increased motor activity on the dimensions and the staining properties of the neuron soma. *J. Comp. Neurol.* 107:295–304.

Eken, T., and Kiehn, O. 1989. Bistable firing properties of soleus motor units in unrestrained rats. *Acta Physiol. Scand.* 136:383–394.

Ennion, S., Sant'ana Pereira, J., Sargeant, A.J., Young, A., and Goldspink, G. 1995. Characterization of human skeletal muscle fibres according to the myosin heavy chains they express. *J. Muscle Res. Cell Motil.* 16:35–43.

Essen, B., Jansson, E., Henriksson, J., Taylor, A.W., and Saltin, B. 1975. Metabolic characteristics of fiber types in human skeletal muscle. *Acta Physiol. Scand.* 95:153–165.

Evertsen, F., Medbo, J., Jebens, E., and Gjovaag, T. 1999. Effect of training on the activity of five muscle enzymes studies on elite cross-country skiers. *Acta Physiol. Scand.* 167:247–257.

Ewing, J.L., Wolfe, D.R., Rogers, M.A., Amundson, M.L., and Stull, G.A. 1990. Effects of velocity of isokinetic training on strength, power, and quadriceps muscle fibre characteristics. *Eur. J. Appl. Physiol.* 61:159–162.

Fahim, M.A. 1989. Rapid neuromuscular remodeling following limb immobilization. *Anat. Rec.* 224:102–109.

Fahim, M.A., and Robbins, N. 1986. Remodeling of the neuromuscular junction after subtotal disuse. *Brain Res.* 383:353–356.

Fallentin, N., Jorgensen, K., and Simonsen, E. 1993. Motor unit recruitment during prolonged isometric contractions. *Eur. J. Appl. Physiol.* 67:335–341.

Farrell, P., Fedele, M., Vary, T., Kimball, S., Lang, C., and Jefferson, L. 1999. Regulation of protein synthesis after acute resistance in diabetic rats. *Am. J. Physiol.* 276:E721–E727.

Fell, R.D., Steffen, J.M., and Musacchia, X.J. 1985. Effect of hypokinesia-hypodynamia on rat muscle oxidative capacity and glucose uptake. *Am. J. Physiol. Regul. Integr. Comp. Physiol.* 249 (18): R308–R312.

Fernandez, H.L., and Donoso, J.A. 1988. Exercise selectively increases G_4 AChE activity in fast-twitch muscle. *J. Appl. Physiol.* 65:2245–2252.

Fernandez, H.L., and Hodges-Savola, C.A. 1996. Physiological regulation of G_4 AChE in fast-twitch muscle: Effects of exercise and CGRP. *J. Appl. Physiol.* 80:357–362.

Ferrando, A.A., Lane, H.W., Stuart, C.A., Davis-Street, J., and Wolfe, R.R. 1996. Prolonged bed rest decreases skeletal muscle and whole body protein synthesis. *Am. J. Physiol. Endocrinol. Metab.* 270:E627–E633.

Fitts, R.H. 1994. Cellular mechanisms of muscle fatigue. *Physiol. Rev.* 74:49–94.

Fitts, R.H., Bodine, S.C., Romatowski, J.G., and Widrick, J.J. 1998. Velocity, force, power, and Ca^{2+} sensitivity of fast and slow monkey skeletal muscle fibers. *J. Appl. Physiol.* 84:1776–1787.

Fitts, R.H., Brimmer, C., Heywood-Cooksey, A., and Timmerman, R. 1989. Single muscle fiber enzyme shifts with hindlimb suspension and immobilization. *Am. J. Physiol. Cell Physiol.* 25:C1082–C1091.

Fitts, R.H., Costill, D.L., and Gardetto, P.R. 1989. Effect of swim exercise training on human muscle fiber function. *J. Appl. Physiol.* 66:465–475.

Foehring, R.C., and Munson, J.B. 1990. Motoneuron and muscle-unit properties after long-term direct innervation of soleus muscle by medial gastrocnemius nerve in cat. *J. Neurophysiol.* 64:847–861.

Foehring, R.C., Sypert, G., and Munson, J.B. 1986. Properties of self-reinnervated motor units of medial gastrocnemius of cat: I. Long-term reinnervation. *J. Neurophysiol.* 55:931–946.

Foley, J., Jayaraman, R., Prior, B., Pivarnik, J., and Meter, R. 1999. MR measurements of muscle damage and adaptation after eccentric exercise. *J. Appl. Physiol.* 87:2311–2318.

Foster, C., Costill, D.L., Daniels, J.T., and Fink, W.J. 1978. Skeletal muscle enzyme activity, fiber composition and VO_2 max in relation to distance running performance. *Eur. J. Appl. Physiol.* 39:73–80.

Fournier, M., Roy, R., Perham, H., Simard, C., and Edgerton, V. 1983. Is limb immobilization a model of muscle disuse? *Exp. Neurol.* 80:147–156.

Freyssenet, D., Connor, M., Takahashi, M., and Hood, D. 1999. Cytochrome c transcriptional activation and mRNA stability during contractile activity in skeletal muscle. *Am. J. Physiol.* 277:E26–E32.

Freyssenet, D., Di Carlo, M., and Hood, D. 1999. Calcium-dependent regulation of cytochrome c gene expression in skeletal muscle cells: Identification of a protein kinase C–dependent pathway. *J. Biol. Chem.* 274:9305–9311.

Fridén, J. 1984. Changes in human skeletal muscle induced by long-term eccentric exercise. *Cell Tissue Res.* 236:365–372.

Fridén, J., and Lieber, R.L. 1992. Structural and mechanical basis of exercise-induced muscle injury. *Med. Sci. Sports Exerc.* 24:521–530.

Fridén, J., and Lieber, R.L. 1996. Ultrastructural evidence for loss of calcium homeostasis in exercised skeletal muscle. *Acta Physiol. Scand.* 158:381–382.

Fridén, J., and Lieber, R.L. 1998. Segmental muscle fiber lesions after repetitive eccentric contractions. *Cell Tissue Res.* 293:165–171.

Fridén, J., Seger, J., and Ekblom, B. 1998. Sublethal muscle fibre injuries after high-tension anaerobic exercise. *Eur. J. Appl. Physiol.* 57:360–368.

Froese, E.A., and Houston, M.E. 1987. Performance during the Wingate anaerobic test and muscle morphology in males and females. *Int. J. Sports Med.* 8:35–39.

Fuglevand, A.J., Zackowski, K.M., Huey, K.A., and Enoka, R.M. 1993. Impairment of neuromuscular propagation during human fatiguing contractions at submaximal forces. *J. Physiol.* (Lond.) 460:549–572.

Funakoshi, H., Belluardo, N., Arenas, E., Yamamoto, Y., Casabona, A., Persson, H., and Ibáñez, C.F. 1995. Muscle-derived neurotrophin-4 as an activity-dependent trophic signal for adult motor neurons. *Science* 268:1495–1499.

Gallego, R., Kuno, M., Nunez, R., and Snider, W.D. 1979a. Dependence of motoneurone properties on the length of immobilized muscle. *J. Physiol.* (Lond.) 291:179–189.

Gallego, R., Kuno, M., Nunez, R., and Snider, W.D. 1979b. Disuse enhances synaptic efficacy in spinal motoneurones. *J. Physiol.* 291:191–205.

Galler, S., Hilber, K., Gohlsch, B., and Pette, D. 1997. Two functionally distinct myosin heavy chain isoforms in slow skeletal muscle fibres. *FEBS Lett.* 410:150–152.

Galler, S., Schmitt, T.L., Hilber, K., and Pette, D. 1997. Stretch activation and isoforms of myosin heavy chain and troponin-T of rat skeletal muscle fibres. *J. Muscle Res. Cell Motil.* 18:555–561.

Galler, S., Schmitt, T.L., and Pette, D. 1994. Stretch activation, unloaded shortening velocity, and myosin heavy chain isoforms of rat skeletal muscle fibres. *J. Physiol.* (Lond.) 478:513–521.

Galvas, P.E., Neaves, W.B., and Gonyea, W.J. 1982. Direct correlation of histochemical profile to the ultrastructure of single myofibers and their neuromuscular junctions from a mixed muscle. *Anat. Rec.* 203:1–17.

Gamrin, L., Berg, H.E., Essén, P., Tesch, P.A., Hultman, E., Garlick, P.J., McNurlan, M.A., and Wernerman, J. 1998. The effect of unloading on protein synthesis in human skeletal muscle. *Acta Physiol. Scand.* 163:369–377.

Gandevia, S.C. 1998. Neural control in human muscle fatigue: Changes in muscle afferents, motoneurones and motocortical drive. *Acta Physiol. Scand.* 162:275–283.

Gandevia, S.C., Allen, G.M., Butler, J.E., and Taylor, J.L. 1996. Supraspinal factors in human muscle fatigue: Evidence for suboptimal output from the motor cortex. *J. Physiol.* 490:529–536.

Gandevia, S.C., Petersen, N., Butler, J.E., and Taylor, J.L. 1999. Impaired response of human motoneurones to corticospinal stimulation after voluntary exercise. *J. Physiol.* (Lond.) 521:749–759.

Gandevia, S.C., and Rothwell, J.C. 1987. Knowledge of motor commands and the recruitment of human motoneurons. *Brain* 110:1117–1130.

Gardetto, P.R., Schluter, J.M., and Fitts, R.H. 1989. Contractile function of single muscle fibers after hindlimb suspension. *J. Appl. Physiol.* 66:2739–2749.

Gardiner, P.F. 1993. Physiological properties of motoneurons innervating different muscle unit types in rat gastrocnemius. *J. Neurophysiol.* 69:1160–1170.

Gardiner, P.F., Favron, M., and Corriveau, P. 1992. Histochemical and contractile responses of rat medial gastrocnemius to 2 weeks of complete disuse. *Can. J. Physiol. Pharmacol.* 70:1075–1081.

Gardiner, P.F., and Lapointe, M. 1982. Daily in vivo neuromuscular stimulation effects on immobilized rat hindlimb muscles. *J. Appl. Physiol.* 53:960–966.

Gardiner, P.F., Michel, R., and Iadeluca, G. 1984. Previous exercise training influences functional sprouting of rat hindlimb motoneurons in response to partial denervation. *Neurosci. Lett.* 45:123–127.

Gardiner, P.F., and Olha, A.E. 1987. Contractile and electromyographic characteristics of rat plantaris motor unit types during fatigue *in situ. J. Physiol.* 385:13–34.

Gardiner, P.F., and Seburn, K.L. 1997. The effects of tetrodotoxin-induced muscle paralysis on the physiological properties of muscle units and their innervating motoneurons in rat. *J. Physiol.* (Lond.) 499:207–216.

Garland, S.J., Enoka, R., Serrano, L., and Robinson, G. 1994. Behavior of motor units in human biceps brachii during a submaximal fatiguing contraction. *J. Appl. Physiol.* 76:2411–2419.

Garland, S.J., Garner, S.H., and McComas, A.J. 1988. Reduced voluntary electromyographic activity after fatiguing stimulation of human muscle. *J. Physiol.* 401:547–556.

Garland, S.J., Griffin, L., and Ivanova, T. 1997. Motor unit discharge rate is not associated with muscle relaxation time in sustained submaximal contractions in humans. *Neurosci. Lett.* 239:25–28.

Garland, S.J., and McComas, A.J. 1990. Reflex inhibition of human soleus muscle during fatigue. *J. Physiol.* (Lond.) 429:17–27.

Garnett, R.A.F., O'Donovan, M.J., Stephens, J.A., and Taylor, A. 1979. Motor unit organization of human medial gastrocnemius. *J. Physiol.* 287:33–43.

Gerchman, L.B., Edgerton, V.R., and Carrow, R.E. 1975. Effects of physical training on the histochemistry and morphology of ventral motor neurons. *Exp. Neurol.* 49:790–801.

Gerrits, H., De Haan, A., Hopman, M., Van Der Woude, L., Jones, D., and Sargeant, A. 1999. Contractile properties of the quadriceps muscle in individuals with spinal cord injury. *Muscle & Nerve* 22:1249–1256.

Gertler, R.A., and Robbins, N. 1978. Differences in neuromuscular transmission in red and white muscles. *Brain Res.* 142:160–164.

Gharakhanlou, R., Chadan, S., and Gardiner, P.F. 1999. Increased activity in the form of endurance training increases calcitonin gene-related peptide content in lumbar motoneuron cell bodies and in sciatic nerve in the rat. *Neuroscience* 89:1229–1239.

Gibala, M.J., MacDougall, J.D., Tarnopolsky, M.A., Stauber, W.T., and Elorriaga, A. 1995. Changes in human skeletal muscle ultrastructure and force production after acute resistance exercise. *J. Appl. Physiol.* 78:702–708.

Gibson, J., Halliday, D., Morrison, W., Stoward, P.J., Hornsby, G.A., Watt, P., Murdoch, G., and Rennie, M. 1987. Decrease in human quadriceps muscle protein turnover consequent upon leg immobilization. *Clin. Sci.* 72:503–509.

Giddings, C.J., and Gonyea, W.J. 1992. Morphological observations supporting muscle fiber hyperplasia following weight-lifting exercise in cats. *Anat. Rec.* 233:178–195.

Giddings, C.J., Neaves, W.B., and Gonyea, W.J. 1985. Muscle fiber necrosis and regeneration induced by prolonged weight-lifting exercise in the cat. *Anat. Rec.* 211:133–141.

Giniatullin, R.A., Bal'tser, S.K., Nikol'skii, E.E., and Magazanik, L.G. 1986. Postsynaptic potentiation and desensitization at the frog neuromuscular junction produced by repeated stimulation of the motor nerve. *Neirofiziologiya* 18:645–654.

Gisiger, V., Bélisle, M., and Gardiner, P.F. 1994. Acetylcholinesterase adaptation to voluntary wheel running is proportional to the volume of activity in fast, but not slow, rat hindlimb muscles. *Eur. J. Neurosci.* 6:673–680.

Gisiger, V., Sherker, S., and Gardiner, P.F. 1991. Swimming training increases the G_4 acetylcholinesterase content of both fast ankle extensors and flexors. *FEBS Lett.* 278:271–273.

Gisiger, V., and Stephens, H.R. 1988. Localization of the pool of G_4 acetylcholinesterase characterizing fast muscles and its alteration in murine muscular dystrophy. *J. Neurosci. Res.* 19:62–78.

Goldspink, D.F. 1977a. The influence of activity on muscle size and protein turnover. *J. Physiol.* 264:283–296.

Goldspink, D.F. 1977b. The influence of immobilization and stretch on protein turnover of rat skeletal muscle. *J. Physiol.* 264:267–282.

Goldspink, D.F., Cox, V.M., Smith, S.K., Eaves, L.A., Osbaldeston, N.J., Lee, D.M., and Mantle, D. 1995. Muscle growth in response to mechanical stimuli. *Am. J. Physiol. Endocrinol. Metab.* 268:E288–E297.

Goldspink, D.F., Garlick, P.J., and McNurlan, M.A. 1983. Protein turnover measured *in vivo* and *in vitro* in muscles undergoing compensatory growth and subsequent denervation atrophy. *Biochem. J.* 210:89–98.

Gonyea, W.J., Sale, D.G., Gonyea, F.B., and Mikesky, A. 1986. Exercise induced increases in muscle fiber number. *Eur. J. Appl. Physiol.* 55:137–141.

Goodyear, L.J., Chang, P.-Y., Sherwood, D.J., Dufresne, S.D., and Moller, D.E. 1996. Effects of exercise and insulin on mitogen-activated protein kinase signaling pathways in rat skeletal muscle. *Am. J. Physiol. Endocrinol. Metab.* 34:E403–E408.

Gorassini, M.A., Bennett, D.J., and Yang, J.F. 1998. Self-sustained firing of human motor units. *Neurosci. Lett.* 247:13–16.

Goslow, G.E., Jr., Cameron, W.E., and Stuart, D.G. 1977. The fast twitch motor units of cat ankle flexor: 1. Tripartite classification on basis of fatigability. *Brain Res.* 134:35–46.

Graham, S.C., Roy, R.R., Navarro, C., Jiang, B., Pierotti, D., Bodine-Fowler, S., and Edgerton, V.R. 1992. Enzyme and size profiles in chronically inactive cat soleus muscle fibers. *Muscle & Nerve* 15:27–36.

Graham, S.C., Roy, R.R., West, S.P., Thomason, D., and Baldwin, K.M. 1989. Exercise effects on the size and metabolic properties of soleus fibers in hindlimb-suspended rats. *Aviat. Space Environ. Med.* 60:226–234.

Grana, E.A., Chiou-Tan, F., and Jaweed, M.M. 1996. Endplate dysfunction in healthy muscle following a period of disuse. *Muscle & Nerve* 19:989–993.

Granit, R., Kernell, D., and Shortess, G.K. 1963a. The behaviour of mammalian motoneurones during long-lasting orthodromic, antidromic and transmembrane stimulation. *J. Physiol* 169:743–754.

Granit, R., Kernell, D., and Shortess, G.K. 1963b. Quantitative aspects of repetitive firing of mammalian motoneurones, caused by injected currents. *J. Physiol.* 168:911–931.

Greaser, M., Moss, R., and Reiser, P. 1988. Variations in contractile properties of rabbit single muscle fibres in relation to troponin T isoforms and myosin light chains. *J. Physiol.* (Lond.) 406:85–98.

Green, H.J., Ball-Burnett, M., Chin, E., and Pette, D. 1992. Time-dependent increases in Na$^+$,K$^+$-ATPase content of low-frequency-stimulated rabbit muscle. *FEBS Lett.* 310:129–131.

Green, H.J., Düsterhöft, S., Dux, L., and Pette, D. 1992. Metabolite patterns related to exhaustion, recovery and transformation of chronically stimulated rabbit fast-twitch muscle. *Pflugers Arch.* 420:359–366.

Gregor, R.J., Edgerton, V.R., Perrine, J.J., Campion, D.S., and DeBus, C. 1979. Torque–velocity relationships and muscle fiber composition in elite female athletes. *J. Appl. Physiol. Respir. Environ. Exerc. Physiol.* 47:388–392.

Griffin, L., Ivanova, T., and Garland, S. 2000. Role of limb movement in the modulation of motor unit discharge rate during fatiguing contractions. *Exp. Brain Res.* 130:392–400.

Grimby, G., Broberg, C., Krotkiewska, I., and Krotkiewski, M. 1976. Muscle fiber composition in patients with traumatic cord lesion. *Scand. J. Rehab. Med.* 8:37–42.

Grimby, L., Hannerz, J., and Hedman, B. 1981. The fatigue and voluntary discharge properties of single motor units in man. *J. Physiol.* (Lond.) 316:545–554.

Grossman, E.J., Roy, R.R., Talmadge, R.J., Zhong, H., and Edgerton, V.R. 1998. Effects of inactivity on myosin heavy chain composition and size of rat soleus fibers. *Muscle & Nerve* 21:375–389.

Guertin, P.A., and Hounsgaard, J. 1999. Non-volatile general anaesthetics reduce spinal activity by suppressing plateau potentials. *Neuroscience* 88:353–358.

Gustafsson, B., Katz, R., and Malmsten, J. 1982. Effects of chronic partial deafferentation on the electrical properties of lumbar alpha-motoneurones in the cat. *Brain Res.* 246:23–33.

Gustafsson, B., and Pinter, M.J. 1984a. An investigation of threshold properties among cat spinal alpha-motoneurons. *J. Physiol.* (Lond.) 357:453–483.

Gustafsson, B., and Pinter, M.J. 1984b. Relations among passive electrical properties of lumbar alpha-motoneurones of the cat. *J. Physiol.* 356:401–431.

Gustafsson, B., and Pinter, M.J. 1985. Factors determining the variation of the afterhyperpolarization duration in cat lumbar alpha-motoneurons. *Brain Res.* 326:392–395.

Gydikov, A., Dimitrov, G., Kosarov, D., and Dimitrova, N. 1976. Functional differentiation of motor units in human opponens pollicis muscle. *Exp. Neurol.* 50:36–47.

Haddad, F., Herrick, R.E., Adams, G.R., and Baldwin, K.M. 1993. Myosin heavy chain expression in rodent skeletal muscle: Effects of exposure to zero gravity. *J. Appl. Physiol.* 75:2471–2477.

Haddad, F., Qin, A.X., Zeng, M., McCue, S.A., and Baldwin, K.M. 1998. Interaction of hyperthyroidism and hindlimb suspension on skeletal myosin heavy chain expression. *J. Appl. Physiol.* 85:2227–2236.

Hagbarth, K.-E., Bongiovanni, L.G., and Nordin, M. 1995. Reduced servo-control of fatigued human finger extensor and flexor muscles. *J. Physiol.* (Lond.) 485:865–872.

Hagbarth, K.-E., Kunesch, E., Nordin, M., Schmidt, R., and Wallin, E. 1986. Gamma loop contributing to maximal voluntary contractions in man. *J. Physiol.* (Lond.) 380:575–591.

Hainaut, K., and Duchateau, J. 1989. Muscle fatigue, effects of training and disuse. *Muscle & Nerve* 12:660–669.

Häkkinen, K., Kallinen, M., Izquierdo, M., Jokelainen, K., Lassila, H., Mälkiä, E., Kraemer, W.J., Newton, R.U., and Alén, M. 1998. Changes in agonist–antagonist EMG, muscle CSA, and force during strength training in middle-aged and older people. *J. Appl. Physiol.* 84:1341–1349.

Häkkinen, K., Kallinen, M., Linnamo, V., Pastinen, U.M., Newton, R.U., and Kraemer, W.J. 1996. Neuromuscular adaptations during bilateral versus unilateral strength training in middle-aged and elderly men and women. *Acta Physiol. Scand.* 158:77–88.

Häkkinen, K., and Komi, P.V. 1983a. Alterations of mechanical characteristics of human skeletal muscle during strength training. *Eur. J. Appl. Physiol.* 50:161–172.

Häkkinen, K., and Komi, P.V. 1983b. Electromyographic and mechanical characteristics of human skeletal muscle during fatigue under voluntary and reflex conditions. *Electroencephalogr. Clin. Neurophysiol.* 55:436–444.

Häkkinen, K., and Komi, P.V. 1986. Effects of fatigue and recovery on electromyographic and isometric force- and relaxation-time characteristics of human skeletal muscle. *Eur. J. Appl. Physiol.* 55:588–596.

Häkkinen, K., Komi, P.V., and Alén, M. 1985. Effect of explosive type strength training on isometric force- and relaxation-time, electromyographic and muscle fibre characteristics of leg extensor muscles. *Acta Physiol. Scand.* 125:587–600.

Häkkinen, K., Newton, R.U., Gordon, S.E., McCormick, M., Volek, J.S., Nindl, B.C., Gotshalk, L.A., Campbell, W.W., Evans, W.J., Häkkinen, A., Humphries, B.J., and Kraemer, W.J. 1998. Changes in muscle morphology, electromyographic activity, and force production characteristics during progressive strength training in young and older men. *J. Gerontol. A* 53A:B415–B423.

Haller, H., Lindschau, C., and Luft, F.C. 1994. Role of protein kinase C in intracellular signaling. *Ann. NY Acad. Sci.* 733:313–324.

Halter, J.A., Carp, J.S., and Wolpaw, J.R. 1995. Operantly conditioned motoneuron plasticity: Possible role of sodium channels. *J. Neurophysiol.* 73:867–871.

Hämäläinen, N., and Pette, D. 1995. Patterns of myosin isoforms in mammalian skeletal muscle fibres. *Microscopy Res. Technique* 30:381–389.

Hamm, T.M., Nemeth, P.M., Solanki, L., Gordon, D.A., Reinking, R.M., and Stuart, D.G. 1988. Association between biochemical and physiological properties in single motor units. *Muscle & Nerve* 11:245–254.

Harridge, S.D.R., Bottinelli, R., Canepari, M., Pellegrino, M.A., Reggiani, C., Esbjörnsson, M., Balsom, P.D., and Saltin, B. 1998. Sprint training, *in vitro* and *in vivo* muscle function, and myosin heavy chain expression. *J. Appl. Physiol.* 84:442–449.

Harridge, S.D.R., Bottinelli, R., Canepari, M., Pellegrino, M.A., Reggiani, C., Esbjörnsson, M., and Saltin, B. 1996. Whole-muscle and single-fibre contractile properties and myosin heavy chain isoforms in humans. *Pflügers Arch.* 432:913–920.

Hather, B.M., Adams, G.R., Tesch, P.A., and Dudley, G.A. 1992. Skeletal muscle responses to lower limb suspension in humans. *J. Appl. Physiol.* 72:1493–1498.

Hather, B.M., Tesch, P.A., Buchanan, P., and Dudley, G.A. 1991. Influence of eccentric actions on skeletal muscle adaptations to resistance training. *Acta Physiol. Scand.* 143:177–185.

Hauschka, E.O., Roy, R.R., and Edgerton, V.R. 1987. Size and metabolic properties of single muscle fibers in rat soleus after hindlimb suspension. *J. Appl. Physiol.* 62:2338–2347.

Hayward, L., Breitbach, D., and Rymer, W.Z. 1988. Increased inhibitory effects on close synergists during muscle fatigue in the decerebrate cat. *Brain Res.* 440:199–203.

Hayward, L., Wesselmann, U., and Rymer, W.Z. 1991. Effects of muscle fatigue on mechanically sensitive afferents of slow conduction velocity in the cat triceps surae. *J. Neurophysiol.* 65:360–370.

Heckman, C.J., and Binder, M.D. 1991. Computer simulation of the steady-state input-output function of the cat medial gastrocnemius motoneuron pool. *J. Neurophysiol.* 65:952–967.

Heckman, C.J., and Binder, M.D. 1993. Computer simulations of motoneuron firing rate modulation. *J. Neurophysiol.* 69:1005–1008.

Heilmann, C., and Pette, D. 1979. Molecular transformations in sarcoplasmic reticulum of fast-twitch muscle by electro-stimulation. *Eur. J. Biochem.* 93:437–446.

Heiner, L., Domonkos, J., Motika, D., and Vargha, M. 1984. Role of the nervous system in regulation of the sarcoplasmic membrane function in different muscle fibres. *Acta Physiol. Hung.* 64:129–133.

Henneman, E. 1981. Recruitment of motor units: The size principle. In *Motor unit types, recruitment and plasticity in health and disease,* ed. J.E. Desmedt, 26-60. New York: Karger.

Henneman, E., Somjen, G., and Carpenter, D.O. 1965. Excitability and inhibitility of motoneurons of different sizes. *J. Neurophysiol.* 28:599–620.

Henriksson, J., Chi, M.M.-Y., Hintz, C.S., Young, D.A., Kaiser, K.K., Salmons, S., and Lowry, O.H. 1986. Chronic stimulation of mammalian muscle: Changes in enzymes of six metabolic pathways. *Am. J. Physiol. Cell Physiol.* 251:C614–C632.

Hensbergen, E., and Kernell, D. 1992. Task-related differences in distribution of electromyographic activity within peroneus longus muscle of spontaneously moving cats. *Exp. Brain Res.* 89:682–685.

Herbert, M.E., Roy, R.R., and Edgerton, V.R. 1988. Influence of one-week hindlimb suspension and intermittent high load exercise on rat muscles. *Exp. Neurol.* 102:190–198.

Herbert, R.D., Dean, C., and Gandevia, S.C. 1998. Effects of real and imagined training on voluntary muscle activation during maximal isometric contractions. *Acta Physiol. Scand.* 163:361–368.

Herbert, R.D., and Gandevia, S.C. 1996. Muscle activation in unilateral and bilateral efforts assessed by motor nerve and cortical stimulation. *J. Appl. Physiol.* 80:1351–1356.

Hesketh, J.E., and Whitelaw, P.F. 1992. The role of cellular oncogenes in myogenesis and muscle cell hypertrophy. *Int. J. Biochem.* 24:193–203.

Hicks, A., Ohlendieck, K., Göpel, S.O., and Pette, D. 1997. Early functional and biochemical adaptations to low-frequency stimulation of rabbit fast-twitch muscle. *Am. J. Physiol. Cell Physiol.* 273:C297–C305.

Higbie, E.J., Cureton, K.J., Warren, G.L., III, and Prior, B.M. 1996. Effects of concentric and eccentric training on muscle strength, cross-sectional area, and neural activation. *J. Appl. Physiol.* 81:2173–2181.

Hikida, R.S., Gollnick, P.D., Dudley, G.A., Convertino, V.A., and Buchanan, P. 1989. Structural and metabolic characteristics of human skeletal muscle following 30 days of simulated microgravity. *Aviat. Space Environ. Med.* 60:664–670.

Hill, J.M. 2000. Discharge of group IV phrenic afferent fibers increases during diaphragmatic fatigue. *Brain Res.* 856:240–244.

Hintz, C.S., Chi, M.M.-Y., and Lowry, O.H. 1984. Heterogeneity in regard to enzymes and metabolites within individual muscle fibers. *Am. J. Physiol. Cell Physiol.* 246:C288–C292.

Hintz, C.S., Coyle, E.F., Kaiser, K.K., Chi, M.M.-Y., and Lowry, O.H. 1984. Comparison of muscle fiber typing by quantitative enzyme assays and by myosin ATPase staining. *J. Histochem. Cytochem.* 32:655–660.

Hintz, C.S., Lowry, C.V., Kaiser, K.K., McKee, D., and Lowry, O.H. 1980. Enzyme levels in individual rat muscle fibers. *Am. J. Physiol. Cell Physiol.* 239:C58–C65.

Hochman, S., and McCrea, D.A. 1994a. Effects of chronic spinalization on ankle extensor motoneurons: I. Composite monosynaptic Ia EPSPs in four motoneuron pools. *J. Neurophysiol.* 71:1452–1467.

Hochman, S., and McCrea, D.A. 1994b. Effects of chronic spinalization on ankle extensor motoneurons: II. Motoneuron electrical properties. *J. Neurophysiol.* 71:1468–1479.

Hodgson, J.A., Roy, R.R., De Leon, R., Dobkin, B., and Edgerton, V.R. 1994. Can the mammalian lumbar spinal cord learn a motor task? *Med. Sci. Sports Exerc.* 26:1491–1497.

Hofmann, P.A., Metzger, J.M., Greaser, M.L., and Moss, R.L. 1990. Effects of partial extraction of light chain 2 on the Ca^{2+} sensitivities of isometric tension, stiffness, and velocity of shortening in skinned skeletal muscle fibers. *J. Gen. Physiol.* 95:477–498.

Hong, S.J., and Lnenicka, G.A. 1993. Long-term changes in the neuromuscular synapses of a crayfish motoneuron produced by calcium influx. *Brain Res.* 605:121–127.

Hood, D.A., and Parent, G. 1991. Metabolic and contractile responses of rat fast-twitch muscle to 10-Hz stimulation. *Am. J. Physiol. Cell Physiol.* 260:C832–C840.

Horowitz, J.F., Sidossis, L.S., and Coyle, E.F. 1994. High efficiency of type I fibers improves performance. *Int. J. Sports Med.* 15:152–157.

Hortobágyi, T., Barrier, J., Beard, D., Braspennincx, J., Koens, P., Devita, P., Dempsey, L., and Lambert, N.J. 1996. Greater initial adaptations to submaximal muscle lengthening than maximal shortening. *J. Appl. Physiol.* 81:1677–1682.

Hortobágyi, T., Hill, J.P., Houmard, J.A., Fraser, D.D., Lambert, N.J., and Israel, R.G. 1996. Adaptive responses to muscle lengthening and shortening in humans. *J. Appl. Physiol.* 80:765–772.

Houston, M.E., Norman, R.W., and Froese, E.A. 1988. Mechanical measures during maximal velocity knee extension exercise and their relation to fibre composition of the human vastus lateralis muscle. *Eur. J. Appl. Physiol.* 58:1–7.

Howald, H., Hoppeler, H., Claassen, H., Mathieu, O., and Straub, R. 1985. Influences of endurance training on the ultrastructural composition of the different muscle fiber types in humans. *Pflugers Arch.* 403:369–376.

Howard, G., Steffen, J.M., and Geoghegan, T.E. 1989. Transcriptional regulation of decreased protein synthesis during skeletal muscle unloading. *J. Appl. Physiol.* 66:1093–1098.

Howard, J.D., and Enoka, R.M. 1991. Maximum bilateral contractions are modified by neurally mediated interlimb effects. *J. Appl. Physiol.* 70:306–316.

Howell, J.N., Chleboun, G., and Conatser, R. 1993. Muscle stiffness, strength loss, swelling and soreness following exercise-induced injury in humans. *J. Physiol. (Lond.)* 464:183–196.

Howell, J.N., Fuglevand, A.J., Walsh, M., and Bigland-Ritchie, B.R. 1995. Motor unit activity during isometric and concentric-eccentric contractions of the human first dorsal interosseus muscle. *J. Neurophysiol.* 74:901–904.

Howell, S., Zhan, W.Z., and Sieck, G.C. 1997. Diaphragm disuse reduces Ca^{2+} uptake capacity of sarcoplasmic reticulum. *J. Appl. Physiol.* 82:164–171.

Hsiao, C.F., Trueblood, P.R., Levine, M.S., and Chandler, S.H. 1997. Multiple effects of serotonin on membrane properties of trigeminal motoneurons *in vitro*. *J. Neurophysiol.* 77:2910–2924.

Hu, P., Yin, C., Zhang, K.M., Wright, L.D., Nixon, T.E., Wechsler, A.S., Spratt, J.A., and Briggs, F.N. 1995. Transcriptional regulation of phospholamban gene and translational regulation of SERCA2 gene produces coordinate expression of these two sarcoplasmic reticulum proteins during skeletal muscle phenotype switching. *J. Biol. Chem.* 270:11619–11622.

Hubatsch, D.A., and Jasmin, B.J. 1997. Mechanical stimulation increases expression of acetylcholinesterase in cultured myotubes. *Am. J. Physiol. Cell Physiol.* 273:C2002–C2009.

Huber, B., and Pette, D. 1996. Dynamics of parvalbumin expression in low-frequency-stimulated fast-twitch rat muscle. *Eur. J. Biochem.* 236:814–819.

Hultborn, H., Katz, R., and Mackel, R. 1988. Distribution of recurrent inhibition within a motor nucleus: II. Amount of recurrent inhibition in motoneurones to fast and slow units. *Acta Physiol. Scand.* 134:363–374.

Hultborn, H., and Kiehn, O. 1992. Neuromodulation of vertebrate motor neuron membrane properties. *Curr. Opin. Neurobiol.* 2:770–775.

Hultborn, H., Lipski, J., Mackel, R., and Wigström, H. 1988. Distribution of recurrent inhibition within a motor nucleus: I. Contribution from slow and fast motor units to the excitation of Renshaw cells. *Acta Physiol. Scand.* 134:347–361.

Hultborn, H., Meunier, S., Morin, C., and Pierrot-Deseilligny, E. 1987. Assessing changes in presynaptic inhibition of Ia fibres: A study in man and the cat. *J. Physiol.* (Lond.) 389:729–756.

Hultborn, H., Meunier, S., Pierrot-Deseilligny, E., and Shindo, M. 1987. Changes in presynaptic inhibition of Ia fibres at the onset of voluntary contraction in man. *J. Physiol.* (Lond.) 389:757–772.

Hulten, B., and Karlsson, J. 1974. Relationship between isometric endurance and muscle fiber type composition. *Acta Physiol. Scand.* 91:A46–A47.

Hutton, R.S., and Nelson, D.L. 1986. Stretch sensitivity of Golgi tendon organs in fatigued gastrocnemius muscle. *Med. Sci. Sports Exerc.* 18:69–74.

Inbar, O., Kaiser, P., and Tesch, P. 1981. Relationships between leg muscle fiber type distribution and leg exercise performance. *Int. J. Sports Med.* 2:154–159.

Ishihara, A., Ohira, Y., Roy, R.R., Nagaoka, S., Sekiguchi, C., Hinds, W.E., and Edgerton, V.R. 1996. Influence of spaceflight on succinate dehydrogenase activity and soma size of rat ventral horn neurons. *Acta Anat.* (Basel) 157:303–308.

Ishihara, A., Oishi, Y., Roy, R.R., and Edgerton, V.R. 1997. Influence of two weeks of non–weight bearing on rat soleus motoneurons and muscle fibers. *Aviat. Space Environ. Med.* 68:421–425.

Ishihara, A., Roy, R.R., and Edgerton, V.R. 1995. Succinate dehydrogenase activity and soma size of motoneurons innervating different portions of the rat tibialis anterior. *Neuroscience* 68:813–822.

Ivanova, T., Garland, S.J., and Miller, K.J. 1997. Motor unit recruitment and discharge behavior in movements and isometric contractions. *Muscle & Nerve* 20:867–874.

Ivy, J.L., Withers, R.T., Brose, G., Maxwell, B.D., and Costill, D.L. 1981. Isokinetic contractile properties of the quadriceps with relation to fiber type. *Eur. J. Appl. Physiol.* 47:247–255.

Jacobs-El, J., Ashley, W., and Russell, B. 1993. IIx and slow myosin expression follow mitochondrial in-

creases in transforming muscle fibers. *Am. J. Physiol. Cell Physiol.* 265:C79–C84.

Jacobs-El, J., Zhou, M.-Y., and Russell, B. 1995. MRF-4, Myf-5, and myogenin mRNAs in the adaptive responses of mature rat muscle. *Am. J. Physiol. Cell Physiol.* 268:C1045–C1052.

Jain, N., Florence, S.L., and Kaas, J.H. 1998. Reorganization of somatosensory cortex after nerve and spinal cord injury. *News Physiol. Sci.* 13:143–149.

Jakab, G., Dux, L., Tabith, K., and Guba, F. 1987. Effects of disuse on the function of fragmented sarcoplasmic reticulum of rabbit m. gastrocnemius. *Gen. Physiol. Biophys.* 6:127–135.

Jami, L., Murthy, K.S.K., Petit, J., and Zytnicki, D. 1983. After-effects of repetitive stimulation at low frequency on fast-contracting motor units of cat muscle. *J. Physiol.* 340:129–143.

Jänkälä, H., Harjola, V.P., Petersen, N.E., and Härkönen, M. 1997. Myosin heavy chain mRNA transform to faster isoforms in immobilized skeletal muscle: A quantitative PCR study. *J. Appl. Physiol.* 82:977–982.

Jansson, E., and Hedberg, G. 1991. Skeletal muscle fiber types in teenagers: Relationship to physical performance and activity. *Scand. J. Med. Sci. Sports* 1:31–44.

Jansson, E., Sylvén, C., Arvidsson, I., and Eriksson, E. 1988. Increase in myoglobin content and decrease in oxidative enzyme activities by leg muscle immobilization in man. *Acta Physiol. Scand.* 132:515–517.

Jaschinski, F., Schuler, M., Peuker, H., and Pette, D. 1998. Changes in myosin heavy chain mRNA and protein isoforms of rat muscle during forced contractile activity. *Am. J. Physiol. Cell Physiol.* 274:C365–C370.

Jasmin, B.J., Gardiner, P.F., and Gisiger, V. 1991. Muscle acetylcholinesterase adapts to compensatory overload by a general increase in its molecular forms. *J. Appl. Physiol.* 70:2485–2489.

Jasmin, B.J., and Gisiger, V. 1990. Regulation by exercise of the pool of G_4 acetylcholinesterase characterizing fast muscles: Opposite effect of running training in antagonist muscles. *J. Neurosci.* 10:1444–1454.

Jasmin, B., Lavoie, P., and Gardiner, P.F. 1988. Fast axonal transport of labeled proteins in motoneurons of exercise-trained rats. *Am. J. Physiol. Cell Physiol.* 255:C731–C736.

Jiang, B., Ohira, Y., Roy, R.R., Nguyen, Q., Ilyina-Kakueva, E.I., Oganov, V., and Edgerton, V.R. 1992. Adaptation of fibers in fast-twitch muscles of rats to spaceflight and hindlimb suspension. *J. Appl. Physiol.* 73 suppl.: 58S–65S.

Jiang, B., Roy, R.R., and Edgerton, V.R. 1990. Enzymatic plasticity of medial gastrocnemius fibers in the adult chronic spinal cat. *Am. J. Physiol. Cell Physiol.* 259:C507–C514.

Jiang, B., Roy, R.R., Navarro, C., Nguyen, Q., Pierotti, D., and Edgerton, V.R. 1991. Enzymatic responses of cat medial gastrocnemius fibers to chronic inactivity. *J. Appl. Physiol.* 70:231–239.

Johansson, C., Lorentzon, R., Sjöström, M., Fagerlund, M., and Fugl-Meyer, A.R. 1987. Sprinters and marathon runners: Does isokinetic knee extensor performance reflect muscle size and structure? *Acta Physiol. Scand.* 130:663–669.

Johnson, B.D., and Sieck, G.C. 1993. Differential susceptibility of diaphragm muscle fibers to neuromuscular transmission failure. *J. Appl. Physiol.* 75:341–348.

Johnson, L.D., Jiang, Y., and Rall, J.A. 1999. Intracellular EDTA mimics parvalbumin in the promotion of skeletal muscle relaxation. *Biophys. J.* 76:1514–1522.

Jokl, P., and Konstadt, S. 1983. The effect of limb immobilization on muscle function and protein composition. *Clin. Orthop. Rel. Res.* 174:222–229.

Jones, D.A., and Rutherford, O.M. 1987. Human muscle strength training: The effects of three different regimes and the nature of the resultant changes. *J. Physiol* 391:1–11.

Jones, D.A., Rutherford, O.M., and Parker, D.F. 1989. Physiological changes in skeletal muscle as a result of strength training. *Q. J. Exp. Physiol.* 74:233–256.

Jones, K.E., Bawa, P., and McMillan, A.S. 1993. Recruitment of motor units in human flexor carpi ulnaris. *Brain Res.* 602:354–356.

Jones, K.E., Lyons, M., Bawa, P., and Lemon, R.N. 1994. Recruitment order of motoneurons during functional tasks. *Exp. Brain Res.* 100:503–508.

Jones, K.J. 1993. Gonadal steroids and neuronal regeneration: A therapeutic role. *Adv. Neurol.* 59:227–240.

Jones, T., Chu, C., Grande, L., and Gregory, A. 1999. Motor skills training enhances lesion-induced structural plasticity in the motor cortex of adult rats. *J. Neurosci.* 19:10153–10163.

Jones, T.A., Kleim, J.A., and Greenough, W.T. 1996. Synaptogenesis and dendritic growth in the cortex opposite unilateral sensorimotor cortex damage in adult rats: A quantitative electron microscopic examination. *Brain Res.* 733:142–148.

Jozsa, L., Kannus, P., Thoring, J., Reffy, A., Järvinen, M., and Kvist, M. 1990. The effect of tenotomy and immobilisation on intramuscular connective tissue. *J. Bone Joint Surg.* 72B:293–297.

Julian, F.J., Moss, R.L., and Waller, G.S. 1981. Mechanical properties and myosin light chain composition of skinned muscle fibers from adult and new-born rabbits. *J. Physiol.* (Lond.) 311:201–218.

Jürimäe, J., Abernethy, P.J., Quigley, B.M., Blake, K., and McEniery, M.T. 1997. Differences in muscle contractile characteristics among bodybuilders, endurance trainers and control subjects. *Eur. J. Appl. Physiol. Occup. Physiol.* 75:357–362.

Jürimäe, J., Blake, K., Abernethy, P.J., and McEniery, M.T. 1996. Changes in the myosin heavy chain isoform profile of the triceps brachii muscle following 12 weeks of resistance training. *Eur. J. Appl. Physiol. Occup. Physiol.* 74:287–292.

Kaczkowski, W., Montgomery, D.L., Taylor, A.W., and Klissouras, V. 1982. The relationship between muscle fiber composition and maximal anaerobic power and capacity. *J. Sports Med. Phys. Fitness* 22:407–413.

Kadi, F., and Thornell, L. 1999. Training affects myosin heavy chain phenotype in the trapezius muscle of women. *Histochem. Cell. Biol.* 112:73–78.

Kadi, F., and Thornell, L. 2000. Concomitant increases in myonuclear and satellite cell content in female trapezius muscle following strength training. *Histochem. Cell Biol.* 113:99–103.

Kanda, K., Burke, R.E., and Walmsley, B. 1977. Differential control of fast and slow twitch motor units in the decerebrate cat. *Exp. Brain Res.* 29:57–74.

Kang, C.-M., Lavoie, P.-A., and Gardiner, P.F. 1995. Chronic exercise increases SNAP-25 abundance in fast-transported proteins of rat motoneurones. *Neuroreport* 6:549–553.

Karlsson, J., and Jacobs, F. 1982. Onset of blood lactate accumulation during muscular exercise as a threshold concept. *Int. J. Sports Med.* 3:190–201.

Kaufman, M.P., Longhurst, J.C., Rybicki, K.J., Wallach, J.H., and Mitchell, J.H. 1983. Effects of static muscular contraction on impulse activity of groups III and IV afferents in cats. *J. Appl. Physiol. Respir. Environ. Exerc. Physiol.* 55:105–112.

Kawakami, Y., Abe, T., Kuno, S.Y., and Fukunaga, T. 1995. Training-induced changes in muscle architecture and specific tension. *Eur. J. Appl. Physiol.* 72:37–43.

Kelley, G. 1996. Mechanical overload and skeletal muscle fiber hyperplasia: A meta-analysis. *J. Appl. Physiol.* 81:1584–1588.

Kent-Braun, J.A., and Le Blanc, R. 1996. Quantitation of central activation failure during maximal voluntary contractions in humans. *Muscle & Nerve* 19:861–869.

Kent-Braun, J.A., Ng, A.V., Castro, M., Weiner, M.W., Gelinas, D., Dudley, G.A., and Miller, R.G. 1997. Strength, skeletal muscle composition, and enzyme activity in multiple sclerosis. *J. Appl. Physiol.* 83:1998–2004.

Kernell, D. 1965a. The adaptation and the relation between discharge frequency and current strength of cat lumbosacral motoneurones stimulated by long-lasting injected currents. *Acta Physiol. Scand.* 65:65–73.

Kernell, D. 1965b. High-frequency repetitive firing of cat lumbosacral motoneurones stimulated by long-lasting injected currents. *Acta Physiol. Scand.* 65:74–86.

Kernell, D. 1965c. The limits of firing frequency in cat lumbosacral motoneurons possessing different time course of afterhyperpolarization. *Acta Physiol. Scand.* 65:87–100.

Kernell, D. 1979. Rhythmic properties of motoneurones innervating muscle fibres of different speed in m. gastrocnemius medialis of the cat. *Brain Res.* 160:159–162.

Kernell, D. 1983. Functional properties of spinal motoneurons and gradation of muscle force. In *Motor control mechanisms in health and disease,* ed. J.E. Desmedt, 213-226. New York: Raven Press.

Kernell, D. 1984. The meaning of discharge rate: Excitation-to-frequency transduction as studied in spinal motoneurons. *Arch. Ital. Biol.* 122:5–15.

Kernell, D. 1992. Organized variability in the neuromuscular system: A survey of task-related adaptations. *Arch. Ital. Biol.* 130:19–66.

Kernell, D., Eerbeek, O., Verhey, B., and Donselaar, Y. 1987. Effects of physiological amounts of high- and low-rate chronic stimulation on fast-twitch muscle of the cat hindlimb: I. Speed- and force-related properties. *J. Neurophysiol.* 58:598–612.

Kernell, D., and Monster, A. 1981. Threshold current for repetitive impulse firing in motoneurones innervating muscle fibers of different fatigue sensitivity in the cat. *Brain Res.* 229:193–196.

Kernell, D., and Zwaagstra, B. 1981. Input conductance, axonal conduction velocity and cell size among hindlimb motoneurones of the cat. *Brain Res.* 204:311–326.

Kernell, D., and Zwaagstra, B. 1989. Dendrites of cat's spinal motoneurones: Relationship between stem diameter and predicted input conductance. *J. Physiol.* 413:255–269.

Kiehn, O., Erdal, J., Eken, T., and Bruhn, T. 1996. Selective depletion of spinal monoamines changes the rat soleus EMG from a tonic to a more phasic pattern. *J. Physiol.* 492:173–184.

Kilgour, R.D., Gariepy, P., and Rehel, R. 1991. Facial cooling does not benefit cardiac dynamics during recovery from exercise hyperthermia. *Aviat. Space Environ. Med.* 62:849–854.

Kim, D.H., Witzmann, F.A., and Fitts, R.H. 1982. Effect of disuse on sarcoplasmic reticulum in fast and slow skeletal muscle. *Am. J. Physiol. Cell Physiol.* 243 (12): C156–C160.

Kirby, C.R., Ryan, M.J., and Booth, F.W. 1992. Eccentric exercise training as a countermeasure to non-weight-bearing soleus muscle atrophy. *J. Appl. Physiol.* 73:1894–1899.

Kirschbaum, B.J., Schneider, S., Izumo, S., Mahdavi, V., Nadal-Ginard, B., and Pette, D. 1990. Rapid and reversible changes in myosin heavy chain expression in response to increased neuromuscular activity of rat fast-twitch muscle. *FEBS Lett.* 268:75–78.

Kleim, J.A., Barbay, S., and Nudo, R.J. 1998. Functional reorganization of the rat motor cortex following motor skill learning. *J. Neurophysiol.* 80:3321–3325.

Klitgaard, H., Bergman, O., Betto, R., Salviati, G., Schiaffino, S., Clausen, T., and Saltin, B. 1990. Co-existence of myosin heavy chain I and IIa isoforms in human skeletal muscle fibres with endurance training. *Pflugers Arch.* 416:470–472.

Klitgaard, H., Zhou, M., and Richter, E.A. 1990. Myosin heavy chain composition of single fibers from m. biceps brachii of male body builders. *Acta Physiol. Scand.* 140:175–180.

Kniffki, K.-D., Schomburg, E.D., and Steffens, H. 1981. Convergence in segmental reflex pathways from fine muscle afferents and cutaneous or group II muscle afferents to alpha-motoneurones. *Brain Res.* 218:342–346.

Komi, P.V., Rusko, H., Vos, J., and Vhiko, V. 1977. Anaerobic performance capacity in athletes. *Acta. Physiol. Scand.* 100:107–114.

Komulainen, J., Takala, T.E.S., Kuipers, H., and Hesselink, M.K.C. 1998. The disruption of myofibre structures in rat skeletal muscle after forced lengthening contractions. *Pflugers Arch.* 436:735–741.

Koryak, Y. 1998. Electromyographic study of the contractile and electrical properties of the human triceps surae muscle in a simulated microgravity environment. *J. Physiol.* 510:287–295.

Koryak, Y. 1999. The effects of long-term simulated microgravity on neuromuscular performance in men and women. *Eur. J. Appl. Physiol. Occup. Physiol.* 79:168–175.

Kossev, A., and Christova, P. 1998. Discharge pattern of human motor units during dynamic concentric and eccentric contractions. *Electroencephalogr. Clin. Neurophysiol. Electromyogr. Motor Control* 109:245–255.

Krippendorf, B., and Riley, D. 1994. Temporal changes in sarcomere lesions of rat adductor longus muscles during hindlimb reloading. *Anat. Rec.* 238:304–310.

Krnjevic, K., and Miledi, R. 1959. Presynaptic failure of neuromuscular propagation in rats. *J. Physiol.* 149:1–22.

Kudina, L.P., and Alexeeva, N.L. 1992. After-potentials and control of repetitive firing in human motoneurones. *Electroencephalogr. Clin. Neurophysiol.* 85:345–353.

Kudina, L.P., and Churikova, L.I. 1990. Testing excitability of human motoneurones capable of firing double discharges. *Electroencephalogr. Clin. Neurophysiol.* 75:334–341.

Kuei, J.H., Shadmehr, R., and Sieck, G.C. 1990. Relative contribution of neurotransmission failure to diaphragm fatigue. *J. Appl. Physiol.* 68:174–180.

Kugelberg, E., and Lindegren, B. 1979. Transmission and contraction fatigue of rat motor units in relation to succinate dehydrogenase activity of motor unit fibres. *J. Physiol.* (Lond.) 288:285–300.

Kukulka, C.G., and Clamann, H.P. 1981. Comparison of the recruitment and discharge properties of motor units in human brachial biceps and adductor pollicis during isometric contractions. *Brain Res.* 219:45–55.

Kukulka, C.G., Moore, M.A., and Russel, A.G. 1986. Changes in human α-motoneuron excitability during sustained maximum isometric contractions. *Neurosci. Lett.* 68:327–333.

Lafleur, J., Zytnicki, D., Horcholle-Bossavit, G., and Jami, L. 1992. Depolarization of Ib afferent axons in the cat spinal cord during homonymous muscle contraction. *J. Physiol.* 445:345–354.

Lapier, T.K., Burton, H.W., Almon, R., and Cerny, F. 1995. Alterations in intramuscular connective tissue after limb casting affect contraction-induced muscle injury. *J. Appl. Physiol.* 78:1065–1069.

Larsson, L. 1992. Is the motor unit uniform? *Acta Physiol. Scand.* 144:143–154.

Larsson, L., Ansved, T., Edström, L., Gorza, L., and Schiaffino, S. 1991. Effects of age on physiological, immunohistochemical and biochemical properties of fast-twitch single motor units in the rat. *J. Physiol.* (Lond.) 443:257–275.

Larsson, L., Edström, L., Lindegren, B., Gorza, L., and Schiaffino, S. 1991. MHC composition and enzyme-histochemical and physiological properties of a novel fast-twitch motor unit type. *Am. J. Physiol. Cell Physiol.* 261:C93–C101.

Larsson, L., Li, X.P., Berg, H.E., and Frontera, W.R. 1996. Effects of removal of weight-bearing function on contractility and myosin isoform composition in single human skeletal muscle cells. *Pflugers Arch.* 432:320–328.

Larsson, L., and Moss, R.L. 1993. Maximum velocity of shortening in relation to myosin isoform composition in single fibres from human skeletal muscles. *J. Physiol.* (Lond.) 472:595–614.

Larsson, L., and Tesch, P.A. 1986. Motor unit fibre density in extremely hypertrophied skeletal muscles in man. *Eur. J. Appl. Physiol.* 55:130–136.

Lavoie, P., Collier, B., and Tenenhouse, A. 1976. Comparison of alpha-bungarotoxin binding to skeletal muscles after inactivity or denervation. *Nature* 260:349–350.

Leberer, E., Härtner, K.-T., Brandl, C.J., Fujii, J., Tada, M., MacLennan, D.H., and Pette, D. 1989. Slow/cardiac sarcoplasmic reticulum Ca^{2+}-ATPase and phospholamban mRNAs are expressed in chronically stimulated rabbit fast-twitch muscle. *Eur. J. Biochem.* 185:51–54.

Leblanc, A.D., Schneider, V.S., Evans, H.J., Pientok, C., Rowe, R., and Spector, E. 1992. Regional changes in muscle mass following 17 weeks of bed rest. *J. Appl. Physiol.* 73:2172–2178.

Lee, R.H., and Heckman, C.J. 1998a. Bistability in spinal motoneurons *in vivo:* Systematic variations in persistent inward currents. *J. Neurophysiol.* 80:583–593.

Lee, R.H., and Heckman, C.J. 1998b. Bistability in spinal motoneurons *in vivo:* Systematic variations in rhythmic firing patterns. *J. Neurophysiol.* 80:572–582.

Lee, Y.S., Ondrias, K., Duhl, A.J., Ehrlich, B.E., and Kim, D.H. 1991. Comparison of calcium release from sarcoplasmic reticulum of slow and fast twitch muscles. *J. Membr. Biol.* 122:155–163.

Leterme, D., Cordonnier, C., Mounier, Y., and Falempin, M. 1994. Influence of chronic stretching upon rat soleus muscle during non-weight-bearing conditions. *Pflugers Arch.* 429:274–279.

Leterme, D., and Falempin, M. 1994. Compensatory effects of chronic electrostimulation on unweighted rat soleus muscle. *Pflugers Arch.* 426:155–160.

Lexell, J., Henriksson-Larsen, K., Winblad, B., and Sjöström, M. 1983. Distribution of different fiber types in human skeletal muscles: Effects of aging studied in whole muscle cross sections. *Muscle & Nerve* 6:588–595.

Lexell, J., Jarvis, J., Downham, D., and Salmons, S. 1992. Quantitative morphology of stimulation-induced damage in rabbit fast-twitch skeletal muscles. *Cell Tissue Res.* 269:195–204.

Li, X.P., and Larsson, L. 1996. Maximum shortening velocity and myosin isoforms in single muscle fibers from young and old rats. *Am. J. Physiol. Cell Physiol.* 270:C352–C360.

Lieber, R.L., Fridén, J.O., Hargens, A.R., Danzig, L.A., and Gershuni, D.H. 1988. Differential response of the dog quadriceps muscle to external skeletal fixation of the knee. *Muscle & Nerve* 11:193–201.

Lieber, R.L., Fridén, J.O., Hargens, A.R., and Feringa, E.R. 1986. Long-term effects of spinal cord transection on fast and slow rat skeletal muscle: 2. Morphometric properties. *Exp. Neurol.* 91:435–448.

Lieber, R.L., Johansson, C.B., Vahlsing, H.L., Hargens, A.R., and Feringa, E.R. 1986. Long-term effects of spinal cord transection on fast and slow rat skeletal muscle: 1. Contractile properties. *Exp. Neurol.* 91:423–434.

Liepert, J., Miltner, W.H.R., Bauder, H., Sommer, M., Dettmers, C., Taub, E., and Weiller, C. 1998. Motor cortex plasticity during constraint-induced movement therapy in stroke patients. *Neurosci. Lett.* 250:5–8.

Lind, A., and Kernell, D. 1991. Myofibrillar ATPase histochemistry of rat skeletal muscles: A "two-dimensional" quantitative approach. *J. Histochem. Cytochem.* 39:589–597.

Linderman, J.K., Gosselink, K.L., Booth, F.W., Mukku, V.R., and Grindeland, R.E. 1994. Resistance exercise and growth hormone as countermeasures for skeletal muscle atrophy in hindlimb-suspended rats. *Am. J. Physiol. Regul. Integr. Comp. Physiol.* 267:R365–R371.

Linderman, J.K., Whittall, J.B., Gosselink, K.L., Wang, T.J., Mukku, V.R., Booth, F.W., and Grindeland, R.E. 1995. Stimulation of myofibrillar protein synthesis in hindlimb suspended rats by resistance exercise and growth hormone. *Life Sci.* 57:755–762.

Liu, Y.W., and Schneider, M.F. 1998. Fibre type–specific gene expression activated by chronic electrical stimulation of adult mouse skeletal muscle fibres in culture. *J. Physiol.* 512:337–344.

Ljubisavljevic, M., Jovanovic, K., and Anastasijevic, R. 1994. Fusimotor responses to fatiguing muscle contractions in non-denervated hindlimb of decerebrate cats. *Neuroscience* 61:683–689.

Ljubisavljevic, M., Milanovic, S., Radovanovic, S., Vukcevic, I., Kostic, V., and Anastasijevic, R. 1996. Central changes in muscle fatigue during sustained submaximal isometric voluntary contraction as revealed by transcranial magnetic stimulation. *Electroencephalogr. Clin. Neurophysiol.* 101:281–288.

Ljubisavljevic, M., Radovanovic, S., Vukcevic, I., and Anastasijevic, R. 1995. Fusimotor outflow to pretibial flexors during fatiguing contractions of the triceps surae in decerebrate cats. *Brain Res.* 691:99–105.

Lnenicka, G.A., and Atwood, H.L. 1986. Impulse activity of a crayfish motoneuron regulates its neuromuscular synaptic properties. *J. Neurophysiol.* 61:91–96.

Lnenicka, G.A., and Atwood, H.L. 1988. Long-term changes in neuromuscular synapses with altered sensory input to a crayfish motoneuron. *Exp. Neurol.* 100:437–447.

Locke, M., and Noble, E.G. 1995. Stress proteins: The exercise response. *Can. J. Appl. Physiol.* 20:155–167.

Loeb, G.E. 1987. Hard lessons in motor control from the mammalian spinal cord. *Trends Neurosci.* 10:108–113.

Longhurst, C.M., and Jennings, L.K. 1998. Integrin-mediated signal transduction. *Cell. Mol. Life Sci.* 54:514–526.

Löscher, W.N., Cresswell, A.G., and Thorstensson, A. 1996a. Central fatigue during a long-lasting submaximal contraction of the triceps surae. *Exp. Brain Res.* 108:305–314.

Löscher, W.N., Cresswell, A.G., and Thorstensson, A. 1996b. Excitatory drive to the motoneuron pool during a fatiguing submaximal contraction in man. *J. Physiol.* (Lond.) 491:271–280.

Loughna, P.T., Goldspink, D.F., and Goldspink, G. 1987. Effects of hypokinesia and hypodynamia upon protein turnover in hindlimb muscles of the rat. *Aviat. Space Environ. Med.* 58:A133–A138.

Lovely, R.G., Gregor, R.J., Roy, R.R., and Edgerton, V.R. 1986. Effects of training on the recovery of full-weight-bearing stepping in the adult spinal cat. *Exp. Neurol.* 92:421–435.

Lovely, R.G., Gregor, R.J., Roy, R.R., and Edgerton, V.R. 1990. Weight-bearing hindlimb stepping in treadmill-exercised adult spinal cats. *Brain Res.* 514:206–218.

Lowe, D.A., Warren, G.L., Ingalls, C.P., Boorstein, D.B., and Armstrong, R.B. 1995. Muscle function and protein metabolism after initiation of eccentric contraction–induced injury. *J. Appl. Physiol.* 79:1260–1270.

Lowey, S., Waller, G.S., and Trybus, K.M. 1993a. Function of skeletal muscle myosin heavy and light chain isoforms by an in vitro motility assay. *J. Biol. Chem.* 268:20414–20418.

Lowey, S., Waller, G.S., and Trybus, K.M. 1993b. Skeletal muscle myosin light chains are essential for physiological speeds of shortening. *Nature* 365:454–456.

MacDougall, J.D., Green, H.J., Sutton, J.R., Coates, G., Cymerman, A., Young, P., and Houston, C.S. 1991. Operation Everest II: Structural adaptations in skeletal muscle in response to extreme simulated altitude. *Acta Physiol. Scand.* 142:421–427.

MacDougall, J.D., Elder, G., Sale, D.G., Moroz, J., and Sutton, J.R. 1980. Effects of strength training and immobilization on human muscle fibres. *Eur. J. Appl. Physiol.* 43:25–34.

MacDougall, J.D., Sale, D.G., Alway, S.E., and Sutton, J.R. 1984. Muscle fiber number in biceps brachii in bodybuilders and control subjects. *J. Appl. Physiol. Respir. Environ. Exerc. Physiol.* 57:1399–1403.

MacDougall, J.D., Tarnopolsky, M.A., Chesley, A., and Atkinson, S.A. 1992. Changes in muscle protein synthesis following heavy resistance exercise in humans: A pilot study. *Acta Physiol. Scand.* 146:403–404.

Macefield, V.G., Hagbarth, K.-E., Gorman, R.B., Gandevia, S.C., and Burke, D. 1991. Decline in spindle support to alpha-motoneurones during sustained voluntary contractions. *J. Physiol.* 440:497–512.

Macefield, V.G., Gandevia, S.C., Bigland-Ritchie, B., Gorman, R.B., and Burke, D. 1993. The firing rates of human motoneurons voluntarily activated in the absence of muscle afferent feedback. *J. Physiol.* (Lond.) 471:429–443.

MacIntosh, B.R., Herzog, W., Suter, E., Wiley, J.P., and Sokolosky, J. 1993. Human skeletal muscle fibre types and force: Velocity properties. *Eur. J. Appl. Physiol.* 67:499–506.

Magleby, K.L., and Pallotta, B.S. 1981. A study of desensitization of acetylcholine receptors using nerve-released transmitter in the frog. *J. Physiol.* 316:225–250.

Maier, A., Gorza, L., Schiaffino, S., and Pette, D. 1988. A combined histochemical and immunohistochemical study on the dynamics of fast-to-slow fiber transformation in chronically stimulated rabbit muscle. *Cell Tissue Res.* 254:59–68.

Maier, A., and Pette, D. 1987. The time course of glycogen depletion in single fibers of chronically stimulated rabbit fast-twitch muscle. *Pflugers Arch.* 408:338–342.

Malathi, S., and Batmanabane, M. 1983. Alterations in the morphology of the neuromuscular junctions following experimental immobilization in cats. *Experientia* 39:547–549.

Mambrito, B., and De Luca, C.J. 1983. Acquisition and decomposition of the EMG signal. *Prog. Clin. Neurophysiol.* 10:52–72.

Marsden, C.D., Meadows, J.C., and Merton, P. 1983. "Muscular wisdom" that minimizes fatigue during prolonged effort in man: Peak rates of motoneuron discharge and slowing of discharge during fatigue. In *Motor control mechanisms in health and disease,* ed. J.E. Desmedt, 169-211. New York: Raven Press.

Martin, L., Cometti, G., Pousson, M., and Morlon, B. 1993. Effect of electrical stimulation training on the contractile characteristics of the triceps surae muscle. *Eur. J. Appl. Physiol.* 67:457–461.

Martin, T.P., Edgerton, V.R., and Grindeland, R.E. 1988. Influence of spaceflight on rat skeletal muscle. *J. Appl. Physiol.* 65:2318–2325.

Martin, T.P., Stein, R.B., Hoeppner, P.H., and Reid, D.C. 1992. Influence of electrical stimulation on the morphological and metabolic properties of paralyzed muscle. *J. Appl. Physiol.* 72:1401–1406.

Martineau, L., and Gardiner, P.F. 1999. Static stretch induces MAPK activation in skeletal muscle [abstract]. *FASEB J.* 13:A4101999.

McDonagh, J.C., Binder, M.C., Reinking, R.M., and Stuart, D.G. 1980. A commentary on muscle unit properties in cat hindlimb muscles. *J. Morphol.* 166:217–230.

McDonagh, M.J.N., Hayward, C.M., and Davies, C.T.M. 1983. Isometric training in human elbow flexor muscles: The effects on voluntary and electrically evoked forces. *J. Bone Joint Surg.* 65B:355–358.

McDonald, K.S., Blaser, C.A., and Fitts, R.H. 1994. Force-velocity and power characteristics of rat soleus muscle fibers after hindlimb suspension. *J. Appl. Physiol.* 77:1609–1616.

McDonald, K.S., Delp, M.D., and Fitts, R.H. 1992. Fatigability and blood flow in the rat gastrocnemius-plantaris-soleus after hindlimb suspension. *J. Appl. Physiol.* 73:1135–1140.

McDonald, K.S., and Fitts, R.H. 1993. Effect of hindlimb unweighting on single soleus fiber maximal shortening velocity and ATPase activity. *J. Appl. Physiol.* 74:2949–2957.

McDonald, K.S., and Fitts, R.H. 1995. Effect of hindlimb unloading on rat soleus fiber force, stiffness, and calcium sensitivity. *J. Appl. Physiol.* 79:1796–1802.

McKay, W.B., Stokic, D.S., Sherwood, A.M., Vrbova, G., and Dimitrijevic, M.R. 1996. Effect of fatiguing maximal voluntary contraction on excitatory and inhibitory responses elicited by transcranial magnetic motor cortex stimulation. *Muscle & Nerve* 19:1017–1024.

McKay, W.B., Tuel, S., Sherwood, A.M., Stokic, D.S., and Dimitrijevic, M.R. 1995. Focal depression of cortical excitability induced by fatiguing muscle contraction: A transcranial magnetic stimulation study. *Exp. Brain Res.* 105:276–282.

Meissner, J., Kubis, H.-P., Scheibe, R., and Gros, G. 2000. Reversible Ca^{2+}-induced fast-to-slow transition in primary skeletal muscle culture cells at the mRNA level. *J. Physiol.* (Lond.) 523:19–28.

Melloni, E., and Pontremoli, S. 1989. The calpains. *Trends Neurosci.* 12:438–444.

Mendell, L.M., Cope, T.C., and Nelson, S.G. 1982. Plasticity of the group Ia fiber pathway to motoneurons. In *Changing concepts of the nervous system,* ed. A.R. Morrison and P.L. Strick, 69–78. New York: Academic Press.

Mercier, A.J., Bradacs, H., and Atwood, H.L. 1992. Long-term adaptation of crayfish neurons depends on the frequency and number of impulses. *Brain Res.* 598:221–224.

Michel, J., Ordway, G.A., Richardson, J.A., and Williams, R.S. 1994. Biphasic induction of immediate early gene expression accompanies activity-dependent angiogenesis and myofiber remodeling of rabbit skeletal muscle. *J. Clin. Invest.* 94:277–285.

Michel, R.N., Cowper, G., Chi, M.M.-Y., Manchester, J.K., Falter, H., and Lowry, O.H. 1994. Effects of tetrodotoxin-induced neural inactivation on single muscle fiber metabolic enzymes. *Am. J. Physiol. Cell Physiol.* 267:C55–C66.

Michel, R.N., and Gardiner, P.F. 1990. To what extent is hindlimb suspension a model of disuse? *Muscle & Nerve* 13:646–653.

Michel, R.N., Parry, D.J., and Dunn, S.E. 1996. Regulation of myosin heavy chain expression in adult rat hindlimb muscles during short-term paralysis: Comparison of denervation and tetrodotoxin-induced neural inactivation. *FEBS Lett.* 391:39–44.

Miles, M.P., Clarkson, P.M., Bean, M., Ambach, K., Mulroy, J., and Vincent, K. 1994. Muscle function at the wrist following 9 d of immobilization and suspension. *Med. Sci. Sports Exerc.* 26:615–623.

Miller, K.J., Garland, S.J., Ivanova, T., and Ohtsuki, T. 1996. Motor-unit behavior in humans during fatiguing arm movements. *J. Neurophysiol.* 75:1629–1636.

Mills, K.R., and Thomson, C.C.B. 1995. Human muscle fatigue investigated by transcranial magnetic stimulation. *Neuroreport* 6:1966–1968.

Milner-Brown, H.S., Stein, R.B., and Lee, R.G. 1975. Synchronization of human motor units: Possible roles of exercise and supraspinal reflexes. *Electroencephalogr. Clin. Neurophysiol.* 38:245–254.

Milner-Brown, H.S., Stein, R.B., and Yemm, R. 1973a. The contractile properties of human motor units during voluntary isometric contractions. *J. Physiol.* (Lond.) 228:285–306.

Milner-Brown, H.S., Stein, R.B., and Yemm, R. 1973b. The orderly recruitment of human motor units during voluntary isometric contractions. *J. Physiol.* 230:359–370.

Monster, A.W., and Chan, H. 1977. Isometric force production by motor units of extensor digitorum communis muscle in man. *J. Neurophysiol.* 40:1432–1443.

Moore, R.L., and Stull, J.T. 1984. Myosin light chain phosphorylation in fast and slow skeletal muscles *in situ. Am. J. Physiol. Cell Physiol.* 247:C462–C471.

Moritani, T., and De Vries, H.A. 1979. Neural factors versus hypertrophy in the time course of muscle strength gain. *Am. J. Phys. Med.* 58:115–131.

Moritani, T., Muramatsu, S., and Muro, M. 1988. Activity of motor units during concentric and eccentric contractions. *Am. J. Phys. Med.* 66:338–350.

Morrison, P.R., Montgomery, J.A., Wong, T.S., and Booth, F.W. 1987. Cytochrome c protein-synthesis rates and mRNA contents during atrophy and recovery in skeletal muscle. *Biochem. J.* 241:257–263.

Moss, R.L., Diffee, G.M., and Greaser, M.L. 1995. Contractile properties of skeletal muscle fibers in relation to myofibrillar protein isoforms. *Rev. Physiol. Biochem. Pharmacol.* 126:1–63.

Moss, R.L., Reiser, P.J., Greaser, M.L., and Eddinger, T.J. 1990. Varied expression of myosin alkali light chains is associated with altered speed of contraction in rabbit fast twitch skeletal muscles. In *The dynamic state of muscle fibers,* ed. D. Pette, 353-368. Berlin: DeGruyter.

Munson, J.B., Foehring, R.C., Lofton, S.A., Zengel, J.E., and Sypert, G.W. 1986. Plasticity of medial gastrocnemius motor units following cordotomy in the cat. *J. Neurophysiol.* 55:619–634.

Munson, J.B., Foehring, R.C., Mendell, L.M., and Gordon, T. 1997. Fast-to-slow conversion following chronic low-frequency activation of medial gastrocnemius muscle in cats: 2. Motoneuron properties. *J. Neurophysiol.* 77:2605–2615.

Müntener, M., Käser, L., Weber, J., and Berchtold, M.W. 1995. Increase of skeletal muscle relaxation speed by direct injection of parvalbumin cDNA. *Proc. Natl. Acad. Sci. USA* 92:6504–6508.

Naito, H., Powers, S., Demirel, H., Sugiura, T., Dodd, S., and Aoki, J. 2000. Heat stress attenuates skeletal muscle atrophy in hindlimb-unweighted rats. *J. Appl. Physiol.* 88:359–363.

Nakano, H., Masuda, K., Sasaki, S.Y., and Katsuta, S. 1997. Oxidative enzyme activity and soma size in motoneurons innervating the rat slow-twitch and fast-twitch muscles after chronic activity. *Brain Res. Bull.* 43:149–154.

Nakazawa, K., Kawakami, Y., Fukunaga, T., Yano, H., and Miyashita, M. 1993. Differences in activation patterns in elbow flexor muscles during isometric, concentric and eccentric contractions. *Eur. J. Appl. Physiol.* 66:214–220.

Nardone, A., Romano, C., and Schieppati, M. 1989. Selective recruitment of high-threshold human motor units during voluntary isotonic lengthening of active muscles. *J. Physiol.* 409:451–471.

Narici, M.V., Roi, G.S., Landoni, L., Minetti, A.E., and Cerretelli, P. 1989. Changes in force, cross-sectional area and neural activation during strength training and detraining of the human quadriceps. *Eur. J. Appl. Physiol.* 59:310–319.

Naya, F., Mercer, B., Shelton, J., Richardson, J., Williams, R., and Olson, E. 2000. Stimulation of slow skeletal muscle fiber gene expression by calcineurine *in vitro*. *J. Biol. Chem.* 275:4545–4548.

Nelson, D.L., and Hutton, R.S. 1985. Dynamic and static stretch responses in muscle spindle receptors in fatigued muscle. *Med. Sci. Sports Exerc.* 17:445–450.

Nemeth, P.M., and Pette, D. 1981. Succinate dehydrogenase activity in fibres classified by myosin ATPase in three hind limb muscles of rat. *J. Physiol.* 320:73–80.

Nemeth, P.M., Pette, D., and Vrbova, G. 1981. Comparison of enzyme activities among single muscle fibres within defined motor units. *J. Physiol.* 311:489–495.

Neufer, P.D., Ordway, G.A., Hand, G.A., Shelton, J.M., Richardson, J.A., Benjamin, I.J., and Williams, R.S. 1996. Continuous contractile activity induces fiber type specific expression of HSP70 in skeletal muscle. *Am. J. Physiol. Cell Physiol.* 271:C1828–C1837.

Nguyen, P.V., and Atwood, H.L. 1990. Expression of long-term adaptation of synaptic transmission requires a critical period of protein synthesis. *J. Neurosci.* 10:1099–1109.

Nguyen, P.V., and Atwood, H.L. 1992. Maintenance of long-term adaptation of synaptic transmission requires axonal transport following induction in an identified crayfish motoneuron. *Exp. Neurol.* 115:414–422.

Nielsen, J., and Kagamihara, Y. 1993. The regulation of presynaptic inhibition during co-contraction of antagonistic muscles in man. *J. Physiol.* (Lond.) 464:575–593.

Nilsson, J., Tesch, P., and Thorstensson, A. 1977. Fatigue and EMG of repeated fast voluntary contractions in man. *Acta Physiol. Scand.* 101:194–198.

Nordstrom, M.A., Enoka, R.M., Reinking, R.M., Callister, R.C., and Stuart, D.G. 1995. Reduced motor unit activation of muscle spindles and tendon organs in the immobilized cat hindlimb. *J. Appl. Physiol.* 78:901–913.

Nudo, R.J., Wise, B.M., SiFuentes, F., and Milliken, G.W. 1996. Neural substrates for the effects of rehabilitative training on motor recovery after ischemic infarct. *Science* 272:1791–1794.

Ogata, T. 1988. Structure of motor endplates in the different fiber types of vertebrate skeletal muscles. *Arch. Histol. Cytol.* 51:385–424.

Ohira, Y., Jiang, B., Roy, R.R., Oganov, V., Ilyina-Kakueva, E., Marini, J.F., and Edgerton, V.R. 1992. Rat soleus muscle fiber responses to 14 days of spaceflight and hindlimb suspension. *J. Appl. Physiol.* 73 suppl.: 51S–57S.

Ohlendieck, K., Fromming, G., Murray, B., Maguire, P., Liesner, E., Traub, I., and Pette, D. 1999. Effects of chronic low-frequency stimulation on Ca^{2+}-regulatory membrane proteins in rabbit fast muscle. *Pflugers Arch.* 438:700–708.

Oishi, Y., Ishihara, A., and Katsuta, S. 1992. Muscle fibre number following hindlimb immobilization. *Acta Physiol. Scand.* 146:281–282.

Ordway, G., Neufer, P., Chin, E., and DiMartino, G. 2000. Chronic contractile activity upregulates the proteosome system in rabbit skeletal muscle. *J. Appl. Physiol.* 88:1134–1141.

Ornatsky, O.I., Connor, M.K., and Hood, D.A. 1995. Expression of stress proteins and mitochondrial chaperonins in chronically stimulated skeletal muscle. *Biochem. J.* 311:119–123.

Osbaldeston, N.J., Lee, D.M., Cox, V.M., Hesketh, J.E., Morrison, J.F.J., Blair, G.E., and Goldspink, D.F. 1995. The temporal and cellular expression of *c-fos* and *c-jun* in mechanically stimulated rabbit latissimus dorsi muscle. *Biochem. J.* 308:465–471.

Pachter, B., and Eberstein, A. 1984. Neuromuscular plasticity following limb immobilization. *J. Neurocytol.* 13:1013–1025.

Pagala, M.K.D., Namba, T., and Grob, D. 1984. Failure of neuromuscular transmission and contractility during muscle fatigue. *Muscle & Nerve* 7:454–464.

Pagala, M.K.D., and Taylor, S.R. 1998. Imaging caffeine-induced Ca^{2+} transients in individual fast-twitch and slow-twitch rat skeletal muscle fibers. *Am. J. Physiol. Cell Physiol.* 274:C623–C632.

Paintal, A.S. 1960. Functional analysis of group III afferent fibres of mammalian muscles. *J. Physiol.* (Lond.) 152:250–270.

Panenic, R., and Gardiner, P.F. 1998. The case for adaptability of the neuromuscular junction to endurance exercise training. *Can. J. Appl. Physiol.* 23:339–360.

Papadaki, M., and Eskin, S. 1997. Effects of fluid shear stress on gene regulation of vascular cells. *Biotechnol. Prog.* 13:209–221.

Pattullo, M.C., Cotter, M.A., Cameron, N.E., and Barry, J.A. 1992. Effects of lengthened immobilization on functional and histochemical properties of rabbit tibialis anterior muscle. *Exp. Physiol.* 77:433–442.

Pearce, A., Thickbroom, G., Byrnes, M., and Mastaglia, F. 2000. Functional reorganization of the corticomotor projection to the hand in skilled racquet players. *Exp. Brain Res.* 130:238–243.

Pedersen, J., Ljubisavljevic, M., Bergenheim, M., and Johansson, H. 1998. Alterations in information transmission in ensembles of primary muscle spindle afferents after muscle fatigue in heteronymous muscle. *Neuroscience* 84:953–959.

Péréon, Y., Dettbarn, C., Lu, Y., Westlund, K.N., Zhang, J.T., and Palade, P. 1998. Dihydropyridine receptor isoform expression in adult rat skeletal muscle. *Pflugers Arch.* 436:309–314.

Pestronk, A., Drachman, D.B., and Griffin, J.W. 1976. Effect of muscle disuse on acetylcholine receptors. *Nature* 260:352–353.

Peters, E.J.D., and Fuglevand, A.J. 1999. Cessation of human motor unit discharge during sustained maximal voluntary contraction. *Neurosci. Lett.* 274:66–70.

Petit, J., and Gioux, M. 1993. Properties of motor units after immobilization of cat peroneus longus muscle. *J. Appl. Physiol.* 74:1131–1139.

Petrofsky, J.S., and Phillips, C.A. 1985. Discharge characteristics of motor units and the surface EMG during fatiguing isometric contractions at submaximal tensions. *Aviat. Space Environ. Med.* 56:581–586.

Pette, D. 1998. Training effects on the contractile apparatus. *Acta Physiol. Scand.* 162:367–376.

Pette, D., and Düsterhöft, S. 1992. Altered gene expression in fast-twitch muscle induced by chronic low-frequency stimulation. *Am. J. Physiol. Regul. Integr. Comp. Physiol.* 262:R333–R338.

Pette, D., and Staron, R.S. 1993. The molecular diversity of mammalian muscle fibers. *News Physiol. Sci.* 8:153–157.

Pette, D., and Staron, R.S. 1997. Mammalian skeletal muscle fiber type transitions. *Int. Rev. Cytol.* 170:143–223.

Pette, D., and Vrbova, G. 1992. Adaptation of mammalian skeletal muscle fibers to chronic electrical stimulation. *Rev. Physiol. Biochem. Pharmacol.* 120:115–202.

Pette, D., Wimmer, M., and Nemeth, P.M. 1980. Do enzyme activities vary along muscle fibres? *Histochemistry* 67:225–231.

Peuker, H., Conjard, A., and Pette, D. 1998. Alpha-cardiac-like myosin heavy chain as an intermediate between MHCIIa and MHCI beta in transforming rabbit muscle. *Am. J. Physiol. Cell Physiol.* 274:C595–C602.

Peuker, H., and Pette, D. 1997. Quantitative analyses of myosin heavy-chain mRNA and protein isoforms in single fibers reveal a pronounced fiber heterogeneity in normal rabbit muscles. *Eur. J. Biochem.* 247:30–36.

Phelan, J.N., and Gonyea, W.J. 1997. Effect of radiation on satellite cell activity and protein expression in overloaded mammalian skeletal muscle. *Anat. Rec.* 247:179–188.

Phillips, S.M., Tipton, K.D., Aarsland, A., Wolf, S.E., and Wolfe, R.R. 1997. Mixed muscle protein synthesis and breakdown after resistance exercise in humans. *Am. J. Physiol. Endocrinol. Metab.* 273:E99–E107.

Phillips, S.M., Tipton, K.D., Ferrando, A.A., and Wolfe, R.R. 1999. Resistance training reduces the acute exercise-induced increase in muscle protein turnover. *Am. J. Physiol. Endocrinol. Metab.* 276:E118–E124.

Pickar, J.G., Hill, J.M., and Kaufman, M.P. 1994. Dynamic exercise stimulates group III muscle afferents. *J. Neurophysiol.* 71:753–760.

Pierotti, D.J., Roy, R.R., Bodine-Fowler, S.C., Hodgson, J.A., and Edgerton, V.R. 1991. Mechanical and morphological properties of chronically inactive cat tibialis anterior motor units. *J. Physiol.* (Lond.) 444:175–192.

Pierotti, D.J., Roy, R.R., Flores, V., and Edgerton, V.R. 1990. Influence of 7 days of hindlimb suspension and intermittent weight support on rat muscle mechanical properties. *Aviat. Space Environ. Med.* (March):205–210.

Pierotti, D.J., Roy, R.R., Hodgson, J.A., and Edgerton, V.R. 1994. Level of independence of motor unit properties from neuromuscular activity. *Muscle & Nerve* 17:1324–1335.

Ploutz, L.L., Tesch, P.A., Biro, R.L., and Dudley, G.A. 1994. Effect of resistance training on muscle use during exercise. *J. Appl. Physiol.* 76:1675–1681.

Ploutz-Snyder, L.L., Tesch, P.A., Crittenden, D.J., and Dudley, G.A. 1995. Effect of unweighting on skeletal muscle use during exercise. *J. Appl. Physiol.* 79:168–175.

Pluskal, M.G., and Sreter, F.A. 1983. Correlation between protein phenotype and gene expression in adult rabbit fast twitch muscles undergoing a fast to slow fiber transformation in response to electrical stimulation *in vivo. Biochem. Biophys. Res. Commun.* 113:325–331.

Pool, C.W., Moll, H., and Diegenbach, P.C. 1979. Quantitative succinate-dehydrogenase histochemistry. *Histochemistry* 64:273–278.

Powers, S., Ji, L., and Leeuwenburgh, C. 1999. Exercise training–induced alterations in skeletal muscle antioxidant capacity: A brief review. *Med. Sci. Sports Exerc.* 31:987–997.

Prakash, Y.S., Miller, S.M., Huang, M., and Sieck, G.C. 1996. Morphology of diaphragm neuromuscular junctions on different fibre types. *J. Neurocytol.* 25:88–100.

Psek, J.A., and Cafarelli, E. 1993. Behavior of coactive muscles during fatigue. *J. Appl. Physiol.* 74:170–175.

Puntschart, A., Wey, E., Jostarndt, K., Vogt, M., Wittwer, M., Widmer, H., Hoppeler, H., and Billeter, R. 1998. Expression of *fos* and *jun* genes in human skeletal muscle after exercise. *Am. J. Physiol.* 274:C129–C137.

Raj, D.A., Booker, T.S., and Belcastro, A.N. 1998. Striated muscle calcium-stimulated cysteine protease (calpain-like) activity promotes myeloperoxidase activity with exercise. *Pflugers Arch.* 435:804–809.

Rall, J.A. 1996. Role of parvalbumin in skeletal muscle relaxation. *News Physiol. Sci.* 11:249–255.

Rall, W., Burke, R.E., Holmes, W.R., Jack, J.J.B., Redman, S.J., and Segev, I. 1992. Matching dendritic neuron models to experimental data. *Physiol. Rev.* 72 suppl.: S159–S186.

Reggiani, C., Bottinelli, R., and Stienen, G. 2000. Sarcomeric myosin isoforms: Fine tuning of a molecular motor. *News Physiol. Sci.* 15:26–33.

Reichmann, H., and Pette, D. 1982. A comparative microphotometric study of succinate dehydrogenase activity levels in type I, IIA and IIB fibres of mammalian and human muscles. *Histochemistry* 74:27–41.

Reid, B., Slater, C.R., and Bewick, G.S. 1999. Synaptic vesicle dynamics in rat fast and slow motor nerve terminals. *J. Neurosci.* 19:2511–2521.

Reiser, P.J., Moss, R.L., Giulian, G.G., and Greaser, M.L. 1985. Shortening velocity in single fibers from adult rabbit soleus muscles is correlated with myosin heavy chain composition. *J. Biol. Chem.* 260:9077–9080.

Rice, C.L., Cunningham, D.A., Taylor, A.W., and Paterson, D.H. 1988. Comparison of the histochemical and contractile properties of human triceps surae. *Eur. J. Appl. Physiol.* 58:165–170.

Richter, E.A., and Nielsen, N.B.S. 1991. Protein kinase C activity in rat skeletal muscle: Apparent relation to body weight and muscle growth. *FEBS Lett.* 289:83–85.

Riek, S., and Bawa, P. 1992. Recruitment of motor units in human forearm extensors. *J. Neurophysiol.* 68:100–108.

Riley, D.A., Bain, J.L.W., Thompson, J., Fitts, R., Widrick, J., Trappe, S., Trappe, T., and Costill, D. 2000. Decreased thin filament density and length in human atrophic soleus muscle fibers after spaceflight. *J. Appl. Physiol.* 88:567–572.

Riley, D.A., Ellis, S., Giometti, C.S., Hoh, J.F.Y., Ilyina-Kakueva, E.I., Oganov, V.S., Slocum, G.R., Bain, J.L.W., and Sedlak, F.R. 1992. Muscle sarcomere lesions and thrombosis and suspension unloading. *J. Appl. Physiol.* 73 suppl.: 33S–43S.

Riley, D.A., Ellis, S., Slocum, G.R., Sedlak, F.R., Bain, J.L.W., Krippendorf, B., Lehman, C., Macias, M., Thompson, J., Vijayan, K., and De Bruin, J. 1996. In-flight and post-flight changes in skeletal muscles of SLS-1 and SLS-2 spaceflown rats. *J. Appl. Physiol.* 81:133–144.

Rivero, J.-L., Talmadge, R.J., and Edgerton, V.R. 1998. Fibre size and metabolic properties of myosin heavy chain–based fibre types in rat skeletal muscle. *J. Muscle Res. Cell Motil.* 19:733–742.

Robbins, N., and Fischbach, G.D. 1971. Effect of chronic disuse of rat soleus neuromuscular junctions on presynaptic function. *J. Neurophysiol.* 34:570–578.

Robinson, G.A., Enoka, R.M., and Stuart, D.G. 1991. Immobilization-induced changes in motor unit force and fatigability in the cat. *Muscle & Nerve* 14:563–573.

Romaiguère, P., Vedel, J.-P., and Pagni, S. 1993. Comparison of fluctuations of motor unit recruitment and de-recruitment thresholds in man. *Exp. Brain Res.* 95:517–522.

Romaiguère, P., Vedel, J.-P., Pagni, S., and Zenatti, A. 1989. Physiological properties of the motor units of the wrist extensor muscles in man. *Exp. Brain Res.* 78:51–61.

Rome, L.C., Funke, R., McNeill Alexander, R., Lutz, G., Aldridge, H., and Freadman, M. 1988. Why animals have different muscle fibre types. *Nature* 335:824–827.

Rome, L.C., Sosnicki, A.A., and Goble, D.O. 1990. Maximum velocity of shortening of three fibre types from horse soleus muscle: Implications for scaling with body size. *J. Physiol.* (Lond.) 431:173–185.

Rome, L.C., Swank, D., and Corda, D. 1993. How fish power swimming. *Science* 261:340–343.

Rosenblatt, J.D., Yong, D., and Parry, D.J. 1994. Satellite cell activity is required for hypertrophy of overloaded adult rat muscle. *Muscle & Nerve* 17:608–613.

Rothmuller, C., and Cafarelli, E. 1995. Effect of vibration on antagonist muscle coactivation during progressive fatigue in humans. *J. Physiol.* (Lond.) 485:857–864.

Rothwell, J.C., Thompson, P.D., Day, B.L., Boyd, S., and Marsden, C.D. 1991. Stimulation of the motor cortex through the scalp. *Exp. Physiol.* 76:159–200.

Rotto, D.M., and Kaufman, M.P. 1988. Effect of metabolic products of muscular contraction on discharge of group III and IV afferents. *J. Appl. Physiol.* 64:2306–2313.

Round, J.M., Barr, F.M.D., Moffat, B., and Jones, D.A. 1993. Fibre areas and histochemical fibre types in the quadriceps muscle of paraplegic subjects. *J. Neurol. Sci.* 116:207–211.

Roy, R.R., and Acosta, L.J. 1986. Fiber type and fiber size changes in selected thigh muscles six months after low thoracic spinal cord transection in adult cats: Exercise effects. *Exp. Neurol.* 93:675–685.

Roy, R.R., Bello, M.A., Bouissou, P., and Edgerton, V.R. 1987. Size and metabolic properties of fibers in rat fast-twitch muscles after hindlimb suspension. *J. Appl. Physiol.* 62:2348–2357.

Roy, R.R., Eldridge, L., Baldwin, K.M., and Edgerton, V.R. 1996. Neural influence on slow muscle properties: Inactivity with and without cross-reinnervation. *Muscle & Nerve* 19:707–714.

Roy, R.R., Pierotti, D.J., Baldwin, K.M., Zhong, H., Hodgson, J.A., and Edgerton, V.R. 1998. Cyclical passive stretch influences the mechanical properties of the inactive cat soleus. *Exp. Physiol.* 83:377–385.

Roy, R.R., Pierotti, D.J., Flores, V., Rudolph, W., and Edgerton, V.R. 1992. Fibre size and type adaptations to spinal isolation and cyclical passive stretch in cat hindlimb. *J. Anat.* 180:491–499.

Roy, R.R., Sacks, R.D., Baldwin, K.M., Short, M., and Edgerton, V.R. 1984. Interrelationships of contraction time, Vmax, and myosin ATPase after spinal transection. *J. Appl. Physiol. Respir. Environ. Exerc. Physiol.* 56:1594–1601.

Roy, R.R., Talmadge, R.J., Hodgson, J.A., Oishi, Y., Baldwin, K.M., and Edgerton, V.R. 1999. Differential response of fast hindlimb extensor and flexor muscles to exercise in adult spinalized cats. *Muscle & Nerve* 22:230–241.

Roy, R.R., Talmadge, R.J., Hodgson, J.A., Zhong, H., Baldwin, K.M., and Edgerton, V.R. 1998. Training effects on soleus of cats spinal cord transected (T12-13) as adults. *Muscle & Nerve* 21:63–71.

Rube, N., and Secher, N.H. 1990. Effect of training on central factors in fatigue following two- and one-leg static exercise in man. *Acta Physiol. Scand.* 141:87–95.

Ruff, R.L. 1992. Na current density at and away from end plates on rat fast- and slow-twitch skeletal muscle fibers. *Am. J. Physiol. Cell Physiol.* 262:C229–C234.

Ruff, R.L. 1996. Sodium channel slow inactivation and the distribution of sodium channels on skeletal muscle fibres enable the performance properties of different skeletal muscle fibre types. *Acta Physiol. Scand.* 156:159–168.

Ruff, R.L., and Whittlesey, D. 1993. Na currents near and away from endplates on human fast and slow twitch muscle fibers. *Muscle & Nerve* 16:922–929.

Russell, B., Motlagh, D., and Ashley, W. 2000. Form follows function: How muscle shape is regulated by work. *J. Appl. Physiol.* 88:1127–1132.

Ryushi, T., and Fukunaga, T. 1986. Influence of subtypes of fast-twitch fibers on isokinetic strength in untrained men. *Int. J. Sports Med.* 7:250–253.

Sacco, P., Newberry, R., McFadden, L., Brown, T., and McComas, A.J. 1997. Depression of human electromyographic activity by fatigue of a synergistic muscle. *Muscle & Nerve* 20:710–717.

Sacco, P., Thickbroom, G.W., Thompson, M.L., and Mastaglia, F.L. 1997. Changes in corticomotor excitation and inhibition during prolonged submaximal muscle contractions. *Muscle & Nerve* 20:1158–1166.

Sadoshima, J., and Izumo, S. 1993. Mechanical stretch rapidly activates multiple signal transduction pathways in cardiac myocytes: Potential involvement of an autocrine/paracrine mechanism. *EMBO J.* 12:1681–1692.

Sale, D.G., and MacDougall, J.D. 1984. Isokinetic strength in weight-trainers. *Eur. J. Appl. Physiol.* 53:128–132.

Sale, D.G., MacDougall, J.D., Alway, S.E., and Sutton, J.R. 1987. Voluntary strength and muscle characteristics in untrained men and women and male bodybuilders. *J. Physiol.* 62:1786–1793.

Sale, D.G., MacDougall, J.D., Upton, A.R.M., and McComas, A.J. 1983. Effect of strength training upon motoneuron excitability in man. *Med. Sci. Sports Exerc.* 15:57–62.

Sale, D.G., Martin, J.E., and Moroz, D.E. 1992. Hypertrophy without increased isometric strength after weight training. *Eur. J. Appl. Physiol.* 64:51–55.

Sale, D.G., McComas, A.J., MacDougall, J.D., and Upton, A.R.M. 1982. Neuromuscular adaptation in human thenar muscles following strength training and immobilization. *J. Appl. Physiol. Respir. Environ. Exerc. Physiol.* 53:419–424.

Sale, D.G., Upton, A.R.M., McComas, A.J., and MacDougall, J.D. 1983. Neuromuscular function in weight-trainers. *Exp. Neurol.* 82:521–531.

Salviati, G., and Volpe, P. 1988. Ca^{2+}-release from sarcoplasmic reticulum of skinned fast- and slow-twitch muscle fibers. *Am. J. Physiol. Cell Physiol.* 254:C459–C465.

Samii, A., Wassermann, E.M., and Hallett, M. 1997. Post-exercise depression of motor evoked potentials as a function of exercise duration. *Electroencephalogr. Clin. Neurophysiol.* 105:352–356.

Sandercock, T.G., Faulkner, J.A., Albers, J.W., and Abbrecht, P.H. 1985. Single motor unit and fiber action potentials during fatigue. *J. Appl. Physiol.* 58 (4): 1073–1079.

Sant'ana Pereira, J.A.A., De Haan, A., Wessels, A., Moorman, A.F.M., and Sargeant, A.J. 1995. The mATPase histochemical profile of rat type IIX fibres: Correlation with myosin heavy chain immunolabelling. *Histochem. J.* 27:715–722.

Sant'ana Pereira, J.A.A., Ennion, S., Sargeant, A.J., Moorman, A.F.M., and Goldspink, G. 1997. Comparison of the molecular, antigenic and ATPase determinants of fast myosin heavy chains in rat and human: A single-fibre study. *Pflugers Arch.* 435:151–163.

Sant'ana Pereira, J.A.A., Wessels, A., Nijtmans, L., Moorman, A.F.M., and Sargeant, A.J. 1995. New method for the accurate characterization of single human skeletal muscle fibres demonstrates a relation between mATPase and MyHC expression in pure and hybrid fibre types. *J. Muscle Res. Cell Motil.* 16:21–34.

Sargeant, A.J., Davies, C., Edwards, R., Maunder, C., and Young, A. 1977. Functional and structural changes after disuse of human muscle. *Clin. Sci. Mol. Med.* 52:337–342.

Savard, G.K., Richter, E.A., Strange, S., Kiens, B., Christensen, N.J., and Saltin, B. 1989. Norepinephrine spillover from skeletal muscle during exercise in humans: Role of muscle mass. *Am. J. Physiol.* 257:H1812–H1818.

Sawczuk, A., Powers, R.K., and Binder, M.D. 1995. Spike frequency adaptation studied in hypoglossal motoneurons of the rat. *J. Neurophysiol.* 73:1799–1810.

Saxton, J.M., and Donnelly, A.E. 1996. Length-specific impairment of skeletal muscle contractile function after eccentric muscle actions in man. *Clin. Sci.* 90:119–125.

Schachat, F., Diamond, M., and Brandt, P. 1987. Effect of different troponin T–tropomyosin combinations on thin filament activation. *J. Mol. Biol.* 198:551–554.

Schantz, P.G., Moritani, T., Karlson, E., Johansson, E., and Lundh, A. 1989. Maximal voluntary force of bilateral and unilateral leg extension. *Acta Physiol. Scand.* 136:185–192.

Schiaffino, S., and Reggiani, C. 1996. Molecular diversity of myofibrillar proteins: Gene regulation and functional significance. *Physiol. Rev.* 76:371–423.

Schmidt, C., Pommerenke, H., Dürr, F., Nebe, B., and Rychly, J. 1998. Mechanical stressing of integrin receptors induces enhanced tyrosine phosphorylation of cytoskeletally anchored proteins. *J. Biol. Chem.* 273:5081–5085.

Schmied, A., Morin, D., Vedel, J.P., and Pagni, S. 1997. The "size principle" and synaptic effectiveness of muscle afferent projections to human extensor carpi radialis motoneurones during wrist extension. *Exp. Brain Res.* 113:214–229.

Schmitt, T.L., and Pette, D. 1991. Fiber type–specific distribution of parvalbumin in rabbit skeletal muscle: A quantitative microbiochemical and immunohistochemical study. *Histochemistry* 96:459–465.

Schuler, M., and Pette, D. 1996. Fiber transformation and replacement in low-frequency stimulated rabbit fast-twitch muscles. *Cell Tissue Res.* 285:297–303.

Schwaller, B., Dick, J., Dhoot, G., Carroll, S., Vrbova, G., Nicotera, P., Pette, D., Wyss, A., Bluethmann, H., Hunziker, W., and Celio, M.R. 1999. Prolonged contraction-relaxation cycle of fast-twitch muscles in parvalbumin knockout mice. *Am. J. Physiol. Cell Physiol.* 276:C395–C403.

Schwindt, P.C., and Calvin, W.H. 1972. Membrane-potential trajectories between spikes underlying motoneuron firing rates. *J. Neurophysiol.* 35:311–325.

Seburn, K.L., Coicou, C., and Gardiner, P.F. 1994. Effects of altered muscle activation on oxidative enzyme activity in rat alpha-motoneurons. *J. Appl. Physiol.* 77:2269–2274.

Seburn, K.L., and Gardiner, P.F. 1996. Properties of sprouted rat motor units: Effects of period of enlargement and activity level. *Muscle & Nerve* 19:1100–1109.

Seedorf, K. 1995. Intracellular signaling by growth factors. *Metabolism* 44:24–32.

Seger, J.Y., Arvidsson, B., and Thorstensson, A. 1998. Specific effects of eccentric and concentric training on muscle strength and morphology in humans. *Eur. J. Appl. Physiol. Occup. Physiol.* 79:49–57.

Semmler, J.G., Kutzscher, D., and Enoka, R. 1999. Gender differences in the fatigability of human skeletal muscle. *J. Neurophysiol.* 82:3590–3593.

Semmler, J.G., and Nordstrom, M.A. 1998. Motor unit discharge and force tremor in skill- and strength-trained individuals. *Exp. Brain Res.* 119:27–38.

Shields, R.K. 1995. Fatigability, relaxation properties, and electromyographic responses of the human paralyzed soleus muscle. *J. Neurophysiol.* 73:2195–2206.

Siegel, G., Agranoff, B., Albers, R., and Molinoff, P. 1989. *Basic neurochemistry.* New York: Raven Press.

Simard, C.P., Spector, S.A., and Edgerton, V.R. 1982. Contractile properties of rat hind limb muscles immobilized at different lengths. *Exp. Neurol.* 77:467–482.

Simoneau, J.A., and Bouchard, C. 1989. Human variation in skeletal muscle fiber-type proportion and enzyme activities. *Am. J. Physiol. Endocrinol. Metab.* 257 (20): E567–E572.

Simoneau, J.A., and Bouchard, C. 1995. Genetic determinism of fiber type proportion in human skeletal muscle. *FASEB J.* 9:1091–1095.

Sinoway, L.I., Hill, J.M., Pickar, J.G., and Kaufman, M.P. 1993. Effects of contraction and lactic acid on the discharge of group III muscle afferents in cats. *J. Neurophysiol.* 69:1053–1059.

Sjodin, B., Jacobs, I., and Karlsson, J. 1981. Onset of blood lactate accumulation and enzyme activities in M. vastus lateralis muscle. *Int. J. Sports Med.* 2:166–170.

Sketelj, J., Crne-Finderle, N., Strukelj, B., Trontelj, J.V., and Pette, D. 1998. Acetylcholinesterase mRNA level and synaptic activity in rat muscles depend on nerve-induced pattern of muscle activation. *J. Neurosci.* 18:1944–1952.

Skorjanc, D., Traub, I., and Pette, D. 1998. Identical responses of fast muscle to sustained activity by low-frequency stimulation in young and aging rats. *J. Appl. Physiol.* 85:437–441.

Smerdu, V., Karsch-Mizrachi, I., Campione, M., Leinwand, L., and Schiaffino, S. 1994. Type IIx myosin heavy chain transcripts are expressed in type IIb fibers of human skeletal muscle. *Am. J. Physiol. Cell Physiol.* 267:C1723–C1728.

Smith, J.L., Smith, L.A., Zernicke, R.F., and Hoy, M. 1982. Locomotion in exercised and non-exercised cats cordotomized at two or twelve weeks of age. *Exp. Neurol.* 76:393–413.

Smith, L.A., Eldred, E., and Edgerton, V.R. 1993. Effects of age at cordotomy and subsequent exercise on contraction times of motor units in the cat. *J. Appl. Physiol.* 75:2683–2688.

Sogaard, K., Christensen, H., Fallentin, N., Mizuno, M., Quistorff, B., and Sjogaard, G. 1998. Motor unit activation patterns during concentric wrist flexion in humans with different muscle fibre composition. *Eur. J. Appl. Physiol.* 78:411–416.

Sokoloff, A.J., and Cope, T.C. 1996. Recruitment of triceps surae motor units in the decerebrate cat: 2. Heterogeneity among soleus motor units. *J. Neurophysiol.* 75:2005–2016.

Spector, S.A. 1985. Effects of elimination of activity on contractile and histochemical properties of rat soleus muscle. *J. Neurosci.* 5:2177–2188.

Spector, S.A., Simard, C.P., Fournier, M., Sternlicht, E., and Edgerton, V.R. 1982. Architectural alterations of rat hind-limb skeletal muscles immobilized at different lengths. *Exp. Neurol.* 76:94–110.

Spencer, M.J., and Tidball, J.G. 1997. Calpain II expression is increased by changes in mechanical loading of muscle *in vivo*. *J. Cell Biochem.* 64:55–66.

Spielmann, J.M., Laouris, Y., Nordstrom, M.A., Robinson, G.A., Reinking, R.M., and Stuart, D.G. 1993. Adaptation of cat motoneurons to sustained and intermittent extracellular activation. *J. Physiol.* (Lond.) 464:75–120.

Sreter, F., Lopez, J.R., Alamo, L., Mabuchi, K., and Gergely, J. 1987. Changes in intracellular ionized Ca concentration associated with muscle fiber type transformation. *Am. J. Physiol.* 253:C296–C300.

Staron, R.S. 1991. Correlation between myofibrillar ATPase activity and myosin heavy chain composition in single human muscle fibers. *Histochemistry* 96:21–24.

Staron, R.S., Karapondo, D.L., Kraemer, W.J., Fry, A.C., Gordon, S.E., Falkel, J.E., Hagerman, F.C., and Hikida, R.S. 1994. Skeletal muscle adaptations during early phase of heavy-resistance training in men and women. *J. Appl. Physiol.* 76:1247–1255.

Staron, R.S., Malicky, E.S., Leonardi, M.J., Falkel, J.E., Hagerman, F.C., and Dudley, G.A. 1989. Muscle hypertrophy and fast fiber type conversions in heavy resistance–trained women. *Eur. J. Appl. Physiol.* 60:71–79.

Staron, R.S., and Pette, D. 1987. Nonuniform myosin expression along single fibers of chronically stimulated and contralateral rabbit tibialis anterior muscles. *Pflugers Arch.* 409:67–73.

Stein, R.B., Gordon, T., Jefferson, J., Sharfenberger, A., Yang, J.F., De Zepetnek, J.T., and Bélanger, M. 1992. Optimal stimulation of paralyzed muscle after human spinal cord injury. *J. Appl. Physiol.* 72:1393–1400.

Steinbach, J.H., Schubert, D., and Eldridge, L. 1980. Changes in cat muscle contractile proteins after prolonged muscle inactivity. *Exp. Neurol.* 67:655–669.

Stephens, J.A., and Taylor, A. 1972. Fatigue of maintained voluntary muscle contraction in man. *J. Physiol.* 220:1–18.

Stephens, J.A., and Usherwood, T.P. 1977. The mechanical properties of human motor units with special reference to their fatiguability and recruitment threshold. *Brain Res.* 125:91–97.

Sterz, R., Pagala, M., and Peper, K. 1983. Postjunctional characteristics of the endplates in mammalian fast and slow muscles. *Pflugers Arch.* 398:48–54.

Stevens, L., Gohlsch, B., Mounier, Y., and Pette, D. 1999. Changes in myosin heavy chain mRNA and protein isoforms in single fibers of unloaded rat soleus muscle. *FEBS Lett.* 463:15–18.

Stevens, L., Sultan, K., Peuker, H., Gohlsch, B., Mounier, Y., and Pette, D. 1999. Time-dependent changes in myosin heavy chain mRNA and protein isoforms in unloaded soleus muscle of rat. *Am. J. Physiol.* 277:C1044–C1049.

Stienen, G.J.M., Kiers, J.L., Bottinelli, R., and Reggiani, C. 1996. Myofibrillar ATPase activity in skinned human skeletal muscle fibres: Fibre type and temperature dependence. *J. Physiol.* 493:299–307.

St.-Pierre, D., and Gardiner, P.F. 1985. Effects of disuse on mammalian fast-twitch muscle: Joint fixation compared to neurally-applied tetrodotoxin. *Exp. Neurol.* 90:635–651.

St.-Pierre, D., Léonard, D., Houle, R., and Gardiner, P.F. 1988. Recovery of muscle from TTX-induced disuse and the influence of daily exercise: II. Muscle enzymes and fatigue characteristics. *Exp. Neurol.* 101:327–346.

Strojnik, V. 1995. Muscle activation level during maximal voluntary effort. *Eur. J. Appl. Physiol.* 72:144–149.

Stuart, D.G., Hamm, T.M., and Vanden Noven, S. 1988. Partitioning of monosynaptic Ia EPSP connections with motoneurons according to neuromuscular topography: Generality and functional implications. *Prog. Neurobiol.* 30:437–447.

Sugden, P.H., and Clerk, A. 1998. Cellular mechanisms of cardiac hypertrophy. *J. Mol. Med.* 76:725–746.

Sugiura, T., Matoba, H., and Murakamis, N. 1992. Myosin light chain patterns in histochemically typed single fibers of the rat skeletal muscle. *Comp. Biochem. Physiol. [B]* 102B:617–620.

Sutherland, H., Jarvis, J.C., Kwende, M.M.N., Gilroy, S.J., and Salmons, S. 1998. The dose-related response of rabbit fast muscle to long-term low-frequency stimulation. *Muscle & Nerve* 21:1632–1646.

Suzuki, H., Tsuzimoto, H., Ishiko, T., Kasuga, N., Taguchi, S., and Ishihara, A. 1991. Effect of endurance training on the oxidative enzyme activity of soleus motoneurons in rats. *Acta Physiol. Scand.* 143:127–128.

Sweeney, H.L., Bowman, B.F., and Stull, J.T. 1993. Myosin light chain phosphorylation in vertebrate striated muscle: Regulation and function. *Am. J. Physiol. Cell Physiol.* 264 (33): C1085–C1095.

Sweeney, H.L., Kushmerick, M.J., Mabuchi, K., Sreter, F.A., and Gergely, J. 1988. Myosin alkali light chain and heavy chain variations correlate with altered shortening velocity of isolated skeletal muscle fibers. *J. Biol. Chem.* 263:9034–9039.

Tabary, J.C., Tabary, C., Tardieu, C., Tardieu, G., and Goldspink, G. 1972. Physiological and structural changes in the cat's soleus muscle due to immobilization at different lengths by plaster casts. *J. Physiol.* 224:231–244.

Takahashi, M., Chesley, A., Freyssenet, D., and Hood, D.A. 1998. Contractile activity-induced adaptations in the mitochondrial protein import system. *Am. J. Physiol. Cell Physiol.* 274:C1380–C1387.

Takekura, H., and Yoshioka, T. 1987. Determination of metabolic profiles on single muscle fibres of different types. *J. Muscle Res. Cell Motil.* 8:342–348.

Takekura, H., and Yoshioka, T. 1989. Ultrastructural and metabolic profiles on single muscle fibers of different types after hindlimb suspension in rats. *Jap. J. Physiol.* 39:385–396.

Takekura, H., and Yoshioka, T. 1990. Ultrastructural and metabolic characteristics of single muscle fibres belonging to the same type in various muscles in rats. *J. Muscle Res. Cell Motil.* 11:98–104.

Talmadge, R.J., Roy, R.R., and Edgerton, V.R. 1995. Prominence of myosin heavy chain hybrid fibers in soleus muscle of spinal cord–transected rats. *J. Appl. Physiol.* 78:1256–1265.

Talmadge, R.J., Roy, R.R., and Edgerton, V.R. 1996a. Distribution of myosin heavy chain isoforms in non-weight-bearing rat soleus muscle fibers. *J. Appl. Physiol.* 81:2540–2546.

Talmadge, R.J., Roy, R.R., and Edgerton, V.R. 1996b. Myosin heavy chain profile of cat soleus following chronic reduced activity or inactivity. *Muscle & Nerve* 19:980–988.

Talmadge, R.J., Roy, R.R., and Edgerton, V.R. 1999. Persistence of hybrid fibers in rat soleus after spinal cord transection. *Anat. Rec.* 255:188–201.

Tamaki, H., Kitada, K., Akamine, T., Murata, F., Sakou, T., and Kurata, H. 1998. Alternating activity in the synergistic muscles during prolonged low-level contractions. *J. Appl. Physiol.* 84:1943–1951.

Tax, A.A.M., Van der Gon, J.J.D., Gielen, C.C.A.M., and Kleyne, M. 1990. Differences in central control of m. biceps brachii in movement tasks and force tasks. *Exp. Brain Res.* 79:138–142.

Tax, A.A.M., Van der Gon, J.J.D., Gielen, C.C.A.M., and van den Tempel, C.M.M. 1989. Differences in the activation of m. biceps brachii in the control of slow isotonic movements and isometric contractions. *Exp. Brain Res.* 76:55–63.

Taylor, J.L., Butler, J.E., Allen, G.M., and Gandevia, S.C. 1996. Changes in motor cortical excitability during human muscle fatigue. *J. Physiol.* (Lond.) 490:519–528.

Termin, A., and Pette, D. 1991. Myosin heavy-chain-based isomyosins in developing, adult fast-twitch and slow-twitch muscles. *Eur. J. Biochem.* 195:577–584.

Termin, A., and Pette, D. 1992. Changes in myosin heavy-chain isoform synthesis of chronically stimulated rat fast-twitch muscle. *Eur. J. Biochem.* 204:569–573.

Tesch, P.A. 1980. Fatigue pattern in subtypes of human skeletal muscle fibers. *Int. J. Sports Med.* 1:79–81.

Tesch, P.A., and Karlsson, J. 1978. Isometric strength performance and muscle fiber type distribution in man. *Acta Physiol. Scand.* 103:47–51.

Tesch, P.A., Thorsson, A., and Colliander, E.B. 1990. Effects of eccentric and concentric resistance training on skeletal muscle substrates, enzyme activities and capillary supply. *Acta Physiol. Scand.* 140:575–580.

Tesch, P.A., Thorsson, A., and Essen-Gustavsson, B. 1989. Enzyme activities of FT and ST muscle fibers in heavy-resistance trained athletes. *J. Appl. Physiol.* 67:83–87.

Thesleff, S. 1959. Motor end-plate desensitization by repetitive nerve stimuli. *J. Physiol.* (Lond.) 148:659–664.

Thickbroom, G., Phillips, B., Morris, I., Byrnes, M., Sacco, P., and Mastaglia, F. 1999. Differences in functional magnetic resonance imaging of sensorimotor cortex during static and dynamic finger flexion. *Exp. Brain Res.* 126:431–438.

Thomas, C.K., Bigland-Ritchie, B., Westling, G., and Johansson, R. S. 1990. A comparison of human thenar motor-unit properties studied by intraneural motor-axon stimulation and spike-triggered averaging. *J. Neurophysiol.* 64:1347–1351.

Thomas, C.K., Ross, B.H., and Calancie, B. 1987. Human motor-unit recruitment during isometric contractions and repeated dynamic movements. *J. Neurophysiol.* 57:311–324.

Thomas, C.K., Woods, J.J., and Bigland-Ritchie, B. 1989. Impulse propagation and muscle activation in long maximal voluntary contractions. *J. Appl. Physiol.* 67:1835–1842.

Thomason, D.B. 1998. Translational control of gene expression in muscle. *Exerc. Sport Sci. Rev.* 26:165–190.

Thomason, D.B., Biggs, R.B., and Booth, F.W. 1989. Protein metabolism and B-myosin heavy-chain mRNA in unweighted soleus muscle. *Am. J. Physiol. Regul. Integr. Comp. Physiol.* 257:R300–R305.

Thomason, D.B., Herrick, R.E., and Baldwin, K.M. 1987. Activity influences on soleus muscle myosin during rodent hindlimb suspension. *J. Appl. Physiol.* 63:138–144.

Thomason, D.B., Morrison, P.R., Oganov, V., Ilyina-Kakueva, E., Booth, F.W., and Baldwin, K.M. 1992. Altered actin and myosin expression in muscle during exposure to microgravity. *J. Appl. Physiol.* 73 suppl.: 90S–93S.

Thompson, H.S., and Scordilis, S.P. 1994. Ubiquitin changes in human biceps muscle following exercise-induced damage. *Biochem. Biophys. Res. Commun.* 204:1193–1198.

Thompson, M.G., and Palmer, R.M. 1998. Signalling pathways regulating protein turnover in skeletal muscle. *Cell. Signal.* 10:1–11.

Thorstensson, A., Grimby, G., and Karlsson, J. 1976. Force–velocity relations and fiber composition in human knee extensor muscles. *J. Appl. Physiol.* 40:12–16.

Thorstensson, A., Hulten, B., Von Dolbeln, W., and Karlsson, J. 1976. Effect of strength training on enzyme activities and fibre characteristics in human skeletal muscle. *Acta Physiol. Scand.* 96:392–398.

Thorstensson, A., and Karlsson, J. 1976. Fatiguability and fibre composition of human skeletal muscle. *Acta Physiol. Scand.* 98:318–322.

Thorstensson, A., Larsson, L., and Karlsson, J. 1977. Muscle strength and fiber composition in athletes and sedentary men. *Med. Sci. Sports Exerc.* 9:26–30.

Thorstensson, A., Sjodin, B., and Karlsson, J. 1975. Enzyme activities and muscle strength after sprint training in man. *Acta Physiol. Scand.* 94:313–318.

Tidball, J.G. 1995. Inflammatory cell response to acute muscle injury. *Med. Sci. Sports Exerc.* 27:1022–1032.

Tihanyi, J., Apor, P., and Fekete, G. 1982. Force-velocity-power characteristics and fiber composition in human knee extensor muscles. *Eur. J. Appl. Physiol.* 48:331–343.

Toursel, T., Stevens, L., and Mounier, Y. 1999. Evolution of contractile and elastic properties of rat soleus muscle fibres under unloading conditions. *Exp. Physiol.* 84:93–107.

Troiani, D., Filippi, G.M., and Bassi, F. 1999. Nonlinear tension summation of different combinations of motor units in the anesthetized cat peroneus longus muscle. *J. Neurophysiol.* 81:771–780.

Tsuboi, T., Sato, T., Egawa, K., and Miyazaki, M. 1995. The effect of fatigue caused by electrical induction or voluntary contraction on Ia inhibition in human soleus muscle. *Neurosci. Lett.* 197:72–74.

Turcotte, R., Panenic, R., and Gardiner, P.F. 1991. TTX-induced muscle disuse alters Ca^{2+} activation characteristics of myofibril ATPase. *Comp. Biochem. Physiol. [A]* 100A:183–186.

Van Bolhuis, B.M., and Gielen, C.C.A.M. 1997. The relative activation of elbow-flexor muscles in isometric flexion and in flexion/extension movements. *J. Biomech.* 30:803–811.

Van Bolhuis, B.M., Medendorp, W.P., and Gielen, C.C.A.M. 1997. Motor unit firing behavior in human arm flexor muscles during sinusoidal isometric contractions and movements. *Exp. Brain Res.* 117:120–130.

Van Cutsem, M., Duchateau, J., and Hainaut, K. 1998. Changes in single motor unit behaviour contribute to the increase in contraction speed after dynamic training in humans. *J. Physiol.* 513:295–305.

Van Cutsem, M., Feiereisen, P., Duchateau, J., and Hainaut, K. 1997. Mechanical properties and behaviour of motor units in the tibialis anterior during voluntary contractions. *Can. J. Appl. Physiol.* 22:585–597.

Vandenborne, K., Elliott, M.A., Walter, G.A., Abdus, S., Okereke, E., Shaffer, M., Tahernia, D., and Esterhai, J.L. 1998. Longitudinal study of skeletal muscle adaptations during immobilization and rehabilitation. *Muscle & Nerve* 21:1006–1012.

Vandenburgh, H.H., Hatfaludy, S., Sohar, I., and Shansky, J. 1990. Stretch-induced prostaglandins and protein turnover in cultured skeletal muscle. *Am. J. Physiol. Cell Physiol.* 259:C232–C240.

Vandenburgh, H.H., Karlisch, P., Shansky, J., and Feldstein, R. 1991. Insulin and IGF-I induce pronounced hypertrophy of skeletal myofibers in tissue culture. *Am. J. Physiol. Cell Physiol.* 260:C475–C484.

Vandenburgh, H.H., Shansky, J., Karlisch, P., and Solerssi, R.L. 1993. Mechanical stimulation of skeletal muscle generates lipid-related second messengers by phospholipase activation. *J. Cell. Physiol.* 155:63–71.

Vandenburgh, H.H., Shansky, J., Solerssi, R.L., and Chromiak, J. 1995. Mechanical stimulation of skeletal muscle increases prostaglandin F_2 production, cyclo-oxygenase activity, and cell growth by a pertussis toxin sensitive mechanism. *J. Cell. Physiol.* 163:285–294.

Vandenburgh, H.H., Swasdison, S., and Karlisch, P. 1991. Computer-aided mechanogenesis of skeletal muscle organs from single cells *in vitro. FASEB J.* 5:2860–2867.

Vandervoort, A., Sale, D., and Moroz, J. 1984. Comparison of motor unit activation during unilateral and bilateral leg extension. *J. Appl. Physiol.* 56:46–51.

Van Lunteren, E., and Moyer, M. 1996. Effects of DAP on diaphragm force and fatigue, including fatigue due to neurotransmission failure. *J. Appl. Physiol.* 81:2214–2220.

Van Praag, H., Christie, B., Sejnowski, T., and Gage, F. 1999. Running enhances neurogenesis, learning, and long-term potentiation in mice. *Proc. Natl. Acad. Sci. USA* 96:13427–13431.

Vary, T., Jefferson, L., and Kimball, S. 2000. Role of eIF4E in stimulation of protein synthesis by IGF-I in perfused rat skeletal muscle. *Am. J. Physiol.* 278:E58–E64.

Viitasalo, J.H.T., Häkkinen, K., and Komi, P.V. 1981. Isometric and dynamic force production and muscle fiber type composition in man. *J. Hum. Mov. Stud.* 7:199–209.

Viitasalo, J.H.T., and Komi, P.V. 1978. Force-time characteristics and fiber composition in human leg extensor muscles. *Eur. J. Appl. Physiol.* 40:7–15.

Viitasalo, J.H.T., and Komi, P.V. 1981. Interrelationships between electromyographic, mechanical, muscle structure and reflex time measurements in man. *Acta Physiol. Scand.* 111:97–103.

Vijayan, K., Thompson, J., and Riley, D. 1998. Sarcomere lesion damage occurs mainly in slow fibers of reloaded rat adductor longus muscles. *J. Appl. Physiol.* 85:1017–1023.

Vollestad, N.K. 1997. Measurement of human muscle fatigue. *J. Neurosci. Methods* 74:219–227.

Vollestad, N.K., and Blom, P.C.S. 1985. Effect of varying exercise intensity on glycogen depletion in human muscle fibres. *Acta Physiol. Scand.* 125:395–405.

Vollestad, N.K., Tabata, I., and Medbo, J.I. 1992. Glycogen breakdown in different human muscle fibre types during exhaustive exercise of short duration. *Acta Physiol. Scand.* 144:135–141.

Vollestad, N., Vaage, O., and Hermansen, L. 1984. Muscle glycogen depletion patterns in type I and subgroups of type II fibres during prolonged severe exercise in man. *Acta Physiol. Scand.* 122:433–441.

Wada, M., Hämäläinen, N., and Pette, D. 1995. Isomyosin patterns of single type IIB, IID and IIA fibres from rabbit skeletal muscle. *J. Muscle Res. Cell Motil.* 16:237–242.

Wada, M., Okumoto, T., Toro, K., Masuda, K., Fukubayashi, T., Kikuchi, K., Niihata, S., and Katsuta, S. 1996. Expression of hybrid isomyosins in human skeletal muscle. *Am. J. Physiol. Cell Physiol.* 271:C1250–C1255.

Wada, M., and Pette, D. 1993. Relationships between alkali light-chain complement and myosin heavy-chain isoforms in single fast-twitch fibers of rat and rabbit. *Eur. J. Biochem.* 214:157–161.

Waerhaug, O., and Lomo, T. 1994. Factors causing different properties at neuromuscular junctions in fast and slow rat skeletal muscles. *Anat. Embryol.* (Berl.) 190:113–125.

Warren, G.L., III, Lowe, D.A., Hayes, D.A., Farmer, M.A., and Armstrong, R.B. 1995. Redistribution of cell membrane probes following contraction-induced injury of mouse soleus muscle. *Cell Tissue Res.* 282:311–320.

Watson, P.A. 1991. Function follows form: Generation of intracellular signals by cell deformation. *FASEB J.* 5:2013–2019.

Watson, P.A., Stein, J., and Booth, F. 1984. Changes in actin synthesis and alpha-actin-mRNA content in rat muscle during immobilization. *Am. J. Physiol. Cell Physiol.* 247:C39–C44.

Wehring, M., Cal, B., and Tidball, J. 2000. Modulation of myostatin expression during modified muscle use. *FASEB J.* 14:103–110.

Weir, J.P., Keefe, D.A., Eaton, J.F., Augustine, R.T., and Tobin, D.M. 1998. Effect of fatigue on hamstring coactivation during isokinetic knee extensions. *Eur. J. Appl. Physiol.* 78:555–559.

Westerblad, H., Allen, D.G., Bruton, J.D., Andrade, F.H., and Lannergren, J. 1998. Mechanisms underlying the reduction of isometric force in skeletal muscle fatigue. *Acta Physiol. Scand.* 162:253–260.

Westgaard, R.H., and De Luca, C.J. 1999. Motor unit substitution in long-duration contractions of the human trapezius muscle. *J. Neurophysiol.* 82:501–504.

Westing, S.H., Cresswell, A.G., and Thorstensson, A. 1991. Muscle activation during maximal voluntary eccentric and concentric knee extension. *Eur. J. Appl. Physiol.* 62:104–108.

Westing, S.H., Seger, J.Y., and Thorstensson, A. 1990. Effects of electrical stimulation on eccentric and concentric torque–velocity relationships during knee extension in man. *Acta Physiol. Scand.* 140:17–22.

Wheeler, M., Snyder, E., Patterson, M., and Swoap, S. 1999. An E-box within the MHC IIB gene is bound by MyoD and is required for gene expression in fast muscle. *Am. J. Physiol.* 276:C1069–C1078.

White, M., Davies, C., and Brooksby, P. 1984. The effects of short-term voluntary immobilization on the contractile properties of the human triceps surae. *Q. J. Exp. Physiol.* 69:685–691.

Whitehead, N.P., Allen, T.J., Morgan, D.L., and Proske, U. 1998. Damage to human muscle from eccentric exercise after training with concentric exercise. *J. Physiol.* 512:615–620.

Whitelaw, P.F., and Hesketh, J.E. 1992. Expression of *c-myc* and *c-fos* in rat skeletal muscle: Evidence for increased levels of *c-myc* mRNA during hypertrophy. *Biochem. J.* 281:143–147.

Wickham, J.B., and Brown, J.M.M. 1998. Muscles within muscles: The neuromotor control of intra-muscular segments. *Eur. J. Appl. Physiol. Occup. Physiol.* 78:219–225.

Widrick, J.J., and Fitts, R.H. 1997. Peak force and maximal shortening velocity of soleus fibers after non-weight-bearing and resistance exercise. *J. Appl. Physiol.* 82:189–195.

Widrick, J.J., Romatowski, J.G., Bain, J., Trappe, S.W., Trappe, T., Thompson, J., Costill, D.L., Riley, D., and Fitts, R.H. 1997. Effect of 17 days of bed rest on peak isometric force and unloaded shortening velocity of human soleus fibers. *Am. J. Physiol.* 273:C1690–C1699.

Widrick, J.J., Romatowski, J.G., Karhanek, M., and Fitts, R.H. 1997. Contractile properties of rat, rhesus monkey, and human type I muscle fibers. *J. Appl. Physiol.* 41:R34–R47.

Widrick, J.J., Trappe, S.W., Blaser, C.A., Costill, D.L., and Fitts, R.H. 1996. Isometric force and maximal shortening velocity of single muscle fibers from elite master runners. *Am. J. Physiol. Cell Physiol.* 271:C666–C675.

Widrick, J.J., Trappe, S.W., Costill, D.L., and Fitts, R.H. 1996. Force-velocity and force-power properties of single muscle fibers from elite master runners and sedentary men. *Am. J. Physiol. Cell Physiol.* 271:C676–C683.

Williams, P.E., and Goldspink, G. 1984. Connective tissue changes in immobilized muscle. *J. Anat.* 138:343–350.

Williams, P.E., Watt, P., Bicik, V., and Goldspink, G. 1986. Effect of stretch combined with electrical stimulation on the type of sarcomeres produced at the ends of muscle fibers. *Exp. Neurol.* 93:500–509.

Williams, R.S., and Neufer, P.D. 1996. Regulation of gene expression in skeletal muscle by contractile activity. In *Handbook of physiology: 12. Exercise, regulation and integration of multiple systems,* ed. L.B. Rowell and J.T. Shepherd, 1124-1150. New York: Oxford University Press.

Windhorst, U., Kirmayer, D., Soibelman, F., Misri, A., and Rose, R. 1997. Effects of neurochemically excited group III-IV muscle afferents on motoneuron after-hyperpolarization. *Neuroscience* 76:915–929.

Windhorst, U., and Kokkoroyiannis, T. 1991. Interaction of recurrent inhibitory and muscle spindle afferent feedback during muscle fatigue. *Neuroscience* 43:249–259.

Windisch, A., Gundersen, K., Szabolcs, M.J., Gruber, H., and Lomo, T. 1998. Fast to slow transformation of denervated and electrically stimulated rat muscle. *J. Physiol.* 510:623–632.

Winiarski, A., Roy, R., Alford, E., Chiang, P., and Edgerton, V. 1987. Mechanical properties of rat skeletal muscle after hind limb suspension. *Exp. Neurol.* 96:650–660.

Witzmann, F.A., Kim, D., and Fitts, R.H. 1982. Hindlimb immobilization: Length–tension and contractile properties of skeletal muscle. *J. Appl. Physiol.* 53:335–345.

Witzmann, F.A., Kim, D., and Fitts, R.H. 1983. Effect of hindlimb immobilization on the fatigability of skeletal muscle. *J. Appl. Physiol. Respir. Environ. Exerc. Physiol.* 54:1242–1248.

Witzmann, F.A., Troup, J.P., and Fitts, R.H. 1982. Acid phosphatase and protease activities in immobilized rat skeletal muscles. *Can. J. Physiol. Pharmacol.* 60:1732–1736.

Wohlfart, B., and Edman, K.A.P. 1994. Rectangular hyperbola fitted to muscle force-velocity data using three-dimensional regression analysis. *Exp. Physiol.* 79:235–239.

Wolpaw, J.R., and Carp, J.S. 1990. Memory traces in spinal cord. *Trends Neurosci.* 13:137–142.

Wolpaw, J.R., and Lee, C.L. 1989. Memory traces in primate spinal cord produced by operant conditioning of H-reflex. *J. Neurophysiol.* 61:563–572.

Wong, T.S., and Booth, F.W. 1990a. Protein metabolism in rat gastrocnemius muscle after stimulated chronic concentric exercise. *J. Appl. Physiol.* 69:1709–1717.

Wong, T.S., and Booth, F.W. 1990b. Protein metabolism in rat tibialis anterior muscle after stimulated chronic eccentric exercise. *J. Appl. Physiol.* 69:1718–1724.

Wood, S.J., and Slater, C.R. 1997. The contribution of postsynaptic folds to the safety factor for neuromuscular transmission in rat fast- and slow-twitch muscles. *J. Physiol.* 500:165–176.

Woods, J., Furbush, F., and Bigland-Ritchie, B. 1987. Evidence for a fatigue-induced reflex inhibition of motoneuron firing rates. *J. Neurophysiol.* 58:125–137.

Wu, L.G., and Betz, W.J. 1998. Kinetics of synaptic depression and vesicle recycling after tetanic stimulation of frog motor nerve terminals. *Biophys. J.* 74:3003–3009.

Yan, Z., Biggs, R.B., and Booth, F.W. 1993. Insulin-like growth factor immunoreactivity increases in muscle after acute eccentric contractions. *J. Appl. Physiol.* 74:410–414.

Yang, S.Y., Alnaqeeb, M., Simpson, H., and Goldspink, G. 1996. Cloning and characterization of an IGF-1 isoform expressed in skeletal muscle subjected to stretch. *J. Muscle Res. Cell Motil.* 17:487–495.

Yaspelkis, B.B., III, Castle, A.L., Ding, Z., and Ivy, J.L. 1999. Attenuating the decline in ATP arrests the exercise training–induced increases in muscle GLUT4 protein and citrate synthase activity. *Acta Physiol. Scand.* 165:71–79.

Yasuda, T., Sakamoto, K., Nosaka, K., Wada, M., and Katsuta, S. 1997. Loss of sarcoplasmic reticulum membrane integrity after eccentric contractions. *Acta Physiol. Scand.* 161:581–582.

Yue, G., and Cole, K. 1992. Strength increases from the motor program: Comparison of training with maximal voluntary and imagined muscle contractions. *J. Neurophysiol.* 67:1114–1123.

Zardini, D.M., and Parry, D.J. 1998. Physiological characteristics of identified motor units in the mouse extensor digitorum longus muscle: An *in vitro* approach. *Can. J. Physiol. Pharmacol.* 76:68–71.

Zemková, H., Teisinger, J., Almon, R.R., Vejsada, R., Hník, P., and Vyskocil, F. 1990. Immobilization atrophy and membrane properties in rat skeletal muscle fibres. *Pflugers Arch.* 416:126–129.

Zengel, J., Reid, S., Sypert, G., and Munson, J. 1985. Membrane electrical properties and prediction of motor-unit type of medial gastrocnemius motoneurons in the cat. *J. Neurophysiol.* 53:1323–1344.

Zhou, M.-Y., Klitgaard, H., Saltin, B., Roy, R.R., Edgerton, V.R., and Gollnick, P.D. 1995. Myosin heavy chain isoforms of human muscle after short-term spaceflight. *J. Appl. Physiol.* 78:1740–1744.

Zijdewind, I., Kernell, D., and Kukulka, C.G. 1995. Spatial differences in fatigue-associated electromyographic behaviour of the human first dorsal interosseus muscle. *J. Physiol.* (Lond.) 483:499–510.

Zijdewind, I., Zwarts, M.J., and Kernell, D. 1999. Fatigue-associated changes in the electromyogram of the human first dorsal interosseous muscle. *Muscle & Nerve* 22:1432–1436.

Zytnicki, D., Lafleur, J., Horcholle-Bossavit, G., Lamy, C., and Jami, L. 1990. Reduction of Ib autogenic inhibition in motoneurones during contractions of an ankle extensor muscle in the cat. *J. Neurophysiol.* 64:1380–1389.

Index

Note: The letters *f* and *t* following page numbers refer to figures and tables, respectively.

A

acetylcholine receptor–inducing activity (ARIA) 131
acetylcholine receptors (AChR) 131. *See also* receptors
    desensitization of, as a possible mechanism during fatigue 88, 130
    response to endurance training 131
acetylcholinesterase (AChE) 130–131
    effects of endurance training 130–131, 131*f*
    molecular forms 130, 131*f*
AChE. *See* acetylcholinesterase
AChR. *See* acetylcholine receptors
afferents
    groups III and IV; role in fatigue 108–109
    onto motoneurons 91, 95*f*
afterhyperpolarization, motoneuronal
    definition 55, 56*f*
    determinants 55
    response to chronic muscle stimulation 135, 136*f*
    response to neuromuscular inactivity 178–179
    role in determining motoneuron firing frequency limits 56–57, 56*f*
    variation among motor unit types 49*t*, 55
alpha-actinin 147
a/P$_o$. *See also* force-velocity relationship of muscle; fiber types, muscle
    definition 13
    differences among fiber types 14, 14*f*
AP-1 transcription factor 126
ARIA. *See* acetylcholine receptor inducing activity
autogenic inhibition 108
axonal transport 133
    definition 133
    effects of endurance training 133, 134*f*
axotomy 138

B

ballistic contractions
    recruitment of motor units during 76, 169
    training effects on recruitment of motor units during 169, 169*f*
basic fibroblast growth factor (bFGF) 154
bedrest, effects on neuromuscular system 193–195, 193*t*, 194*f*, 195*f*, 196*f*
beta-GPA. *See* beta-guanidinoproprionic acid
beta-guanidinoproprionic acid (beta-GPA) 125
bFGF. *See* basic fibroblast growth factor
bilateral contractions 78

biopsy technique 113
bistability 60–62. *See also* plateau potentials
branch-block failure 86

C

calcineurin 126, 127*f*
    possible role during muscle hypertrophy 147*f*
calcitonin gene-related peptide
    effects of endurance training 133, 134*f*
    in motoneurons 133, 134*f*
calcium
    intracellular, as a signal for gene expression 125–126, 127*f*, 147*f*
    regulation of, in different fiber types 28
    and the troponin–tropomyosin complex in muscle fibers 23
calcium-activated neutral proteases. *See* calpains
calpains 152–153
calsequestrin 28, 115
capacitance, motoneuronal 50
central fatigue 98, 99
c-Fos 126. *See also* protooncogenes
    responses to muscle stretch and electrical stimulation 148, 148*f*
CGRP. *See* calcitonin gene-related peptide
chronic muscle electrical stimulation 113–128
    as a model, compared to endurance training 113–115
c-Jun 126. *See also* protooncogenes
    responses to muscle stretch and electrical stimulation 150, 151*f*
c-Jun NH$_2$-terminal kinase (JNK) 126
    response to muscle stretch, acute exercise 148, 148*f*
clenbuterol 125
c-Myc 150. *See also* protooncogenes
    response to muscle compensatory overload 150
coactivation of agonists, antagonists
    defined 77
    during fatigue 104–105, 105*f*
    effects of strength training 167, 168*f*
    possible role of presynaptic inhibition 95
cocontraction of agonists, antagonists. *See* coactivation of agonists, antagonists
compartmentalization of motor units 68
concentric training
    effects, compared to eccentric training 154, 165*f*
connective tissue
    effects of strength training 157
    effects of muscle atrophy caused by joint immobilization 197–198

contractile properties
    of muscles following decreased activity 177, 180*t*,
        180*f*, 182, 184, 189, 194*f*, 195*f*, 198, 199*f*
    of muscles following strength training 158–160
    of single muscle fibers of different types 8–17, 10*t*,
        11*f*, 12*f*, 13*f*, 14*f*, 15*f*, 16*f*, 17*f*, 18*f*
contralateral effects of training 165–167
cooperativity of the troponin–tropomyosin complex
    24, 24*f*, 25*f*
cortex
    excitability changes during fatigue 97, 98*f*, 99*f*,
        100*f*, 105
countermeasures to muscle atrophy 182, 186, 189,
    190*t*
current–frequency relationship of motoneurons 58–59,
    58*f*
cytooxygenase 150
cytoskeleton 147

D

damage, muscular
    during chronic electrical stimulation 127
    effects on protein synthesis, degradation 153–154
    following eccentric exercise 153–154, 160–161
degeneration, muscular 127
dendrites 53
derecruitment of motor units 72, 72*f*
desensitization of postsynaptic membrane as a
    possible source of fatigue 88, 128
dihydropyridine receptors (DHPR) 115. *See also*
    receptors
doublets 57

E

eccentric contractions
    and muscle damage 153–154, 160–161
    recruitment of motor units during 76–77, 76*f*
    specific effects of training 160–161, 165*f*
economy
    of contraction, determinants 17
    variation among fiber types 18, 19*f*
eIF. *See* initiating factors of protein synthesis
electrophoresis. *See* polyacrylamide gel
    electrophoresis
end-plate potential (EPP) 86
    response to endurance training 131, 132*f*
enzymes of energy metabolism in muscle
    differences among fiber types 29–30, 30*t*
    effects of chronic low-frequency electrical muscle
        stimulation 114*t*, 121, 121*f*, 122*f*
    effects of decreased activity 177, 178*f*, 179, 181,
        183, 198
    effects of strength training 157
    related to muscle fatigue resistance 42, 43*f*
EPP. *See* end-plate potential
Erg-1 126
ERK. *See* extracellular signal-related kinase
essential light chains. *See* myosin light chains

excitability
    of motoneurons; determinants 50
    of motor cortex during fatigue 97, 98*f*, 99*f*, 100*f*,
        105–106
extracellular signal-related kinase (ERK)
    activation by contractile activity 148, 148*f*

F

fatigue
    central vs. peripheral 84
    defined 84
    of the neuromuscular junction 41, 85
    relationship to mitochondrial enzymes in muscle
        42, 43*f*
fatigue resistance
    of muscles atrophied following joint
        immobilization 199
    as a property to classify muscle units 39–43, 40*f*
    response to strength training 160
    relationship to mitochondrial enzymes in muscle
        42, 43*f*
fiber types, muscle
    differences among species 4–5, 8, 10, 11*f*
    differences in a/P$_o$ and V$_{opt}$ 15, 16*f*
    differences in calcium-regulatory systems 4*t*, 28–
        29
    differences in enzymes of energy metabolism 29–
        30, 30*t*, 31*f*
    differences in shortening velocity 9–14, 10*t*, 11*f*,
        12*f*, 13*f*, 14*f*
    differences in specific tension 15–17, 17*f*
    as distinguished according to myosin heavy chain
        composition 3, 4*t*, 6*f*
    as distinguished using myofibrillar ATPase
        histochemistry 5–6, 6*f*
    effects of decreased activity 177, 179*f*, 181, 183,
        183*f*, 184, 185*f*, 188, 189*f*, 198
    effects of strength training 155
    hybrids 5, 7*f*, 10
    nomenclature 5
    related to athletic performance 32–36, 33*t*
    in various human muscles 32, 32*f*
fine-wire electrodes 66–68
firing frequency of motoneuron
    decrease during fatigue 88, 89*f*
    determinants of limits 55–60
firing level of motoneuron 55
fMRI. *See* functional magnetic resonance imaging
focal adhesions 147
force-velocity relationship of muscle 12–13, 13*f*
    following strength training 159
functional magnetic resonance imaging (fMRI) 73
fusimotor neurons 91–93, 108
    possible role in fatigue 91–93, 92*f*, 102–103, 109–110

G

gamma-motoneurons. *See* fusimotor neurons
gene sequencing 7

glucose transporters (GLUT) 125
GLUT. *See* glucose transporters
glycogen depletion
   to estimate recruitment of fibers during voluntary
      exercise 78–79, 79*f*, 80*f*
   to visualize activated fibers 42–43, 43f, 46
Golgi tendon organs
   changes in sensitivity during fatigue 104, 108
G proteins 147, 150

H

heat shock proteins (HSP) 120
heavy chains. *See* myosin heavy chains
hindlimb suspension 187–191, 190*t*
Hofmann reflex. *See* H-reflex
H-reflex
   adaptive response to operant conditioning 137, 138*f*
   during prolonged submaximal exercise 99, 99*f*
   at fatigue 104
   measurement of 91
HSP. *See* heat shock proteins
hybrid fibers. *See* fiber types, muscle
hyperplasia
   of muscle fibers as a possible response to strength
      training 155–157
hypertrophy
   of muscle fibers as a result of strength training 154
hypoxia
   as a possible signal for altered gene expression
      126, 127*f*

I

IEGs. *See* immediate early genes
IGF-1. *See* insulinlike growth factor 1
immediate early genes (IEGs) 126
immobilization, effects on neuromuscular system
      194–202, 195*f*, 198*f*, 199*f*, 200*f*, 201*f*, 202*t*
immunohistochemistry of myosin heavy chains, for
      fiber typing 6
inflammation of muscle following contractile activity
      153, 154*f*
inhibition, in spinal cord during fatigue 108
   of motoneurons during fatigue 93, 104
initiating factors of protein synthesis (eIF) 149, 152
injury, muscular. *See* damage, muscular
input resistance, motoneuronal 50–51
insulinlike growth factor 1 (IGF-1) 148–149
integrin 147
intramuscular microstimulation 47
intraneural microstimulation 47
isoforms
   of myosin heavy chains 3, 4, 4*t*
   of myosin light chains 4*t*, 18
   of sarcoplasmic reticulum proteins 4*t*, 28–29
   of troponin–tropomyosin 4*t*, 23–26
isometric contractions
   recruitment of motor units 69–73
isomyosins 26, 26*f*, 28, 28*t*

J

JNK. *See* c-Jun NH_2-terminal kinase

L

late adaptation
   in motoneuron firing, defined 59
   variation among motoneurons innervating
      different muscle unit types 59–60, 60*f*
lengthening contractions. *See* eccentric contractions
light chains. *See* myosin light chains
long-term adaptation (LTA) at neuromuscular junction
      129–130, 129*f*
LTA. *See* long-term adaptation at neuromuscular
      junction
lysosomal proteases 153

M

magnetic resonance imaging (MRI) 161
MAPK. *See* mitogen-activated protein kinase
maximal shortening velocity (V_{max}) of single muscle
      fibers
   comparison with maximal unloaded shortening
      velocity (V_o) 13–14, 14*f*
   measurement 12
MEP. *See* motor evoked potentials
mesencephalic locomotor region (MLR) 57, 108
mitogen-activated protein kinase (MAPK) 126, 147
MLR. *See* mesencephalic locomotor region
motoneuron
   afterhyperpolarization 55–58, 56*f*
   axon diameter, as an estimate of cell size 50, 51*f*
   effect of exhaustive exercise on morphology of 133
   effects of spinal cord transection on properties of
      186–187
   electrophysiological properties
   excitability – determinants 51
   firing frequencies, limits of 55–59
   input resistance 50–51
   late adaptation of firing frequency 59–60, 60*f*
   membrane bistability 60–62, 61*f*, 62*f*
   properties, related to properties of the innervated
      muscle fibers 47–62, 49*t*
   rheobase 53
   size, as estimated from cell capacitance 50
   specific membrane resistivity 51
   synaptic inputs 49, 50*f*, 95*f*
   threshold depolarization 53, 55*f*
motor evoked potentials (MEP) 97, 98*f*, 99
motor unit
   definition 38
   properties 39–47
   types, differences in properties among 39–47
   types, in muscles of humans vs. other species 47,
      48*f*
   uniformity of muscle fibers in 46
MRF. *See* muscle regulatory factors
MRI. *See* magnetic resonance imaging
mRNA stability 115

muscle regulatory factors (MRF) 151
muscle unit. *See also* motor unit
    definition 38
"muscle wisdom" 88
M wave
    response during fatigue 85, 90
myofibrillar ATPase
    and the economy of contraction 17
    quantitative difference among fiber types 17, 18*f*, 19*f*
    relationship to muscle fiber shortening velocity 17–18, 18*f*
    usefulness in fiber-typing 5–8
myonuclei
    increase in number with muscle hypertrophy 149, 156–157, 157*f*
myosin heavy chains
    interspecies differences 3–5, 7–8
    nonuniformity along fiber length 8, 30–32
    relationship to muscle fiber maximal shortening velocity 10
    separation using PAGE 4–5
    types 2, 3, 4*t*, 6*f*
myosin light chains
    essential light chains
    in different muscle fiber types 4*t*, 20
    function 20–22
    location 18–19, 19*f*
    role in modulating muscle fiber contractile speed 20, 20*f*, 21*f*
    regulatory light chains
    functions 22–23
    phosphorylation of 22–23, 22*f*
    types 4*t*, 18

N

neural effects of training 161–169
neuromuscular junction
    differences among muscle fiber types 128
    effects of endurance training 130–133, 132*f*
    effects of reduced neuromuscular activity 175, 181
    failure as a possible cause of fatigue 41, 43, 85–87, 87*f*
neuromuscular propagation failure 41
neuromuscular transmission 85–87
neurotransmitter depletion as a source of fatigue 86
neurotrophin-4 (NT4) 137
NT4. *See* neurotrophin-4
next-neighbor rule of fiber-type transitions 5

O

"onion skin" phenomenon of motor unit recruitment 70–71, 70*f*

P

p38 kinase 126
p70^{s6} kinase 152, 153*f*
PAGE. *See* polyacrylamide gel electrophoresis
parvalbumin 29, 118

PDGF. *See* platelet-derived growth factor
pennateness of muscle, influence on function 2
permeabilized muscle fiber preparation 8
phospholamban 114*t*, 115
phospholipase 149, 150
phosphorylation potential 124
plateau potentials 60–62, 61*f*, 62*f*
platelet-derived growth factor (PDGF) 154
polyacrylamide gel electrophoresis (PAGE)
    as a technique to separate single muscle fiber myosin heavy chains 4, 7, 7*f*
    under non-denaturing conditions, to detect isomyosins 26, 26*f*
postsynaptic membrane
    as a possible fatigue site 88
posttranslational modifications during altered protein synthesis 118, 119*f*
presynaptic inhibition 95–96
    during fatigue 108
pretranslational control of protein synthesis 115, 116*f*
prostaglandins 149–150
protein degradation
    during decreased neuromuscular activity 174, 174*f*
    following an acute session of resistance exercise 145, 146*f*
protein kinase C (PKC) 126
    possible role during muscle hypertrophy 147*f*, 149
protein synthesis
    change in muscle during decreased neuromuscular activity 174, 174*f*
    change in muscle following an acute session of resistance exercise 145, 146*f*
protooncogenes
    possible role in muscle gene expression 150

Q

quantal content 86
quantal release at the neuromuscular junction 128
quantal size 86

R

receptors
    acetylcholine (AChR) 131, 132*f*
    dihydropyridine (DHPR) 115
    transmembrane, role in response of muscle to stretch 147, 147*f*
reciprocal inhibition during fatigue 105
recruitment of motor units
    according to motoneuron size 48, 135
    computer simulation 63–66
    during various tasks 66–81
    effects of strength training 167
    possibility of "rotation" during prolonged contractile activity 107
regeneration of motor axons 133
regeneration of muscle during chronic electrical stimulation 127
regulatory light chains. *See* myosin light chains

Renshaw cells
  influence related to motor unit type 97
  involvement in fatigue 97, 108
resistivity. *See also* motoneuron
  specific, of motoneuronal membrane 51
retrograde
  influence of muscle on motoneuronal properties
    137, 186
reverse transcriptase polymerase chain reaction
    (RTPCR) 7
rheobase. *See also* motoneuron
  defined 53
ribosomes 117
"ribosomal efficiency" 175
RNA activity 145
RTPCR. *See* reverse transcriptase polymerase chain
    reaction
ryanodine calcium release channels 28, 115

S

"safety factor" for neuromuscular transmission 86, 128
sag
  in force during isometric contractions as a
    criterion for motor unit typing 39*f*, 44
  in membrane voltage response to sustained
    current injection in motoneurons 53*f*, 58
sarcalumenin 115
sarcoplasmic/endoplasmic reticulum calcium ATPase
    (SERCA) 115
  isoforms 4*t*, 28
  response to chronic electrical stimulation 114*t*,
    115, 120
sarcoplasmic reticulum
  differences among fiber types 28–29
  effects of chronic electrical stimulation 114*t*, 115
  effects of strength training 157
satellite cells
  and muscle hypertrophy 149, 156
SERCA. *See* sarcoplasmic/endoplasmic reticulum
    calcium ATPase (SERCA)
serotonin, effects on motoneurons 60, 61*f*
size principle of motor unit recruitment 48
slack test, for determination of unloaded shortening
    velocity of single muscle fibers 9, 9*f*
SNAP-25. *See* synaptosome-associated protein of 25
    kilodaltons
sodium channels, at neuromuscular junction 128
spaceflight, effects on neuromuscular system 187–
    192, 189*f*
specific membrane resistivity of motoneurons 51
specific tension of muscle fibers
  measurement of 16–17
  variation among fiber types 17, 17*f*
spike-triggered averaging 47, 68, 69*f*, 72
spinal cord transection 183–187
  effects on neuromuscular system below the
    transection 183–187
  subsequent "training" effects 137, 139*f*, 140*f*, 186, 187*f*

spinal isolation 181–183
spindles
  role in fatigue 91–92, 91*f*, 92*f*, 102, 108, 110
sprouting of motoneurons 133
stimulation of muscle. *See* chronic muscle electrical
    stimulation
strength training
  effects on motor unit recruitment 167, 169*f*
  effects on muscle contractile function 158–160,
    158*f*, 164*f*
  effects on muscle phenotype 154–157
  neural effects 161–169
  role of eccentric contractions 160–161
stretch
  effects on atrophying muscles 173, 198*f*
  of muscle, as a stimulus for altered protein
    synthesis 145
synaptosome-associated protein of 25 kilodaltons
    (SNAP-25) 132*f*, 133, 134*f*
synchronization
  of motor unit firing, effects of strength training
    167–168, 169*f*

T

talin 146
task-partitioning of motor units 68
tendon vibration
  during fatigue 91–93, 92*f*, 93*f*
TES. *See* transcranial electrical stimulation
tetrodotoxin (TTX) 176
TGF. *See* transforming growth factors
threshold depolarization. *See also* motoneuron
  defined 53
TMS. *See* transcranial magnetic stimulation
TNF. *See* tumor necrosis factor
training
  endurance, effects on motoneurons 133–137
  endurance, effects on neuromuscular junction
    128–133, 132*f*
  strength, effects on muscle contractile properties
    158–160, 158*f*, 164*f*
  strength, effects on muscle phenotype 154–157,
    155*f*, 156*f*
  strength, effects on recruitment 167–169, 169*f*
  strength, neural effects 161–167
transcranial electrical stimulation (TES) 97, 105
transcranial magnetic stimulation (TMS) 49, 97–99,
    98*f*, 99*f*, 100*f*, 105
transcription
  changes in muscle during decreased use 174, 199*f*
transforming growth factors (TGF) 154
translation 117
translational control of protein synthesis 117, 118*f*
triadin 115
trophic effect
  of muscle on motoneuron properties 135, 178
  of nerve on muscle during decreased activity 176,
    179

tropomyosin
    function 23–26, 24f, 25f
    isoforms 4t, 25, 25f
troponin C
    isoforms 4t, 24
troponin I
    isoforms 4t
troponin T
    isoforms 4t, 24–25, 25f
troponin–tropomyosin complex 23–26, 24f
    and calcium/force relationships in muscle fibers
        24–25, 24f, 25f
    isoforms 23–24
TTX. *See* tetrodotoxin
tumor necrosis factor (TNF) 154
tungsten electrodes 68
twitch interpolation 79

U

ubiquitin–proteosome system 118–120, 152
ULLS. *See* unilateral lower-limb suspension

unilateral lower-limb suspension (ULLS) 192, 193f,
    193t
unloaded shortening velocity (V$_o$) of single muscle
        fibers
    having different myosin heavy chains 10–12, 10t,
        11f, 12f
    having mixtures of myosin heavy chains 10–11, 12f
    species differences 10–11, 11f

V

vibration
    of muscle tendon during fatigue 91–93, 92f, 93f, 97
V$_o$. *See* unloaded shortening velocity of single muscle
        fibers
V$_{max}$. *See* maximal shortening velocity of single
        muscle fibers
V$_{opt}$ 15, 16f. *See also* fiber types, muscle

About the Author

Phillip Gardiner, PhD, is currently a professor in the department of kinesiology at the Université de Montréal, Québec, Canada. He also is an associate member of the school of physical and occupational therapy at McGill University and associate researcher at the institute de kinesi-therapie at the Free University of Brussels.

For 25 years, Dr. Gardiner has conducted and published research pertaining to the effects of physical activity on the neuromuscular system, and his work has appeared in leading physiology journals. He is president of the Canadian Society for Exercise Physiology.

The former editor of the *Canadian Journal of Applied Physiology*, Dr. Gardiner obtained his doctorate in exercise physiology from the University of Alberta at Edmonton, Alberta, Canada.